Landscaping 6G

At the intersection of cutting-edge technology and environmental stewardship lies the transformative potential of 6G. This book examines the critical relationship between advanced energy-efficient hardware, intelligent infrastructure integration, and environmental conservation, exploring how ultra-fast communication technologies can revolutionize our approach to sustainability while delivering unprecedented connectivity.

From augmented reality applications that promote sustainability to intelligent transport systems, green energy solutions, and environmental conservation support, this book presents real-world case studies that illuminate 6G's practical potential. In the emerging era of generative AI, readers will discover how these technologies can work in concert to address our most pressing ecological challenges while creating new opportunities for innovation and growth.

This essential resource speaks to a diverse audience of technology professionals, environmental scientists, policy makers, and forward-thinking business leaders in the fields of communication technology, environmental science, and sustainable development.

Landscaping 6G

Unlocking the Power of Ultra-Fast Communication

Edited by
Upinder Kaur, Aparna Kumari, Hemant Kumar Saini,
Surbhi B. Khan, and Mariya Ouaissa

CRC Press
Taylor & Francis Group
Boca Raton London New York

CRC Press is an imprint of the
Taylor & Francis Group, an **informa** business

Designed cover image: Shutterstock

First edition published 2026
by CRC Press
4 Park Square, Milton Park, Abingdon, Oxon, OX14 4RN

and by CRC Press
2385 NW Executive Center Drive, Suite 320, Boca Raton FL 33431

CRC Press is an imprint of the Taylor & Francis Group, an informa business

British Library Cataloguing-in-Publication Data
A catalogue record for this book is available from the British Library

ISBN: 9781041162537 (hbk)
ISBN: 9781041162568 (pbk)
ISBN: 9781003683599 (ebk)

DOI: 10.1201/9781003683599

Typeset in Times
by codeMantra

Contents

Preface

This book examines advancements in 6G communication and technology with ultra-fast data transfer speed and reduced latency. The main course of the proposed book is the inclusion of green energy, AR and VR, sustainability, and other cutting-edge technologies that incorporate all the things related to the 6G. This work aims to achieve zero-carbon green transformation solutions, initiate an artistic vision that addresses intersecting sustainability challenges, and highlight the remaining open issues. There are further many more focused areas, such as the technical requirements of 6G for addressing speech challenges in AR/VR 6G solutions, virtual classrooms, AI-based healthcare assistance and surgeries, industrial applications, and global case studies.

This book explores several facets of 6G technology, encompassing the advancement of energy-efficient hardware and the integration of intelligent infrastructure, and analyzes their influence on environmental sustainability. This book works towards inculcating environment-benign practices to stop further deterioration of environmental health. Numerous governing climate-resilient practices seek the prowess of the 6G technology such as AR for environmental sustainability for encouraging sustainability.

Furthermore, this book highlights the significant influence of wireless networks in promoting sustainable behaviors in urban and industrial settings. By harnessing the connectivity and efficiency of 6G networks, individuals may optimize resource utilization, minimize environmental impact, and facilitate positive global transformations. This book covers the challenges in the 6G real-time case studies, intelligent transport, green energy solutions, and sustainable environment conservation support. In the era of generative AI, this book plays a significant role in harnessing the potential of 6G technology to develop sustainable solutions. It inspires and enlightens individuals to collaborate toward a future that is more robust, equitable, and ecologically sensitive.

Each chapter investigates the ecological implications of 6G, providing a comprehensive understanding of its ability to decrease carbon emissions and promote environmental accountability, as well as how 6G will influence the strategies of VR and artificial intelligence software development companies.

Let's take a closer look at the specific themes and contributions of each chapter:

Chapter 1: Traces the historical evolution of mobile networks from 1G to 6G, highlighting the major technological milestones and how 6G is poised to deliver immersive, human-centric, and sustainable connectivity.

Chapter 2: Introduces the key principles, technical enablers, and sustainability aspects of 6G, including spectrum, energy efficiency, and green communication technologies.

Chapter 3: Explores 6G network architectures and their integration with Artificial Intelligence of Things (AIoT), enabling intelligent, autonomous, and highly adaptive applications.

Chapter 4: Examines the legal, regulatory, and ethical dimensions of 6G, focusing on privacy, governance, digital rights, and the balance between innovation and societal protection.

Chapter 5: Presents the cybersecurity challenges and strategies for securing 6G networks, addressing risks like quantum threats, data breaches, and critical infrastructure vulnerabilities.

Chapter 6: Discusses how blockchain can enhance trust, identity management, and access control mechanisms in 6G environments, ensuring decentralized and tamper-resistant security.

Chapter 7: Analyzes the synergy between edge computing, AI, and blockchain for enabling secure, low-latency, real-time surveillance systems in smart cities.

Chapter 8: Focuses on unmanned aerial vehicles (UAVs) in a 6G context, exploring blockchain-based secure communication frameworks, future opportunities, and deployment challenges.

Chapter 9: Proposes a novel framework combining digital twins, blockchain, and federated learning for secure, intelligent, and patient-centric healthcare in the 6G era.

Chapter 10: Provides a broader outlook on healthcare 4.0 powered by 6G, highlighting transformative applications, interoperability challenges, and future research directions.

Chapter 11: Explores how ultra-low latency and massive bandwidth in 6G empower real-time AR/VR ecosystems, enabling immersive experiences in education, industry, and entertainment.

Chapter 12: Investigates the convergence of metaverse technologies, 6G connectivity, and edge intelligence, reshaping digital interactions, economies, and future human experiences.

About the Editors

Upinder Kaur is Head of the Computer Science and Engineering Department at Akal University. Her career in the field of computer science spans more than 20 years, encompassing both academia and research. She has contributed more than 30 research articles, presented at international conferences, and published in reputable journals indexed in renowned databases such as Scopus and SCI. She has also edited three books with reputable publishers. Her research interests are diverse, focusing on artificial intelligence and machine learning, exploring the practical applications of deep learning, and delving into the intricacies of cloud computing. Through Dr. Kaur's innovative perspectives and pioneering research, she continues to advance knowledge within these dynamic and evolving fields.

Aparna Kumari is affiliated with the Department of Computer Science and Engineering, Institute of Technology, Nirma University in Ahmedabad, India. She has over 12 years of experience in research and teaching. Her research interests include smart grid systems, blockchain technology, electric vehicles, and renewable energy sources. In 2023, She was awarded the Shri P. Prahlaad Chabaria Award for Best Woman Professional. She has been listed among the world's top 2% scientists as per the list published by Stanford University in 2021, 2022, and 2023. Two of her novel works on COVID-19 have been approved by the World Health Organization (WHO) and included in their COVID-19 Global literature on coronavirus disease database. She has authored/co-authored 40 publications, including 28 peer-reviewed SCI-indexed international journal articles, 10 papers in top-tier IEEE international conferences, and 2 book chapters with reputed publishers. She is an active member of IEEE and Women in Engineering (WIE) India. Dr. Kumari strives to contribute to society through her outreach efforts and aims to facilitate access to resources for underprivileged students.

Hemant Kumar Saini works as an academic at Lincoln University, Malaysia, and the School of Computer Science Engineering at Bennett University (The Times Group), Greater Noida, India. He earned his Ph.D. in Computer Science and Engineering from Manipal University and has been honored for his research work by the Governor of Rajasthan during his master's studies. He has more than 14 years of teaching experience. Dr. Saini is an active member of ISTE, IEEE, IAENG, and CSI. He has demonstrated his research prowess by publishing articles in renowned SCI and Scopus-indexed journals and conference proceedings. He has published several national and international articles and authored, edited and co-edited 18 books. He has delivered many lectures in the ATAL-sponsored faculty development programs and keynote speeches in National and International Conferences. He is an editorial board member and a reviewer of many international journals of repute. His research interest includes wireless networks, soft computing, deep and machine learning, Internet of Things (IoT), Industry 4.0, and smart contract. His contributions to these fields have been widely recognized, and he continues to push the boundaries of scientific knowledge.

Surbhi B. Khan is working in the School of Science, Engineering and Environment, Data Science Department at the University of Salford, United Kingdom. She is a certified project management professional (PMP) and was ranked among the top 2% of scientists globally by Stanford University for two consecutive years (2023 and 2024). She has published over 200 articles and books in high-indexed outlets with her students and colleagues. She has received several awards, including "best paper", to mention one such example. Surbhi serves as Deputy Editor-in-Chief for *TCE Letters*, Associate Editor for *IEEE Transactions* *on Consumer Electronics*, *HCIS*, and *EAAI* journals and Guest Editor for many other reputed journals. She is also involved in projects funded by external organizations from the UK, India, and Middle Eastern institutes. She has served as a session chair at many IEEE conferences and has been invited as a resource person for international conferences and faculty development programs. Dr. Khan is a member of Women in Data Science Worldwide (WiDS) and was a WiDS Ambassador in 2024. Her research interests include machine learning, deep learning, data science in healthcare, and sentiment analysis.

Mariya Ouaissa is currently a Professor in Cybersecurity and Networks at the Faculty of Sciences Semlalia, Cadi Ayyad University, Marrakech, Morocco. She earned her Ph.D. in Computer Science and Networks from ENSAM-Moulay Ismail University, Meknes, Morocco in 2019. Dr. Ouaissa graduated as a Networks and Telecoms Engineer in 2013 from the National School of Applied Sciences, Khouribga, Morocco. She is a Co-Founder and IT Consultant at IT Support and Consulting Center. She was Visiting Professor at the School of Technology, Meknes, Morocco, between 2013 and 2021. She is a member of the International Association of Engineers and International Association of Online Engineering, and since 2021, she has been an ACM Professional Member. She is an Expert Reviewer for Academic Exchange Information Centre (AEIC) and a Brand Ambassador for Bentham Science. Dr. Ouaissa has served on technical program and organizing committees for several conferences and events and has organized numerous symposiums, workshops, and conferences as General Chair. Her contributions to information security and privacy, IoT security, and wireless and constrained networks security are noteworthy. Her main research topics include IoT, M2M, D2D, WSN, cellular networks, and vehicular networks. Dr. Ouaissa has published over 120 papers, including book chapters, international journal articles, and conference/workshop proceedings, as well as 40 edited books and 10 special issues as Guest Editor.

Contributors

Raghad Jamal Abunasser
Computer Information Systems Department
Jordan University of Science and Technology
Ar-Ramtha, Jordan

Wesam Ahmed
Information Technology Department
Hurghada University
Hurghada, Egypt

Qasem Abu Al-Haija
Cybersecurity Department
Jordan University of Science and Technology
Ar-Ramtha, Jordan

Nada Mohammad Altwaiq
Computer Information Systems Department
Jordan University of Science and Technology
Ar-Ramtha, Jordan

Hella Kaffel Ben Ayed
Computer Science Department
University of Tunis El Manar
Tunis, Tunisia

Jakir Hussain Jareena Begam
Department of Biomedical Engineering
SNS College of Technology
Coimbatore, India

Kishor Kumar Reddy Chinthala
Department of Computer Science Engineering
Stanley College of Engineering and Technology for Women
Hyderabad, India

Muralitharan Deenathayalan
Director of Solution Architecture & Technology
Ryval-X
Bangalore, Karnataka, India

Krishnan Ravichandran Dhanya
Department of Computer Science with Data Analytics
Dr. N.G.P. Arts and Science College
Coimbatore, India

Mohamed El Ghazouani
Department of Computer Science
Chouaib Doukkali University
El Jadida, Morocco

Zakaria Abou El Houda
Institut National de la Recherche Scientifique
University of Quebec
Quebec, Canada

Yassine Eljakani
Computer Systems and Vision Laboratory, Faculty of Sciences
Ibn Zohr University
Agadir, Morocco

My Ahmed El Kiram
Department of Computer Science, Faculty of Sciences Semlalia
University Cadi Ayyad
Marrakesh, Morocco

Latifa Er-Rajy
Department of Computer Science, Faculty of Sciences Semlalia
University Cadi Ayyad
Marrakesh, Morocco

Khadija Rammeh Houerbi
Computer Science Department
Aviation School of Borj el Amri
Borj El Amri, Tunisia

Naser Abbas Hussein
Computer Science Department
University of Tunis El Manar
Tunis, Tunisia

Nabil Idi
Department of Computer Science and Engineering
University of Quebec
Outaouais, Canada

Abdullah Mohammed Kaleem
Department of Electronics and Telecommunication Engineering
MPGI School of Engineering
Nanded, Maharashtra, India

Arunachalam Karthika
Department of Biomedical Engineering
SNS College of Technology
Coimbatore, India

Annadurai Karthikeyan
Department of Electronics and Communication Engineering
SNS College of Technology
Coimbatore, India

Ramandeep Kaur
Department of Computer Science
Baba Farid College of Engineering and Technology
Punjab, India

Subramaniyan Jayaraman Sadheesh Kumar
Department of Electrical and Electronics Engineering
National Institute of Technology Puducherry
Karaikal, India

Oussama Lachihab
Department of Computer Science, Faculty of Sciences Semlalia
University Cadi Ayyad
Marrakesh, Morocco

Amena Mahmoud
Computer Science Department
Kafrelsheikh University
Kafr El Sheikh, Egypt

Hajar Moudoud
Department of Computer Science and Engineering
University of Quebec
Outaouais, Canada

Yassir Moutia
Computer Systems and Vision Laboratory, Faculty of Sciences
Ibn Zohr University
Agadir, Morocco

Vijaishree Dubey Pandey
School of Law
CHRIST University
Bengaluru, India

Shanmugam Pradeep
Department of Electronics and Communication Engineering
SNS College of Technology
Coimbatore, India

Jaganathan Ramkumar
Department of Computer Science
Sri Krishna Arts and Science College
Coimbatore, India

Shubneet
Department of Computer Science
Chandigarh University
Mohali, Punjab, India

Shashank Solanki
School of Law
CHRIST University
Bengaluru, India

Charu Sood
Department of Computer Science
Chandigarh University
Mohali, Punjab, India

Rangasamy Suganthi
Department of Computer Science with Data Analytics
Dr. N.G.P. Arts and Science College
Coimbatore, India

Vijeta Verma
School of Law
CHRIST University
Bengaluru, India

Viswanathan Valarmathi
Department of IT and Cognitive Systems
Sri Krishna Arts and Science College
Coimbatore, India

Anushka Raj Yadav
Department of Computer Science
Chandigarh University
Mohali, Punjab, India

From 1G to 6G

The Evolution of Mobile Communication toward a Sustainable and Immersive Future

1

Latifa Er-Rajy, My Ahmed El Kiram,
Oussama Lachihab, and Mohamed El Ghazouani

1.1 INTRODUCTION

Over the past 40 years, mobile communication has experienced a significant and transformative evolution. Each new generation of mobile technology has been introduced to meet growing demands, such as providing faster internet speeds, improved coverage, and more connected user experiences [1]. While early networks were designed exclusively for voice communication, later generations shifted their focus toward enabling mobile web access, video streaming, and supporting the rapid expansion of smart devices [2].

Today, mobile networks extend far beyond their original purpose. They have become a critical part of modern infrastructure, supporting everything from smart homes and autonomous vehicles to remote surgeries and virtual learning environments [3]. As user expectations continue to rise, this progress brings new challenges—such as the need for efficient energy management, better use of the available spectrum, and the delivery of reliable and immersive services on a global scale [4].

These challenges are essential to consider in the development of 6G. This next generation is not simply a faster extension of its predecessors; it represents a shift in how mobile systems are envisioned [5]. The focus is on ultra-low latency, the integration of AI-driven intelligence directly into the network, and adaptive systems capable of responding in real time to changing conditions. Crucially, 6G is also being designed with sustainability as a central principle. By adopting smarter infrastructure and optimizing energy consumption, the goal is to reduce the environmental impact of these increasingly powerful systems [6].

DOI: 10.1201/9781003683599-1

1.2 THE FIRST STEPS: THE BEGINNING OF 1G NETWORKS

The term 1G refers to the very first technology generation of mobile communication. It appeared in the early 1980s and marked the launch of wireless telephony based on analog radio signals, as shown in Figure 1.1 [2]. This was the first time that people could make phone calls without using a landline, which represented at that time a major leap in communication freedom [7]. Nevertheless, the mobile phones of that era were large and heavy and offered only a few functionalities; still they provided a new level of convenience and mobility that was previously unimaginable [8].

The earliest 1G systems, such as the advanced mobile phone system (AMPS), were designed to support voice communication on the move [9]. These systems made it possible to hold real-time conversations within the limited range of the network, which proved especially beneficial for professionals and emergency services that required mobile accessibility [10].

Despite being a major breakthrough at the time, 1G technology had several notable limitations [11]. Since it relied on analog signals, conversations were frequently affected by interference, and privacy was a significant concern—anyone with the appropriate radio equipment could potentially eavesdrop on calls. Additionally, call quality was generally unreliable due to static and background noise, which were common issues [12].

Signal stability was another major drawback, with frequent drops in coverage, particularly in rural or remote areas. Network congestion further reduced the quality and availability of service. Moreover, the first-generation mobile phones were bulky and had very limited battery life, making them difficult to carry and impractical for everyday use.

1.2.1 Understanding the Working Mechanism of 1G

Let us go back to the early 1980s—the period when 1G, or first-generation mobile technology, first appeared. It was the first time people could make a call without being physically connected to a

FIGURE 1.1 1G generation overview.

telephone cord. At the time, this was considered a groundbreaking innovation, almost like something out of science fiction [13].

However, by today's standards, 1G was quite limited [14]. It operated using analog signals, meaning the voice was converted into a continuous wave and transmitted through the air to the nearest cell tower. This is also where the term *cell phone* comes from—because each tower served a specific "cell" or coverage area. If a user moved during a call, the connection would be handed off from one tower to another, similar to a relay process.

The mobile phones themselves were extremely large and heavy. They came with prominent antennas, bulky casings, and batteries that weighed nearly as much as a modern laptop. Most were not carried in pockets but instead installed in vehicles or transported in suitcase-style cases. These devices were designed for one purpose only—voice calls. There were no screens, no messaging features, no applications, and no additional functions [15].

The quality of the calls was also far from ideal. Audio was often unclear, with interference, static noise, and frequent echoes. In addition, the lack of encryption meant that calls were not private; anyone with a basic radio scanner could listen in, as privacy was not yet a built-in feature [16].

Coverage was another major limitation. While users in urban areas might have had access to a signal, those in rural or remote locations were often left without service [17]. Battery life was also a problem—charging was required frequently, which made the devices far less mobile than their name suggested.

Despite all these drawbacks, 1G marked a major turning point. For the first time, people could communicate while on the move, take business calls outside the office, and contact emergency services without relying on a public payphone. It was the beginning of a new era in connectivity [18].

Nevertheless, the issues with quality, coverage, and security quickly became apparent, and they paved the way for the introduction of 2G technology—which brought significant improvements in digital communication and user experience.

1.2.2 Strengths and Shortcomings of 1G

Despite its limitations, the most significant achievement of 1G was that it introduced the era of mobile communication [19]. Indeed, for the first time, individuals could place calls while driving, working on a construction site, or from nearly any location with signal coverage. It fundamentally transformed how people communicated in their personal lives and in professional contexts, then laid the groundwork for the mobile connectivity we experience today [20].

However, its inadequacies were difficult to manage. In fact, the lack of encryption made calls vulnerable to interception, raising serious concerns about user privacy. Furthermore, the network infrastructure often struggled to support high call volumes, resulting in frequent congestion and dropped connections. Most notably, 1G supported only basic voice communication, without any capability for text messaging, multimedia sharing, or internet access.

These limitations revealed the urgent need for a more advanced system [21]. Users sought clearer audio quality, enhanced security, and the ability to send both messages and data. This growing demand drove the development of a new generation: 2G, which not only enhanced mobile communication but also redefined its possibilities entirely.

1.3 SHIFT TO 2G

The move to 2G in the early 1990s was a major leap forward in mobile communication. Unlike the analog systems of 1G, 2G introduced fully digital networks, and with them, a wave of innovations that changed how people used their phones [22]. One of the biggest game-changers was Short Message Service (SMS), or text messaging. For the first time, users could send short, silent messages instantly, so, no need to call, just type and send. It completely transformed everyday communication [23].

Digital voice transmission in 2G made a big difference in call quality. Calls became much clearer and more reliable, with less background noise and interference, so conversations sounded sharper overall [24]. Security was improved as well, since encryption was added to help prevent calls from being intercepted. This fixed one of the biggest problems users had with 1G. Also, digital networks used radio waves more efficiently, which meant more people could talk at the same time without causing network overload [25].

Another big change that came with 2G was the arrival of mobile data. With the introduction of technologies like GPRS and later EDGE, people could start checking emails, sending picture messages, and even browsing simple websites from their phones [26]. The internet was still very basic on mobile, but it was an important first step. It connected phones to the digital world and prepared the ground for everything that came afterward.

1.3.1 Operation of 2G Networks

When 2G rolled out in the early 1990s, it felt like a huge leap forward compared to 1G. The most important change was the switch from analog to digital signals. That might sound a bit technical—or even boring—but in reality, it completely changed how mobile phones worked. Calls became clearer, more secure, and more reliable [27]. Figure 1.2 gives a simple illustration of how 2G operates and what made it so different from what came before.

Instead of turning the voice into a continuous wave like 1G did, 2G converted it into digital data—basically strings of zeros and ones. This shift made calls a lot more reliable. Conversations became much clearer, the annoying static mostly disappeared, and calls were much harder to listen in on, thanks to the introduction of encryption. At the time, that was a pretty big improvement [28].

What really made 2G stand out for most people, though, was the arrival of text messaging. SMS allowed users to send short, quick messages to friends or family. It was simple, fast, and honestly, kind of addictive.

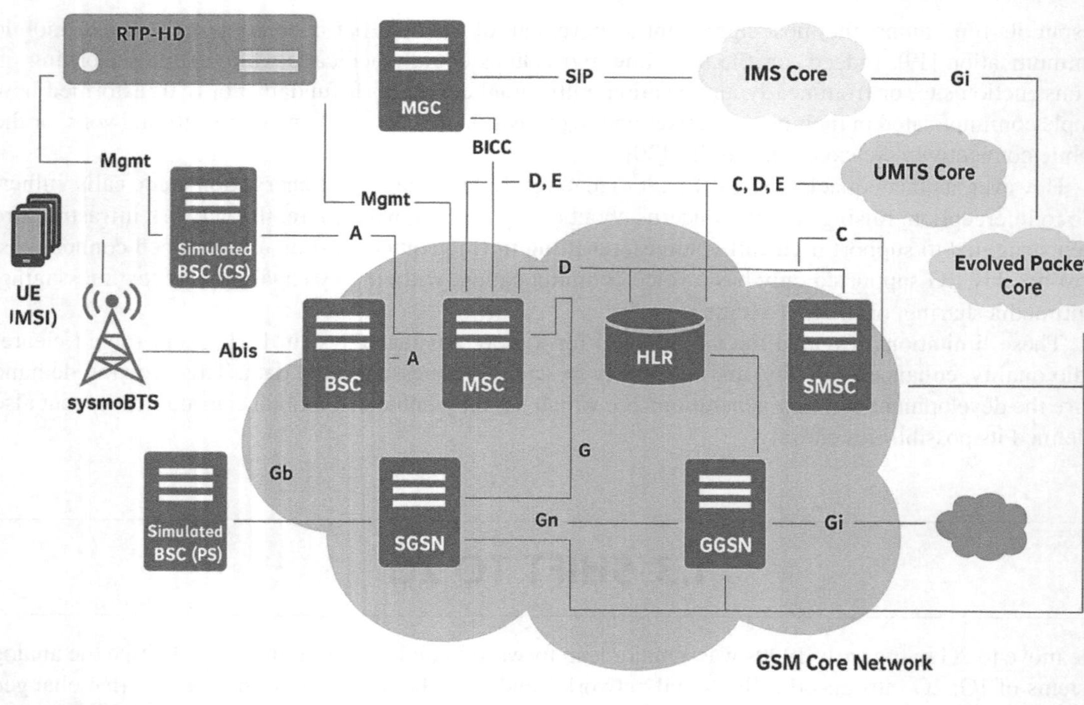

FIGURE 1.2 How 2G works.

If you ever had one of those old Nokia phones, you probably remember pressing the same key over and over just to type out a single word [29].

2G also brought the first version of mobile internet, using technologies like GPRS and later EDGE. Of course, it was very slow by today's standards—loading a webpage could take a long time—but just being able to check an email or read a headline while away from a computer felt almost futuristic back then.

Most 2G networks were built on the GSM standard, which quickly gained popularity around the world. GSM helped make calls and texts more stable and introduced international roaming. That meant people could travel abroad and still use their phones, which made mobile communication even more practical. Phones also got smaller, more affordable, and easier to carry—making them accessible to a lot more people [30].

Still, 2G had its limits. Mobile internet was slow, and things like video streaming or downloading large files just weren't realistic. But compared to the bulky, noisy, and insecure systems of 1G, it was a huge step forward. 2G made mobile phones more than just a luxury—they became everyday tools and set the stage for the smartphones and fast data connections we use today.

1.3.2 Achievements and Limitations of 2G

Table 1.1 provides a comparison between 1G and 2G. The latter brought major improvements to mobile communication. It delivered more secure and reliable voice calls, expanded network coverage, and—perhaps most importantly—introduced text messaging. SMS quickly became a popular way to stay connected: it was fast, simple, and inexpensive. This new form of communication changed the way people interacted, both personally and professionally, and soon became a global habit [31].

TABLE 1.1 Comparison between 1G and 2G

FEATURE	*1G*	*2G*	*DESCRIPTION*
Technology type	Analog	Digital	1G relied on analog transmission, while 2G introduced digital technology, significantly enhancing call quality and reliability.
Launch period	Early 1980s	Early 1990s	1G marked the beginning of mobile communication; about a decade later, 2G emerged with major technological advancements.
Signal type	Analog radio waves	Digital binary signals	Analog signals were continuous and prone to interference; digital signals are more stable and resistant to noise.
Voice quality	Often poor, with static and noise	Clearer, more consistent audio	2G's digital encoding reduced background noise and improved the overall clarity of voice calls.
Security & privacy	No encryption; easily intercepted	Encrypted communications	1G lacked any form of encryption, while 2G introduced basic encryption, enhancing call security and user privacy.
Services supported	Voice calls only	Voice calls, SMS, limited mobile data (GPRS, EDGE)	2G expanded functionality by adding text messaging and basic internet access, unlike 1G which supported only voice.
Coverage	Limited to major urban areas	Broader coverage with international roaming	2G networks expanded geographic coverage and introduced roaming capabilities, allowing use in multiple countries.

(Continued)

TABLE 1.1 (*Continued*) Comparison between 1G and 2G

FEATURE	*1G*	*2G*	*DESCRIPTION*
Device size	Large, heavy, often car-installed	Smaller, lighter, pocket-friendly devices	2G devices became more compact and portable, making mobile phones more practical for daily use.
Battery life	Short, required frequent recharging	Longer battery life	Improved power efficiency in 2G devices led to extended usage times compared to 1G's bulky, power-hungry phones.
Network capacity	Low; few users per cell tower	Higher user capacity	Digital systems in 2G allowed more simultaneous users per cell, improving efficiency and reducing congestion.
Data transfer speed	Not supported	Low-speed data (e.g., GPRS, EDGE)	1G lacked any data capability, while 2G introduced slow-speed mobile internet for basic online functions.
Popularity	Limited by cost and device size	Mass adoption due to affordability and improved functionality	2G democratized mobile technology, making it accessible to a broader population beyond early business users.

However, 2G had its limitations. Data speeds were still very slow—only a few kilobits per second, or slightly better with EDGE. Because of that, it wasn't capable of supporting activities like video streaming or loading complex websites [32]. The mobile internet experience was mostly limited to basic tasks such as checking emails or reading simple, text-based content. As smartphones grew in popularity and users began expecting richer content, social media, and location-based services, it became clear that 2G could no longer keep up.

This growing need for faster data, better internet access, and more advanced mobile features led to the next big step: 3G, a technology that would bring the mobile internet to life in a whole new way (Table 1.1).

1.4 THE REVOLUTION: INTRODUCTION OF 3G

Back in the early 2000s, the arrival of 3G marked a big turning point in mobile communication. Building on the digital foundation laid by 2G, 3G significantly increased data speeds, making it possible to browse the web, check emails, and even stream videos directly from a phone—without all the frustrating delays we were used to [33]. This jump in capability unlocked all kinds of services that hadn't really been possible before. Suddenly, things like video calls, mobile TV, and internet-based apps started becoming part of everyday life, turning the mobile phone into the all-in-one digital tool we now take for granted.

3G networks relied on technologies like UMTS, and later HSPA, which helped boost download speeds from just a few hundred kbps to several Mbps [34]. That made mobile internet feel much closer to what we were used to on a regular computer. Thanks to this improvement, people could actually work, chat, and enjoy online content while on the go—something that was pretty limited before.

But it wasn't just about faster connections. 3G also laid the groundwork for the rise of smartphones. As phones began to feature larger screens, better cameras, and more internal storage, 3G provided the stable connection needed for running apps, accessing cloud services, and chatting in real time. With broader coverage and higher network capacity, more people could get online from more places, making constant connectivity feel normal—even in places where that would've seemed impossible just a few years earlier [35].

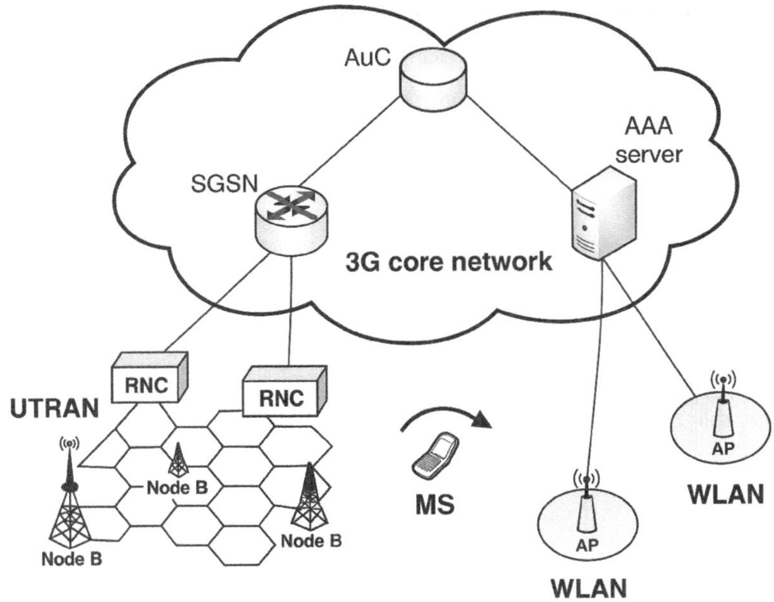

FIGURE 1.3 How 3G works.

1.4.1 The Functioning of 3G Networks

By the early 2000s, phones weren't just for calling or texting anymore. People wanted more—like checking emails, browsing the web, watching videos, and even making video calls. That's where 3G came in. It was the third generation of mobile tech, and it felt like mobile phones were finally catching up with everything we needed [36].

The big game-changer with 3G was that it really focused on data, as shown in Figure 1.3. While 2G was great for texts and basic calls, 3G actually let you go online. It used something called packet-switching—basically, it broke down everything you were trying to send (like a webpage or video) into small chunks called packets. These packets traveled through the network separately and were reassembled on your phone. It was fast and way more efficient.

Behind the scenes, most 3G networks ran on a system called UMTS, which was based on a more technical bit called WCDMA (Wideband Code Division Multiple Access) [37]. Sounds complicated, but the main point is that it allowed lots of people to use the same network at the same time without it crashing. Everyone's data stayed separate and secure, even though they were all sharing the same network.

And this was the moment smartphones really started to shine. Phones stopped being just phones—they turned into mini pocket computers. Apps, emails, maps, games—it all became possible thanks to 3G.

Of course, it wasn't perfect. Speeds could vary depending on your location. If you left a 3G-covered area, your phone would fall back to 2G, and things would get slow again. But even with those ups and downs, 3G was a huge step forward. It set the stage for the fast, always-connected mobile experience we're used to today [38].

1.4.2 Pros and Cons of 3G

3G brought significant improvements to mobile communication, as illustrated in Table 1.2. It made internet access on phones faster and more dependable, introduced features like video streaming and GPS-based

TABLE 1.2 2G vs 3G

FEATURE	*2G*	*3G*	*DESCRIPTION*
Technology type	Digital GSM/CDMA	Digital with WCDMA/UMTS/HSPA	2G used basic digital transmission; 3G introduced more advanced technologies like WCDMA and HSPA for better data handling.
Launch period	Early 1990s	Early 2000s	2G marked the move to digital communication, while 3G brought high-speed internet and multimedia capabilities to mobile devices.
Signal type	Digital binary signals	Digital with packet-switched data	3G used more advanced signal processing to allow faster, more efficient communication, especially for data services.
Voice quality	Clearer than analog, basic call support	Even better voice quality with support for VoIP	3G offered improved voice clarity and allowed voice over IP (VoIP) in some implementations, reducing reliance on traditional circuits.
Security & privacy	Basic encryption	Stronger encryption, improved authentication	3G improved security with stronger encryption and better protection against unauthorized access.
Services supported	Voice calls, SMS, basic mobile internet (GPRS, EDGE)	Voice, high-speed internet, video calls, mobile TV	3G enabled full web browsing, video streaming, real-time video calling, and other rich multimedia services.
Coverage	Broad, with international roaming	Broader, but patchy in rural areas	While 3G expanded coverage, it sometimes offered slower performance in remote regions due to infrastructure limitations.
Device size	Compact and portable phones	Smartphones with larger screens and multimedia functions	3G-era phones became more advanced, with cameras, music players, and app stores, leading to the rise of smartphones.
Battery life	Efficient for calls and basic use	Shorter battery life due to data-heavy tasks	3G devices consumed more power because of high-speed connectivity and constant data use.
Network capacity	Moderate; suitable for basic traffic	Much higher, supports many users with better speeds	3G allowed networks to handle more simultaneous voice and data users with less congestion.
Data transfer speed	30–100 kbps (GPRS/EDGE)	384 kbps – several Mbps (HSPA, UMTS)	3G greatly improved internet speed, making streaming and online applications more practical.
Popularity	Mass adoption for voice and SMS	Widespread adoption with rise of mobile apps and internet	3G fueled the smartphone revolution by supporting rich content and mobile internet usage on a global scale.

apps, and reshaped how people used their devices in everyday life. For businesses, it enabled remote work, mobile payments, and location-based services. For everyday users, 3G offered quick access to social media, online shopping, and on-the-go entertainment [39].

However, 3G had its limitations. Network coverage wasn't consistent, as urban areas generally enjoyed better speeds, while rural regions often faced slower, less reliable connections. And although it was a big step up from 2G, 3G still fell short of meeting the growing demand for HD video streaming and real-time gaming. On top of that, building and maintaining 3G infrastructure came with high costs, making it a major investment for service providers [40].

As smartphones grew more powerful and user expectations kept rising, the limitations of 3G became increasingly apparent. People wanted faster speeds, smoother performance, and instant access to content and services. This growing demand set the stage for the next major leap in mobile technology. Enter 4G, which is a new generation designed to raise the bar and deliver a faster, more seamless mobile experience [41].

1.5 THE EVOLUTION CONTINUES: THE RISE OF 4G

When 4G showed up in the late 2000s, it really took things to a whole new level. It brought way faster speeds, better responsiveness, and a kind of smooth, always-on connection that people just hadn't seen before on mobile. While 3G had built on what came before, 4G was more like a total restart—designed from the ground up to handle heavy stuff like video streaming and real-time apps. Suddenly, your phone wasn't just for calls and light browsing anymore—it could actually handle HD video, online gaming, and smooth video calls without constantly lagging or buffering [42].

The tech behind it—mainly LTE (Long-Term Evolution) and later LTE-Advanced—made the mobile internet feel like real broadband. In fact, in some places, it even felt faster than Wi-Fi at home. It came with lower delays (latency), more stable connections, and download speeds that hit tens of megabits per second—or even over 100 Mbps if you were lucky. And this didn't just apply to phones. Tablets, smartwatches, and other connected gadgets all started working way better thanks to this stronger network [43].

Because of all that, a whole new set of apps and services started popping up. Cloud storage, mobile banking, social media, and even the early Internet of Things (IoT) really took off around this time [44]. People were suddenly doing things on their phones that they'd never even considered before—like editing documents, joining work meetings, or managing money—all from anywhere.

And businesses noticed. With mobile networks finally fast and stable enough, a lot of companies started creating tools and platforms built specifically for mobile. In a way, 4G didn't just change how people used their phones—it kind of changed the way the internet worked altogether [45].

1.5.1 Understanding the Operation of 4G Systems

By the late 2000s, the way people used their mobile phones had evolved significantly. Phones were no longer limited to voice calls or text messages—users were streaming videos, browsing the web, using apps, and actively engaging with social media. As these digital habits grew, the limitations of 3G became increasingly apparent. In response, 4G emerged, not merely as an upgrade, but as a complete transformation of mobile network technology [46].

Fourth-generation mobile networks, or 4G, were designed from the ground up to support high-speed internet access. Rather than building on the structure of 3G, 4G introduced a fundamentally new approach tailored to handle data-intensive services. One of the most notable advantages was speed. With data rates up to ten times faster than 3G, 4G enabled smooth HD video streaming, real-time online gaming, high-quality video calls, and seamless use of cloud-based applications. This performance leap redefined the mobile experience and turned smartphones into true multimedia and productivity tools [47].

Technically speaking, as illustrated in Figure 1.4, 4G departed from older circuit-switched methods used in previous generations. Instead, it adopted a fully IP-based communication model—the same foundation used by the modern internet. The core technology behind this shift was LTE, which allowed all forms of data—whether calls, messages, videos, or apps—to be transmitted as internet packets [48]. This approach greatly improved the efficiency, speed, and reliability of mobile networks.

In short, 4G delivered an internet experience on mobile devices that was comparable to fixed broadband, and it paved the way for new digital services and more connected lifestyles.

This shift fundamentally enhanced the speed and reliability of mobile experiences. Whether users were placing a voice call, streaming a film, or simply scrolling through social media, their devices were continuously exchanging small packets of data at remarkably high speeds. This not only accelerated data transmission but also improved connection stability, reduced latency, and significantly enhanced the quality of voice calls.

However, the impact of 4G extended far beyond improved speed. It unlocked entirely new possibilities [49]. Mobile applications became more sophisticated and responsive, websites loaded almost instantaneously, and social media platforms evolved into mobile-first ecosystems. Moreover, 4G laid the

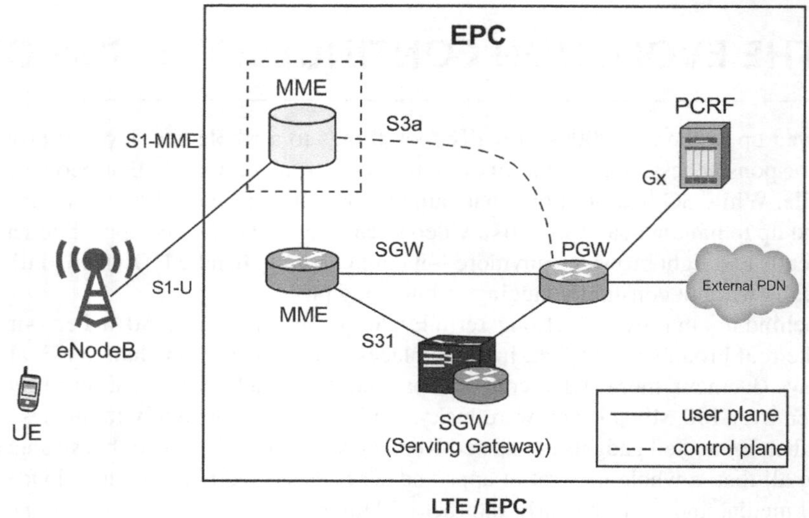

FIGURE 1.4 How 4G works.

technological foundation for services that are now part of everyday life, such as mobile banking, ride-hailing apps, real-time GPS navigation, live video streaming, and the early integration of smart home technologies.

Naturally, the deployment of 4G networks was not without its challenges. Building the necessary infrastructure required substantial investment and time, involving the installation of new cell towers and the modernization of existing ones. In rural and remote regions, consistent 4G coverage remained difficult to achieve. Additionally, as mobile internet speeds improved, users tended to consume more data, often reaching their monthly limits more quickly than before [50].

Despite these hurdles, 4G marked a turning point in mobile communication. It made high-speed, dependable internet access a standard expectation, transforming smartphones from convenient gadgets into indispensable tools woven into the fabric of modern life.

1.5.2 The Pros and Cons of 4G

Table 1.3 highlights the key differences between 3G and 4G technologies. The advantages brought by 4G were substantial. With significantly faster speeds, enhanced security, and more efficient use of the available radio spectrum, networks were able to support a higher number of users with improved service quality [51]. These improvements paved the way for innovation across multiple sectors, including e-commerce, telemedicine, remote work, mobile education, and the development of smarter urban infrastructure. Activities such as streaming high-definition content, navigating with real-time maps, or using cloud-based applications while on the move quickly became part of everyday life.

Of course, 4G was not without its limitations. Deploying the necessary infrastructure—such as new base stations, fiber-optic lines, and upgraded equipment—required substantial investment. Mobile operators faced high costs, some of which were inevitably passed on to consumers, particularly in regions with limited market competition [52]. Moreover, access to 4G was not equally distributed. In many rural or remote areas, coverage often lagged behind that of urban centers, leading to a digital divide. Another emerging challenge was the surge in data consumption. As users took advantage of faster speeds, those with capped data plans frequently encountered higher costs or were forced to closely monitor their usage.

Nonetheless, despite these shortcomings, 4G represented a major technological leap. It fundamentally transformed user expectations—establishing fast, always-connected, and highly capable mobile networks as the new standard. These rising expectations would, in turn, drive the development of the next generation of mobile technology: 5G.

TABLE 1.3 4G vs 3G

FEATURE	*3G*	*4G*	*WHAT IT MEANS*
Technology type	WCDMA, CDMA2000	LTE (Long-Term Evolution), WiMAX	3G combined older voice systems with newer data handling. 4G ditched the old phone tech and went fully digital, built from the ground up for the internet.
Launch period	Early 2000s	Late 2000s	3G made mobile internet possible. 4G made it fast, reliable, and ready for things like streaming and real-time apps.
Data transfer speed	384 Kbps to around 2 Mbps	10–100 Mbps (and even up to 1 Gbps in ideal conditions)	With 3G, you could browse and maybe watch low-res videos. 4G made streaming HD video, fast downloads, and smooth video calls feel normal.
Latency	100–500 ms	30–50 ms or less	4G cuts down the lag. That means better responsiveness for games, video calls, and anything needing real-time feedback.
Voice technology	Traditional calls at first, later added VoIP	VoIP (Voice over LTE or VoLTE)	3G started with older-style calling, but 4G sends calls as data—making them clearer and faster to connect.
Internet experience	Basic browsing and email, low-quality video	HD streaming, real-time video calls, online gaming	4G turned mobile internet into something you could actually enjoy, not just tolerate.
Coverage	Fairly widespread, but patchy in remote areas	Strong in cities, still expanding to rural areas	3G reached many places; 4G brought faster speeds to cities first, with rural areas catching up slowly.
App support	Simple apps like email or messaging	Data-heavy apps like YouTube, Google Maps, cloud services	3G supported early apps. 4G powered the mobile app explosion—with streaming, navigation, live chats, and more.
Network capacity	Moderate	Much higher	4G handles more users at once without the network getting overloaded—perfect for dense urban areas.
Security	Better than 2G, but still limited	Stronger, more advanced security	4G added stronger encryption to keep your data and calls more secure from prying eyes.
Battery usage	Moderate	Higher, but phones got smarter about it	4G can use more power, but modern phones are built to manage it better, so battery life holds up well.
Popularity	Very high—started the smartphone revolution	Extremely high—made smartphones essential	3G brought the internet to your phone. 4G made it fast enough to change how we live, work, and connect on the go.

1.6 THE LEAP FORWARD: THE ARRIVAL OF 5G

The global deployment of 5G—the fifth generation of mobile communication technology—began in the late 2010s, representing a significant advancement in wireless connectivity. In comparison to 4G, 5G delivers dramatically higher data transfer rates—up to 100 times faster—and much lower latency, often reduced to just a few milliseconds. This improvement enables devices to communicate in near real time, making possible a range of applications that were previously too slow or technically unfeasible on earlier mobile networks [53].

A defining feature of 5G is its capacity to support large-scale machine-to-machine (M2M) communication. This capability is essential for emerging domains such as smart cities, autonomous vehicles, industrial automation, and robotics, where vast numbers of connected devices must exchange information simultaneously and with minimal delay [54]. Furthermore, 5G enhances network efficiency and scalability, ensuring consistent performance even as the number of connected devices continues to grow—an increasingly critical requirement in today's interconnected world.

1.6.1 Key Innovations behind 5G Technology

5G has fundamentally transformed how devices connect to the internet. While it continues to rely on radio waves, as did 3G and 4G, the manner in which these frequencies are utilized, combined with the underlying technological infrastructure, sets 5G apart.

Beyond the substantial increase in speed, 5G is specifically engineered to enable real-time communication and to support a massive number of connected devices simultaneously [55]. This allows for seamless operation across diverse applications, even in densely populated areas or complex environments, without compromising performance.

As illustrated in Figure 1.5, 5G operates across three distinct frequency bands, each with its own characteristics and use cases:

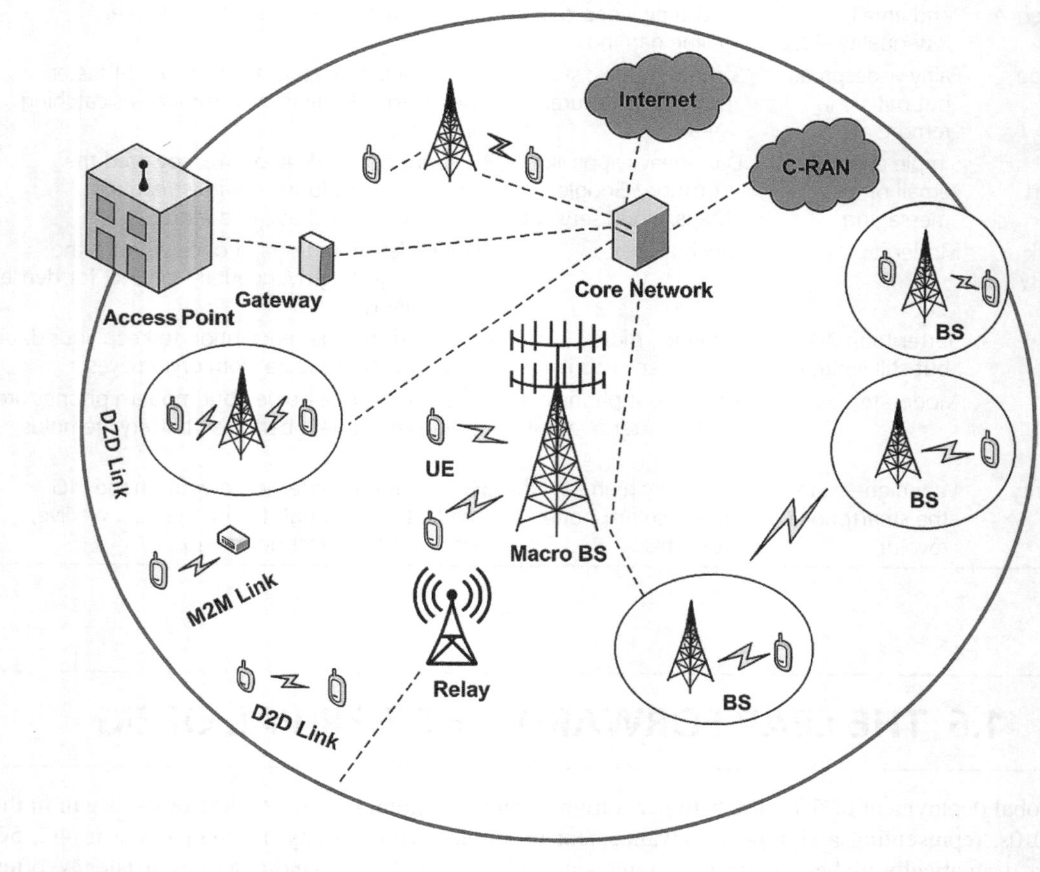

FIGURE 1.5 5G network architecture.

- **Low-Band**: Provides broad coverage and good penetration through obstacles such as walls, but offers only modest speed improvements over 4G.
- **Mid-Band**: Strikes a balance between speed and coverage, making it the backbone of most daily 5G applications.
- **High-Band (or mmWave)**: Delivers extremely high speeds and ultra-low latency, but its range is limited, and it is easily obstructed by buildings, trees, and even weather conditions [56].

Due to the limited range of high-band frequencies, 5G departs from the traditional reliance on large, widely spaced cell towers. Instead, it employs a dense network of small cells—compact antennas installed on rooftops, utility poles, and streetlights—which collectively ensure consistent, high-quality coverage even in high-traffic areas, such as urban centers, stadiums, or transportation hubs.

However, 5G is not solely focused on enhancing mobile phone performance. Its true potential lies in enabling the IoT—from smart home devices and wearable health monitors to industrial automation, autonomous vehicles, and intelligent infrastructure. The network's ability to support millions of simultaneous device connections without sacrificing performance marks a significant leap forward in scalability [57].

Another critical feature of 5G is its ultra-low latency, which enables real-time responsiveness. This is vital for mission-critical applications such as remote surgery, autonomous driving, immersive virtual reality experiences, and interactive cloud-based services.

In essence, 5G achieves its performance through a combination of faster data transmission, intelligent spectrum management, and the deployment of a dense, distributed network of small antennas. While global rollout is still ongoing, and some rural regions may experience delays in access, the transformative potential of 5G spans multiple sectors—including healthcare, education, manufacturing, transportation, and entertainment.

Far more than just an incremental upgrade, 5G serves as the technological foundation for the next generation of connected experiences.

1.6.2 Benefits and Challenges of 5G

Table 1.4 presents a comparison between 4G and 5G technologies. While 5G is commonly associated with significantly faster download speeds, its impact extends far beyond simple data rate improvements. One of its most transformative advantages lies in its ultra-low latency and high reliability, which together enable a new class of advanced applications. These include augmented reality (AR), virtual reality (VR), cloud-based gaming, and large-scale IoT deployments.

Such capabilities are expected to drive substantial innovation across multiple sectors. In healthcare, for example, 5G facilitates real-time remote diagnostics and robotic surgery. In transportation, it enables connected and autonomous vehicles. In entertainment, it supports immersive, low-latency streaming experiences. And in manufacturing, it allows for automated processes, smart factories, and precise remote control of machinery. By supporting real-time data processing and seamless device interconnectivity, 5G is poised to reshape how industries operate and how users experience digital service [58].

However, despite its many advantages, the deployment of 5G is not without significant challenges. Rolling out this technology requires substantial infrastructure investment, including the widespread installation of new antennas and small cells—particularly in densely populated urban areas. At present, 5G coverage remains uneven, with most advancements concentrated in major cities and developed regions, while rural and remote areas often lag [59].

Another growing concern relates to energy consumption. Supporting a vast network of continuously connected devices places increased demands on power infrastructure, raising questions about the long-term sustainability of large-scale 5G networks. Additionally, the expansion of 5G introduces

TABLE 1.4 Comparison between 4G and 5G

FEATURE	*4G*	*5G*	*DESCRIPTION*
Technology type	LTE (long-term evolution)	New radio (NR), including mmWave and sub-6 GHz	5G introduces advanced radio access and a new core network architecture for faster, more efficient, and flexible communication.
Launch period	Late 2000s to early 2010s	Late 2010s onwards	4G brought reliable high-speed mobile internet; 5G began global rollout around 2019 with focus on ultra-low latency and massive IoT.
Maximum speed	Up to 100 Mbps (real-world); peaks near 1 Gbps	1–10 Gbps in real-world use; theoretical peak over 20 Gbps	5G delivers much higher speeds, supporting high-resolution video, real-time cloud access, and large data transfers.
Latency	Around 30–50 ms	As low as 1 ms	5G minimizes delay, enabling applications where real-time response is critical, such as remote surgery, VR, and autonomous driving.
Bandwidth & capacity	Moderate	Significantly higher	5G supports a far greater number of simultaneous connections per square kilometer—ideal for dense urban environments and smart devices.
Frequency bands	Sub-6 GHz	Sub-6 GHz and mmWave (24–100 GHz)	5G adds high-frequency bands (mmWave) for ultra-fast speeds, though these require closely spaced antennas due to limited range.
Network architecture	Centralized; based on large cell towers	Decentralized; relies on small cells and edge computing	5G uses distributed infrastructure to bring processing closer to users, enhancing speed and reducing congestion.
Supported devices	Smartphones, tablets, mobile broadband	Smartphones, IoT, AR/VR, autonomous systems	5G extends support beyond mobile devices to include critical infrastructure, wearables, and industrial equipment.
Energy efficiency	Less energy-efficient	More energy-efficient per bit transmitted	5G is designed for better energy performance, even with more hardware, through techniques like beamforming and power optimization.
Use cases	Mobile browsing, video streaming, VoIP	Smart cities, automation, AR/VR, telemedicine, autonomous vehicles	4G enabled the mobile app ecosystem; 5G unlocks advanced applications that depend on ultra-reliable, low-latency communication.
Coverage (current)	Widely available worldwide	Limited—mostly in urban and high-traffic areas	4G is near-universal; 5G is still expanding and currently concentrated in major metropolitan areas.
Infrastructure cost	High (but mostly deployed)	Very high (due to densification, new antennas, spectrum costs)	5G requires significant investment in new infrastructure and spectrum allocation, especially for mmWave deployments.

complex privacy and cybersecurity risks. With more devices online, the number of potential vulnerabilities grows, increasing the need for robust security protocols to safeguard data and ensure user trust [60].

Nevertheless, despite these hurdles, 5G marks a profound transformation in the way people, systems, and technologies interact. It lays the groundwork for a future defined by smart infrastructure, real-time digital services, and seamlessly connected environments. While the full impact of 5G is still unfolding, its influence is already shaping the next era of digital innovation.

1.7 LOOKING AHEAD: THE VISION FOR 6G

Even though 5G is still undergoing global deployment, researchers, engineers, and industry leaders are already turning their attention to the future: 6G, the sixth generation of wireless communication. Although it is not expected to reach mainstream adoption before 2030 [61], foundational efforts to define, research, and design this next-generation technology are already in progress. Crucially, 6G is not merely seen as a faster iteration of 5G; rather, it represents a fundamental transformation in how we connect, communicate, and engage with digital environments [62].

At the heart of 6G lies the ambition to push the boundaries of what is technologically possible. Projected data transmission rates could reach up to 1 terabit per second (Tbps), with latency reduced to as little as 0.1 ms. These performance enhancements would allow for near-instantaneous communication and enable entirely new categories of applications. Moreover, the integration of artificial intelligence and advanced edge computing will make 6G networks significantly more autonomous, adaptive, and context-aware—capable of self-optimization in real time [63].

Such capabilities could unlock innovations that were, until recently, confined to science fiction: live holographic communication, real-time digital twins that replicate physical systems, and immersive extended reality (XR) environments that fluidly blend augmented, virtual, and mixed realities. These emerging possibilities indicate that 6G will not simply accelerate connectivity—it will redefine the digital experience and pave the way for a more intelligent, responsive, and interconnected world.

1.7.1 Some Key Things That Make 6G Stand Out

While previous generations of mobile networks added artificial intelligence (AI) more as a feature than a foundation, 6G is expected to take a very different approach. AI won't just support the network—it will be part of its core [64]. These future networks will be able to learn from their environment, adjust themselves automatically, and even make decisions in real time without needing human intervention. This could lead to networks that are more efficient, more reliable, and better equipped to handle complex and fast-changing situations.

To keep up with the growing demand for ultra-fast data, 6G will likely use a much higher part of the frequency spectrum known as the terahertz (THz) band. This could open the door to data speeds as high as 1 terabit per second. But working with THz frequencies comes with its own technical challenges—signals don't travel very far, they're easily disrupted, and the equipment needed to make it all work still needs to be developed. Still, if these hurdles can be overcome, the speed and capacity improvements could be transformational [65].

One of the most exciting things about 6G is that it won't just connect devices—it will help them understand their surroundings. Phones, cars, and even wearable devices could be equipped with sensors that "feel" what's happening around them. This could change everything from how we perform medical scans to how autonomous vehicles navigate or how emergency systems detect danger [66].

Another big goal of 6G is to bring high-speed internet to places that have never had reliable access before. This will likely involve combining satellites in low-Earth orbit, high-flying drones or balloons, and a new generation of ground networks. The idea is to remove the concept of "dead zones" entirely—even in the most remote areas. It's a bold vision, but one that could make the digital world truly global [67].

Of course, reaching this future won't be easy. There are still big technical challenges, like building devices that can handle THz signals without overheating or losing performance. On top of that, getting countries and companies around the world to agree on how 6G should work—what standards to follow, what rules to set—takes time and careful coordination.

There's also the issue of energy. With so many devices and antennas working at such high speeds, 6G networks could use a lot of power. That's why making the technology energy-efficient and environmentally sustainable will be just as important as making it fast. And as these networks become smarter and collect more data, protecting people's privacy and ensuring ethical use of AI will be essential. Transparency, user control, and fairness will need to be built in from the start [68–70].

Even with all these challenges, the potential of 6G is incredibly exciting. It's not just about connecting faster—it's about creating a smarter, more responsive, and more inclusive digital world. While we still have a long way to go, the direction is clear, and the possibilities are unlike anything we've seen before.

1.8 CONCLUSION

The evolution of mobile communication from 1G to the anticipated arrival of 6G tells a story that goes far beyond faster speeds or better connections. It reflects a broader transformation in how societies live, interact, and innovate. From 1G's analog voice services in the 1980s to today's digital ecosystems, each generation of mobile technology has responded to growing demands for greater connectivity, capacity, and functionality. While 2G brought digital voice and SMS, 3G opened the door to mobile internet. Then came 4G, which revolutionized daily life through high-speed data, enabling smartphones, cloud computing, and streaming services to become integral to how we work, learn, and communicate.

With 5G, mobile communication entered a new phase—one defined by ultra-fast speeds, ultra-low latency, and the ability to support millions of connected devices at once. This leap in performance has enabled real-time innovations such as smart cities, autonomous transportation, industrial automation, and even remote surgery. It also marked a shift from centralized network designs to more decentralized models, supported by edge computing and small-cell infrastructure. These developments highlight how the evolution of mobile networks is no longer just about performance—it's about enabling new ways of living and working.

Looking ahead, 6G promises to take this transformation even further. It is expected to integrate emerging technologies such as terahertz communication, intelligent edge computing, and AI-powered network automation to build systems that are not just fast but adaptive and context-aware. These networks could unlock immersive digital experiences—like large-scale augmented and VR, real-time holograms, and seamless interaction between humans and machines. Sustainability will also be a major focus, with efforts aimed at improving energy efficiency, reducing environmental impact, and designing infrastructure that supports long-term ecological responsibility.

But realizing this vision won't be without its challenges. Ensuring that the benefits of 6G reach all parts of the world—including rural and underserved communities—will be essential to avoid deepening the digital divide. At the same time, the vast amount of data generated by intelligent systems raises critical concerns around privacy, fairness, and the ethical use of AI. Building trustworthy, secure, and inclusive networks will require thoughtful collaboration across governments, industry, and civil society.

In the end, mobile communication has grown from a luxury into a foundational element of modern life. It supports everything from healthcare and education to agriculture, public services, and economic development. As we move toward the 6G era, the focus is no longer just on connecting devices—it's about empowering people. The networks of tomorrow must not only deliver technical excellence but also serve as engines of progress, inclusion, and human well-being on a global scale.

REFERENCES

1. M. Castells, M. Fernandez-Ardevol, J. L. Qiu, and A. Sey, "Mobile communication and society: A global perspective," *Econ. Geogr.*, vol. 84, no. 1, pp. 119–120, 2008. doi:10.1111/j.1944-8287.2008.tb00398.x.
2. C. Lamberton and A. T. Stephen, "A thematic exploration of digital, social media, and mobile marketing: Research evolution from 2000 to 2015 and an agenda for future inquiry," *J. Mark.*, vol. 80, no. 6, pp. 146–172, 2016. doi:10.1509/jm.15.0415.

3. H. Shakhatreh, A. Sawalmeh, A. H. Alenezi, S. Abdel-Razeq, and A. Al-Fuqaha, "Mobile-IRS assisted next generation UAV communication networks," *Comput. Commun.*, vol. 215, pp. 51–61, 2024. doi:10.1016/j.comcom.2023.12.025.
4. E.-R. Latifa, E. K. My Ahmed, E. G. Mohamed, and O. Mariya, "Foreground-preserving background modification : A deep learning approach," In: *17th International Conference on Signal-Image Technology & Internet-Based Systems (SITIS 2023)*, The Berkeley Hotel Pratunam, Bangkok, Thailand, pp. 261–267, 2023. doi:10.1109/SITIS61268.2023.00047.
5. S. Issa, I. Shayea, U. Tureli, A. A. El-saleh, W. Tashan, and R. Caglar, "Engineering science and technology, an international journal machine learning for handover decision with mobile edge computing in 6G mobile network : A survey," *Eng. Sci. Technol. Int. J.*, vol. 69, p. 102131, 2025. doi:10.1016/j.jestch.2025.102131.
6. S. H. Alsamhi, N. Nasser, Q. Ni, A. Al-Dulaimi, and M. Guizani, "Towards sustainable industry 4.0: A survey on greening IoE in 6G networks," *Ad Hoc Netw.*, vol. 165, p. 103610, 2024. doi:10.1016/j.adhoc.2024.103610.
7. E. Zontou, "Unveiling the evolution of mobile networks: From 1G to 7G," *arXiv* Prepr. arXiv2310.19195, pp. 1–10, 2023. https://arxiv.org/abs/2310.19195.
8. P. Sharma, "Evolution of mobile wireless communication networks-1G to 5G as well as future prospective of next generation communication network," *Int. J. Comput. Sci. Mob. Comput.*, vol. 2, pp. 47–53, 2013. https://www.ijcsmc.com/docs/papers/August2013/V2I8201317.pdf.
9. P. Curwen and J. Whalley, *Fourth Generation Mobile Communication: The Path to Superfast Connectivity*, Edward Elgar Publishing, Cheltenham, 2013. doi:10.4337/9781848444911.00008.
10. K. Lyytinen and V. V. Fomin, "Achieving high momentum in the evolution of wireless infrastructures: The battle over the 1 G solutions," *Telecomm. Policy*, vol. 26, no. 3–4, pp. 149–170, 2002. doi:10.1016/S0308-5961(02)00006-X.
11. S. Wang and M. Higgins, "Limitations of mobile phone learning," In: *Proceedings of the IEEE International Workshop on Wireless and Mobile Technologies in Education (WMTE 2005)*, Tokushima, Japan, pp. 179–181, 2005. doi:10.1109/WMTE.2005.43.
12. G. Y. Koi-Akrofi, M. Kuuboore, D. A. Odai, and A. N. Kotey, "Telecommunications wireless generations: Overview, technological differences, evolutional triggers, and the future," *Int. J. Electron. Telecommun.*, vol. 69, no. 1, pp. 105–114, 2023. doi:10.24425/ijet.2023.144338.
13. D. Vantrease, R. Schreiber, M. Monchiero, M. McLaren, N. P. Jouppi, M. Fiorentino, A. Davis, N. Binkert, R. G. Beausoleil, and J. H. Ahn, "Corona: System implications of emerging nanophotonic technology," In: *Proceedings of the 35th Annual International Symposium on Computer Architecture (ISCA)*, Beijing, China, pp. 153–164, 2008. doi:10.1109/ISCA.2008.35.
14. V. Khare, S. Garg, S. Shukla, and P. Sharma, "Comparative study of 1G, 2G, 3G and 4G," *J. Eng. Comput. Appl. Sci.*, vol. 2, no. 4, pp. 55–63, 2013. https://ptacts.uspto.gov/ptacts/public-informations/petitions/1521258/download-documents?artifactId=LE1BfCcQTT6aYIjgx_DGt3kZbJtiJxjgS76xCAyvgK1TF2FClnbjxXg
15. V. E. Sytnikov, V. S. Vysotsky, I. P. Radchenko, and N. V. Polyakova, "1G versus 2G-comparison from the practical standpoint for HTS power cables use," *J. Phys. Conf. Ser.*, vol. 97, no. 1, 2008. doi:10.1088/1742-6596/97/1/012058.
16. O. T. Eluwole, N. Udon, M. Ojo, C. Okoro, and A. J. Akinyoade, "From_1_G_to_5_G_what_next," *IAENG Int. J. Comput. Sci.*, vol. 45, no. 3, pp. 413–434, 2019.
17. A. Mercian, M. P. McGarry, and M. Reisslein, "Impact of report message scheduling (RMS) in 1G/10G EPON and GPON," *Opt. Switch. Netw.*, vol. 12, pp. 1–13, 2014. doi:10.1016/j.osn.2013.11.004.
18. B. Bakare and E. Bassey, "A comparative study of the evolution of wireless communication technologies from the first generation (1G) to the fourth generation (4G)," *Int. J. Electron. Commun. Comput. Eng.*, vol. 12, no. 3, pp. 73–84, 2021.
19. M. Goyal and Y. Lather, "Advancement of communication technology from 1G to 5G," *Int. J. Adv. Res. IT Eng.*, vol. 4, no. 5, pp. 1–17, 2015. https://www.indianjournals.com/ijor.aspx?target=ijor:ijarie&volume=4&issue=5&article=001
20. G. P. Kaur, J. Birla, and J. Ahlawat, "Generations of wireless technology," *Int. J. Comput. Sci. Manag. Stud.*, vol. 11, no. 2, pp. 176–180, 2011.
21. S. Patel, M. Chauhan, and K. Kapadiya, "5G: Future mobile technology-vision 2020," *Int. J. Comput. Appl.*, vol. 54, no. 17, pp. 6–10, 2012. doi:10.5120/8656-2264.
22. J. R. Sanchez, O. Edfors, and L. Liu, "Positioning for distributed large intelligent surfaces using neural network with probabilistic layer," In: *Proceedings of the 2021 IEEE Globecom Workshops (GC Workshops)*, Madrid, Spain, 2021. doi:10.1109/GCWkshps52748.2021.9682016.
23. P. Gómez-Pérez, M. Crego-García, I. Cuiñas, and R. F. S. Caldeirinha, "Modeling and inferring the attenuation induced by vegetation barriers at 2G/3G/4G cellular bands using artificial neural networks," *Meas. J. Int. Meas. Confed.*, vol. 98, pp. 262–275, 2017. doi:10.1016/j.measurement.2016.12.014.

24. L. R. Rabiner, "Applications of voice processing to telecommunications," *Proc. IEEE,* vol. 82, no. 2, pp. 199–228, 1994. doi:10.1109/5.265347.
25. B. H. Hall, "Innovation and market value," In: R. Barrell, G. Mason, and M. O'Mahony *(Eds.), Productivity, Innovation and Economic Performance*, Cambridge University Press, Cambridge, 2005. doi:10.2139/ssrn.151912.
26. A. K. Salkintzis, C. Fors, and R. Pazhyannur, "WIAN-GPRS integration for next-generation mobile data networks," *IEEE Wirel. Commun.*, vol. 9, no. 5, pp. 112–124, 2002. doi:10.1109/MWC.2002.1043861.
27. O. Osterbo and O. Grondalen, "Benefits of self-organizing networks (SON) for mobile operators," *J. Comput. Networks Commun.*, vol. 2012, Article ID 862527. doi:10.1155/2012/862527.
28. P. Marsch, B. Raaf, A. Szufarska, P. Mogensen, H. Guan, M. Farber, S. Redana, K. Pedersen, and T. Kolding, "Future mobile communication networks: Challenges in the design and operation," *IEEE Veh. Technol. Mag.*, vol. 7, no. 1, pp. 16–23, 2012. doi:10.1109/MVT.2011.2179343.
29. Z. Lalji and J. Good, "Designing new technologies for illiterate populations: A study in mobile phone interface design," *Interact. Comput.*, vol. 20, no. 6, pp. 574–586, 2008. doi:10.1016/j.intcom.2008.09.002.
30. N. Gandal, D. Salant, and L. Waverman, "Standards in wireless telephone networks," *Telecomm. Policy*, vol. 27, no. 5–6, pp. 325–332, 2003. doi:10.1016/S0308-5961(03)00026-0.
31. B. N. Audrey, S. Selian, and R. Edward, "From 2G TO 3G-the evolution of international cellular standards," In: *Proceedings of the 29th Research Conference on Communication, Information and Internet Policy (TPRC 2001)*, Alexandria, Virginia, USA, 2001.
32. S. Gindraux, "From 2 G to 3G: A guide to mobile security," *IEE Conf. Publ.*, vol. 489, pp. 308–311, 2002. doi:10.1049/cp:20020410.
33. S. Guriev, N. Melnikov, and E. Zhuravskaya, "3G internet and confidence in government," *Q. J. Econ.*, vol. 136, no. 4, pp. 2533–2613, 2021. doi:10.1093/qje/qjaa040.
34. N. P. Dharuman, S. Avancha, V. B. R. Bhimanapati, O. Goel, N. Singh, and R. Agarwal, "Multi controller base station architecture for efficient 2G 3G network operations," *Int. J. Res. Mod. Eng. Emerg. Technol.*, vol. 12, no. 10, p. 106, 2024.
35. R. Kumar, R. Geleta, A. Pandey, and D. Sinwar, "Adverse effects of 5th generation mobile technology on flora and fauna: Review study," *IOP Conf. Ser. Mater. Sci. Eng.*, vol. 1099, no. 1, p. 012031, 2021. doi:10.1088/1757-899x/1099/1/012031.
36. W. D. S. Coelho, A. Benhamiche, N. Perrot, and S. Secci, "Network function mapping: From 3G entities to 5G service-based functions decomposition," *IEEE Commun. Stand. Mag.*, vol. 4, no. 3, pp. 46–52, 2020. doi:10.1109/MCOMSTD.001.1900040.
37. R. Mijumbi, J. Serrat, J. L. Gorricho, N. Bouten, F. De Turck, and R. Boutaba, "Network function virtualization: State-of-the-art and research challenges," *IEEE Commun. Surv. Tutorials*, vol. 18, no. 1, pp. 236–262, 2016. doi:10.1109/COMST.2015.2477041.
38. W. L. Tan, F. Lam, and W. C. Lau, "An empirical study on the capacity and performance of 3G networks," *IEEE Trans. Mob. Comput.*, vol. 7, no. 6, pp. 737–750, 2008. doi:10.1109/TMC.2007.70788.
39. H. Ekström, A. Furuskär, J. Karlsson, M. Meyer, S. Parkvall, J. Torsner, and M. Wahlqvist, "Technical solutions for the 3G long-term evolution," *IEEE Commun. Mag.*, vol. 44, no. 3, pp. 38–45, 2006. doi:10.1109/MCOM.2006.1607864.
40. Y. F. Kuo and C. W. Yu, "3G telecommunication operators' challenges and roles: A perspective of mobile commerce value chain," *Technovation*, vol. 26, no. 12, pp. 1347–1356, 2006. doi:10.1016/j.technovation.2005.08.004.
41. A. A. Lazakidou and K. M. Siassiakos, Third generation (3G) cellular networks in telemedicine: Technological overview, applications, and limitations. In: Lazakidou, A. A. (Ed.), *Handbook of Research on Distributed Medical Informatics and E-Health*, IGI Global, Hershey, PA, USA, 2008. doi:10.4018/978-1-60566-002-8.
42. C. A. Teodorescu, A.-N. Ciucu Durnoi, and V. M. Vargas, "The rise of the mobile internet: Tracing the evolution of portable devices," *Proc. Int. Conf. Bus. Excell.*, vol. 17, no. 1, pp. 1645–1654, 2023. doi:10.2478/picbe-2023-0147.
43. M. J. Arshad, A. Farooq, and A. Shah, "Evolution and development towards 4th generation (4G) mobile communication systems," *Proc. Natl. Acad. Sci.*, vol. 3, no. 1, pp. 1–15, 2015.
44. T. Mshvidobadze, "Evolution mobile wireless communication and LTE networks," In: *Proceedings of the 2012 6th International Conference on Application of Information and Communication Technologies (AICT)*, Tbilisi, Georgia, pp. 4–10, 2012. doi:10.1109/ICAICT.2012.6398495.
45. J. Yu, "From 3G to 4G: Technology evolution and path dynamics in China's mobile telecommunication sector," *Technol. Anal. Strateg. Manag.*, vol. 23, no. 10, pp. 1079–1093, 2011. doi:10.1080/09537325.2011.621306.
46. D. Astely, E. Dahlman, A. Furuskar, Y. Jading, M. Lindstrom, and S. Parkvall, "LTE: The evolution of mobile broadband," *IEEE Commun. Mag.*, vol. 47, no. 4, pp. 44–51, 2009.
47. M. Sauter, *From GSM to LTE-Advanced Pro and 5G: An Introduction to Mobile Networks and Mobile Broadband,* 4th edn., Academic Press, Oxford, UK, 2021. doi:10.1002/9781119714712.

48. C. Cox, *An Introduction to LTE: LTE, LTE-Advanced, SAE, VoLTE and 4G Mobile Communications,* 2nd edn., John Wiley & Sons, Ltd, Chichester, UK, 2014. doi:10.1002/9781118818046.
49. K. Gomez, L. Goratti, T. Rasheed, and L. Reynaud, "Enabling disaster-resilient 4G mobile communication networks," *IEEE Commun. Mag.*, vol. 52, no. 12, pp. 66–73, 2014. doi:10.1109/MCOM.2014.6979954.
50. A. Shaik, R. Borgaonkar, N. Asokan, V. Niemi, and J. P. Seifert, "Practical attacks against privacy and availability in 4G/LTE mobile communication systems," In: *Proceedings of the 23rd Annual Network and Distributed System Security Symposium (NDSS 2016)*, San Diego, California, USA, pp. 21–24, 2016. doi:10.14722/ndss.2016.23236.
51. B. G. Evans and K. Baughan, "Visions of 4G," *Electron. Commun. Eng. J.*, vol. 12, no. 6, pp. 293–303, 2000. doi:10.1049/ecej:20000608.
52. J. Z. Sun, J. Sauvola, and D. Howie, "Features in future: 4G visions from a technical perspective," In: *Proceedings of the IEEE Global Telecommunications Conference (GLOBECOM 2001)*, San Antonio, Texas, USA, vol. 6, pp. 3533–3537, 2021. doi:10.1109/glocom.2001.966339.
53. E. Hajlaoui, A. Zaier, A. Khlifi, J. Ghodhbane, M. Ben Hamed, and L. Sbita, "4G and 5G technologies: A comparative study," In: *2020 International Conference on Advanced Technologies for Signal and Image Processing (ATSIP)*, Sousse, Tunisia, 2020. doi:10.1109/ATSIP49331.2020.9231605.
54. M. G. Kachhavay and A. PThakare, "Science and mobile computing 5G technology-evolution and revolution," *Int. J. Comput. Sci. Mob. Comput.*, vol. 3, no. 3, pp. 1080–1087, 2014. www.ijcsmc.com.
55. Y. Yang and K. Hua, "Emerging technologies for 5G-enabled vehicular networks," *IEEE Access*, vol. 7, pp. 181117–181141, 2019. doi:10.1109/ACCESS.2019.2954466.
56. K. E. Georgiou, E. Georgiou, and R. M. Satava, "5G use in healthcare: The future is present," *J. Soc. Laparoendosc. Surg.*, vol. 25, no. 4, 2021. doi:10.4293/JSLS.2021.00064.
57. A. S. George and S. Sagayarajan, "Exploring the potential and limitations of 5G technology : A unique perspective," *Partners Univers. Int. Innov. J.*, vol. 1, pp. 160–174, 2023. doi:10.5281/zenodo.7869011.
58. M. Agiwal, A. Roy, and N. Saxena, "Next generation 5G wireless networks: A comprehensive survey," *IEEE Commun. Surv. Tutorials*, vol. 18, no. 3, pp. 1617–1655, 2016.
59. J. G. Andrews, S. Buzzi, W. Choi, S. V. Hanly, A. Lozano, A. C. K. Soong, and J. C. Zhang, "What will 5G be?" *IEEE J. Sel. Areas Commun.*, vol. 32, no. 6, pp. 1065–1082, 2014. doi:10.1109/JSAC.2014.2328098.
60. R. Dangi, P. Lalwani, G. Choudhary, I. You, and G. Pau, "Study and investigation on 5g technology: A systematic review," *Sensors*, vol. 22, no. 1, pp. 1–32, 2022. doi:10.3390/s22010026.
61. W. Jiang, B. Han, M. A. Habibi, and H. D. Schotten, "The road towards 6G: A comprehensive survey," *IEEE Open J. Commun. Soc.*, vol. 2, pp. 334–366, 2021. doi:10.1109/OJCOMS.2021.3057679.
62. I. F. Akyildiz, A. Kak, and S. Nie, "6G and beyond: The future of wireless communications systems," *IEEE Access*, vol. 8, pp. 133995–134030, 2020. doi:10.1109/ACCESS.2020.3010896.
63. C. De Alwis, A. Kalla, Q.-V. Pham, P. Kumar, K. Dev, W.-J. Hwang, and M. Liyanage, "Survey on 6G frontiers: Trends, applications, requirements, technologies and future research," *IEEE Open J. Commun. Soc.*, vol. 2, pp. 836–886, 2021. doi:10.1109/OJCOMS.2021.3071496.
64. H. Pennanen, T. Hanninen, O. Tervo, A. Tolli, and M. Latva-Aho, "6G: The intelligent network of everything," *IEEE Access*, vol. 13, pp. 1319–1421, 2025. doi:10.1109/ACCESS.2024.3521579.
65. M. Zawish, F. A. Dharejo, S. A. Khowaja, K. Dev, S. Davy, N. M. F. Qureshi, and P. Bellavista, "AI and 6G into the metaverse: Fundamentals, challenges and future research trends," *IEEE Open J. Commun. Soc.*, vol. 5, pp. 730–778, 2024. doi:10.1109/OJCOMS.2024.3349465.
66. P. Ahokangas and A. Aagaard, *The Changing World of Mobile Communications: 5G, 6G and the Future of Digital Services*, Springer Nature, Cham, Switzerland, 2023. doi:10.1007/978-3-031-33191-6.
67. M. A. Jamshed, A. Kaushik, M. Dajer, A. Guidotti, F. Parzysz, E. Lagunas, M. Di Renzo, S. Chatzinotas, and O. A. Dobre. "Non-terrestrial networks for 6G: Integrated, intelligent and ubiquitous connectivity," *IEEE Commun. Stand. Mag.*, 2025. doi:10.1109/mcomstd.2025.3572635.
68. M. Ouaissa, M. Benmoussa, A. Rhattoy, M. Lahmer, and I. Chana, "Performance analysis of random access mechanisms for machine type communications in LTE networks," In: *2016 International Conference on Advanced Communication Systems and Information Security (ACOSIS),* Marrakesh, Morocco, pp. 1–6. IEEE. doi:10.1109/ACOSIS.2016.7843934.
69. M. Ouaissa and A. Rhattoy, "New method based on priority of heterogeneous traffic for scheduling techniques in M2M communications over LTE networks," *Int. J. Intell. Eng. Syst.*, vol. 11, no. 6, 2018. doi:10.22266/ijies2018.1231.21.
70. M. Ouaissa, A. Rhattoy, and M. Lahmer, "New method to control congestion for machine to machine applications in long term evolution system," *Int. J. Commun. Antenna Propag.*, vol. 8, no. 4, pp. 355–363, 2018. doi:10.15866/irecap.v8i4.14481.

Fundamentals of 6G in Sustainable Communication

2

Ramandeep Kaur and Upinder Kaur

2.1 INTRODUCTION

Evolution of fifth-generation (5G) to sixth-generation (6G) was introduced to address the drawbacks of 5G technology and support the arising data-intensive, intellectual Systems [1]. 5G has provided brilliant and advanced services to mobile networks and industries. Besides these, 5G is facing challenges in terms of latency, power efficiency, and enormous device connectivity [2]. Integration of wireless communication along with 5G introduced the platform for 6G technology. International Mobile Communication (IMT) is visualizing the involvement of artificial intelligence (AI) and sensing in network technology till 2030 [3]. Rather than technological betterment, the transition from 5G to 6G confronts many challenges, such as designing the network, terahertz integration, and quantum systems [4]. In order to introduce smarter, faster, and greener communication networks, 6G will launch with transformative inventions, an advanced applications scope, and integration with technologies to sustain emerging use cases since 2030 [5]. 6G of wireless networks brings a revolution in wireless communication, such as holographic telepresence, self-sustaining vehicles, and omnipresent sensing. To achieve this sustainability, 6G networks need high data rates, inferior latency, and ultra-reliable and protected communication [6]. Sustainability can be categorized into three scopes: environmental, social, and financial sustainability [7].

In Green Communication, the goal of 6G networks is to attain high communication speed by minimizing the impact on the environment, low energy consumption, and ensuring no contribution to global climate change [8]. 6G sustainability matrices revolve around energy efficiency (EE), and carbon emission (CC). EE defines the effectiveness of a system to deliver data with low energy consumption. As mentioned in equation 2.1, *R* is the data rate and *P* is the energy consumption in watts [9].

$$EE = R/P_{\text{Total}} \tag{2.1}$$

EE: energy efficiency, *R*: data rate, *p*: energy consumption. CO_2 quantity produced by network is define the term carbon emission as denoted in equation 2.2. *n* is carbon emission produced, *P* is total consumption of power, and *T* is the total no of operations [9].

$$C = n * P_{\text{Total}} * C \tag{2.2}$$

DOI: 10.1201/9781003683599-2

C: carbon emission, *n*: carbon emission produced, *P*: energy consumption, *T*: total no of operations.

This chapter furnishes a complete exploration of the building blocks of 6G in the context of sustainable communication. It contains the necessary principles, boosting technologies, design procedures, and challenges that demonstrate the use of 6G networks. This study includes both technological and systemic enablers of sustainability, targeting energy-efficient communication technologies, optimized network architectures, and emerging paradigms that reduce the environmental impact.

2.2 FUNDAMENTAL PRINCIPLES OF 6G COMMUNICATION SYSTEMS

This section defines the fundamental principles of 6G communication system, viz., Ultra-Low Energy Utilization, Extensive connectivity and scalability, AI-driven Refinement, Security, and Flexibility, as mentioned in Figure 2.1.

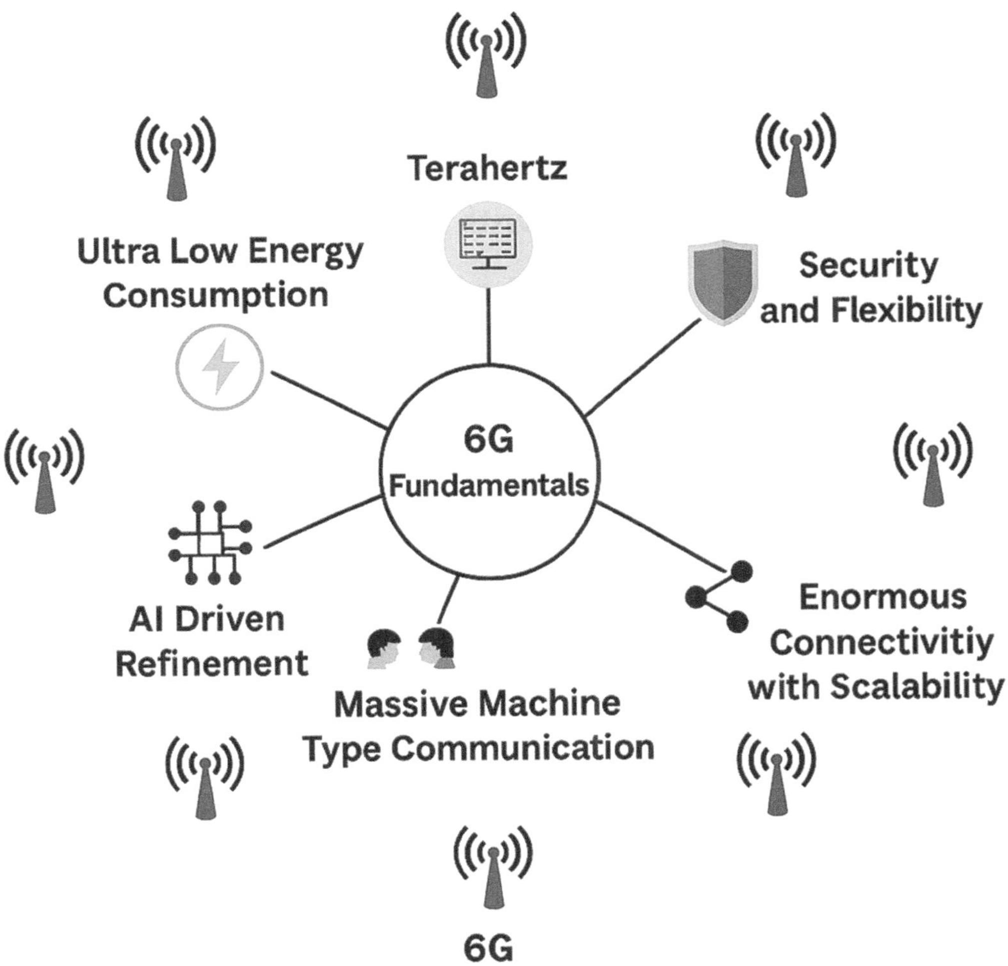

FIGURE 2.1 Fundamental principles of 6G communication systems.

2.2.1 Ultra Low Energy

6G networks aim to achieve EE improvements of 10–100 times over 5G by integrating innovative energy harvesting techniques and dynamic power allocation strategies. High Altitude Platform Stations (HAPS) have shown promise in reducing energy use in radio access networks by providing wide-area coverage with fewer base stations, lowering overall power demands [6]. Ambient sources, viz., thermal, vibration, and RF energy, are also considered to attain sustainable device functioning by offering ultra-low power chip arrangements and zero energy sensors. Collective revolution in expanding these inventions makes sure the future 6G framework reduces carbon emission while holding up enormous connectivity [10]. The ultra-Low energy utilization in 6G network requires strong frameworks. The communication framework is based on rigorous mathematics that balances between transmission power, network architecture, and inference by target energy consumption 1 picojoule per bit, as mentioned in equation 2.3 [11].

$$E_{\text{bit}} \approx 1 \text{ picojoule/bit} \tag{2.3}$$

The second framework, the SINR model based on stochastic geometry, provides core perception in the performance of the system under the practical environment, as mentioned in equation 2.5 [12].

$$\text{SINR} = S/1 + N \tag{2.4}$$

S: required signal power, I: intrusion power from other users, N: noise.

2.2.2 Enormous Connectivity with Scalability

The 6G framework is capable of authorizing 10 million devices per km and also carries heavy ecosystems. 6G is capable of allocating virtual resources dynamically, expanding bandwidth, and enumerating power for various use cases [13]. Edge computing is crucial to reducing latency and distributing computation in ultra-heavy domains [14]. 6G networks connect billions of devices with low latency, with minimum hardware cost and complexity on individual devices, at limitless data access [15]. The expansion in 6G networks does not depend on architectural development but also concerns the interoperability and saving spectrum in the terahertz band, and securing high power consumption [16].

To achieve this scalability, compress sensing for massive random access models is used, which accesses the active devices episodically from a vast pool. It is the pillar for massive machine type communication (mMTC) and allows open access with low complexity. As mentioned in equation 2.5, A is the signature matrix, X is user activity, N is additive white Gaussian noise (AWGN). This approach minimizes the individual scheduling and the grant access overhead [17].

$$Y = AX + N \tag{2.5}$$

where Y: accepted signal matrix, A: signature matrix, X: user activity, N: additive white Gaussian noise matrix.

AI has a drastic contribution to the maximum utilization of network operations in 6G, including traffic routing and reduction in energy consumption. Machine Learning models predict the maximum resource utilization and traffic demands on the basis of history and real-time data [18]. AI improves network flexibility by using prediction algorithms to predict maintenance, problem detection, and outages before they happen [19]. To do the amendments in user demands, AI uses reinforcement learning for real-time adjustments of network chunks [20]. Federated Learning (FL) provides to Internet of Things (IoT) devices, a confidentiality-enhancing distributed model [21]. It has been shown in experimental results that AI can minimize the power consumption by up to 40% without affecting the services [22]. Approximation Message Passing (AMP) is an algorithm used to find out the active devices and signals in vast 6G connectivity. It is helpful to detect the scatter signals, as mentioned in equation 2.6.

$$X^{(t+1)} = \eta\left(X^{(t)} + A^{H} + \left(Y - AX^{(t)}\right)\right) \tag{2.6}$$

where $Y \in C^{M\times T}$ accepted signal matrix, $A \in C^{M\times K}$ is the quantification matrix, $X^{(t)} \in C^{K\times T}$ is the present approximation of the scatter signal, A^H Hermitian transpose A, $\eta(\cdot)$ thresholding function.

2.2.3 Security and Flexibility

Security and flexibility are most important to shield the delicate applications, such as telemedicine and smart grids, from cyber attacks. To maintain long-term security from cyber threats, post-quantum cryptography is being developed. Integration of Zero Trust frameworks and blockchain technology provides a decentralized authentication framework [23]. Communication and sensing collaboratively enable the live monitoring of network connectivity across the tangible and digital realm [24]. The physical layer security mechanism is used to provide secure communication without complex cryptography, as mentioned in equation 2.7 [25].

$$C_s = [C_b - C_e]^+ \tag{2.7}$$

where C_s: confidentiality matrix, C_b: capacity of user 1 channel, C_e: capacity of user 2 channel.

2.3 ENERGY-EFFICIENT 6G FRAMEWORKS

In this section, 6G key energy-efficient frameworks will be discussed, as mentioned in Table 2.1. Visible Light Communication (VLC) has the largest bandwidth, a secure network, and EE over traditional radio frequency communication. Swarm intelligence and genetic algorithms are responsible for the improved functioning of VLC. The basis of this development is to utilize the maximum channel capacity and less noise [21]. Energy harvesting techniques are used in 6G to provide green and better performance in transport systems. The duality theory supports both network throughput and sustainability [22].

In wide and complex networks, Particle Swarm Optimization (PSO) enables 6G to support Signal Annealing (SA) at high speed, maximum accuracy, excellent network performance, and EE. It plays a crucial role in 6G network sustainability [23]. Network slicing makes it possible to assign resources to mobile networks dynamically in different environments. 6G networks require AI-driven frameworks to ensure the sustainability of network slicing [24]. The fundamental requirement of 6G technology is localization. Terahertz ensures the localization in 6G networks. Terahertz also ensures the centimeters- to millimeters-level accuracy with high bandwidth and spectrum resolution [25].

TABLE 2.1 Energy-efficient 6G

REFERENCES	*TECHNIQUE*	*ENERGY-EFFICIENT OFFERING*	*APPLICATION*
[21]	Visible light communication	Largest bandwidth, energy efficiency, security and noise reduction	Indoor location systems
[22]	Energy harvesting	Energy-efficient needs of the network, Improve the throughput of the network, and Enhance the sustainable growth of the network	Intelligent transportation systems
[23]	Particle swarm optimization (PSO)	Increase network performance and energy efficiency in wide environment	Customization energy scatter in wireless network
[24]	Network slicing	Provide on demand energy to network slices	Smart traffic prioritized by network slicing
[25]	Terahertz communication (Thz)	Reduce energy per bit	Wireless indoor access points

TABLE 2.2 Role of AI in 6G sustainability

REF	*TECHNIQUE*	*YEAR*	*CONTRIBUTION*	*CHALLENGE*
[27]	Deep reinforcement learning	2023	Optimal resource allocation	Versatile environment and complex training
[28]	Ant colony optimization	2024	Optimized routing decisions in real-time, Low energy consumption and high data rates	Scalability issues in wide networks
[29]	Deep learning-based uplink integrated sensing and communication (ISAC)	2024	Spectrum and hardware resource sharing	Need of hardware's and complexity of sensing and communication

2.4 ROLE OF AI IN 6G SUSTAINABILITY

AI and 6G technology are bringing drastic change in future data communication and networks with high data rates, accuracy of network, and optimum energy utilization [26]. In this section, Table 2.2 represents the AI techniques in 6G networks.

To improve the EE in Vehicle to Communication (V2c) in 6G network, Deep Reinforcement Learning introduces a framework. This framework uses the Markov Decision Process (MDP) and Proximal Policy Optimization (PPO) algorithms for the resource allocation process and to accept 4 the versatile traffic and communication demands with EE [27]. Ant Colony Optimization (ACO) framework is used in wireless 6G technologies to select the optimum path and efficient energy sustainability. This framework adjusts the network paths as per the real-time conditions and structures. It also ensures the long-term span of the network and throughput in a wide spectrum [28].

Integrating sensing and communication (ISAC) is a deep learning-based framework and an important component of 6G networks. It is responsible for data transmission and communication simultaneously. Neural networks are also used in this framework for complicated mappings. This framework achieves high accuracy over traditional ISAC [29].

2.5 SECURITIES AND RESILIENCE IN 6G

Security and trust are the important and crucial requirements in sustainable 6G. Achieving this on high-demand 6G networks provides security. This section discusses various security mechanisms.

2.5.1 Decentralized Security with Block chain

Decentralized ledger and blockchain ensure the trust and security issues in 6G emerging heterogeneous domains, ensuring the decentralized data sharing between different actors [30, 31]. It ensures the protection of entity management by authorizing in a decentralized ledger [32–34]. It is responsible for the finest resource allocation and network slicing by automatically accepting the contracts without intermediaries [35]. Integration of 6G with blockchain resolves the challenges of latency, EE, and scalability requirements. The future need of 6G with blockchain is to introduce autonomous systems [36]. This collaboration ensures protection from cyber attacks and reduced latency in wide domains [37]. As mentioned in Figure 2.2, the block has a number of blocks, and each block has a header and a Ledger. The header contains the hash of the previous block, Merkle root, timestamp, and nonce. The first block is the Genesis block. Each block is decentralized and is not controlled by central authorization.

2.5.2 Quantum-Safe Cryptography

To protect 6G networks from quantum threats, quantum-safe cryptography plays a crucial role. It is responsible for managing the communication and security [38]. Quantum Key Distribution (QKD) creates secure keys to protect the 6G networks from threats. It makes the 6G networks very complex to break by hackers. The key concept of this technique is cryptography [39]. The practical implementations of quantum-safe cryptography result in the strongest network security in 6G without compromising the performance of the network [40]. As mentioned in Figure 2.3, users are accessing the encrypted signals. Devices or users are not accessing the signals directly from base stations.

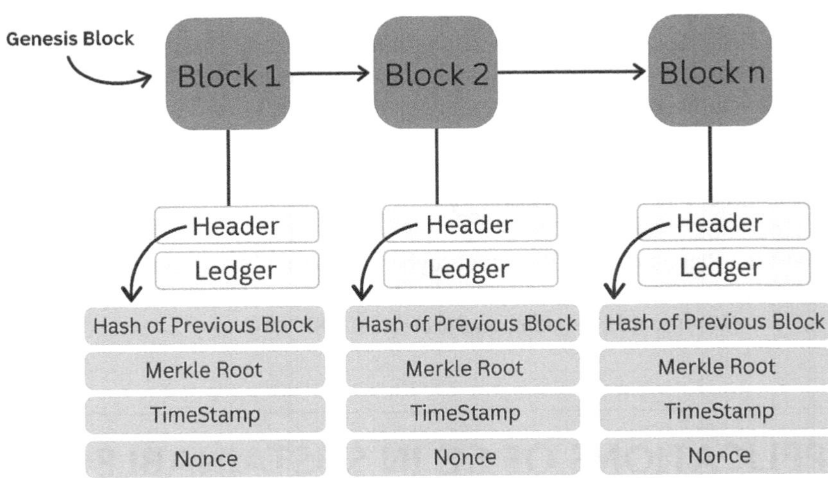

FIGURE 2.2 Decentralized security with blockchain.

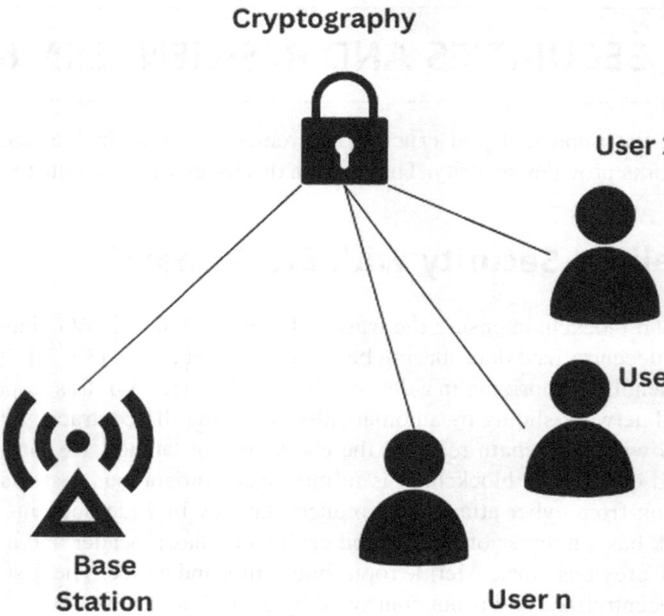

FIGURE 2.3 Quantum-safe cryptography.

2.5.3 Threat Detection with AI

In real-time 6G networks, AI techniques, viz., deep learning, are used to detect cyber threats; the systems have prior knowledge to target the threats. With the increase in connected devices, the 6G networks are able to provide protection and early detection of anomalies [41]. It addresses the key challenges in 6G networks, viz., latency, scalability, and security threats [42]. It introduces robust 6G that automatically detects cyber attacks and handles them in real situations. This system has tremendous results in different spectra and environments. This makes the 6G network strong enough to be easily communicable, reliable, and protected [43]. As mentioned in Figure 2.4, three layers are there. The first layer denotes the expansion of IoT devices such as laptops, mobiles, drones, and automated cars. The second layer is about intrusion attacks, such as unauthorized access and malicious activities. The last and third layer defines the artificial intelligence (AI) algorithms, such as deep learning neural networks, used to detect these attacks priorly.

2.5.4 Definable AI for Security

It introduces robust 6G that automatically detects cyber attacks and handles them in real situations. This system has tremendous results in different spectra and environments. This makes the 6G network strong enough to be easily communicable, reliable, and protected [38].

2.6 APPLICATIONS OF 6G IN SUSTAINABLE SYSTEMS

In this section, various application areas of 6G sustainable systems, viz., Smart Cities, Healthcare, and Environmental Monitoring, are mentioned in Figure 2.5.

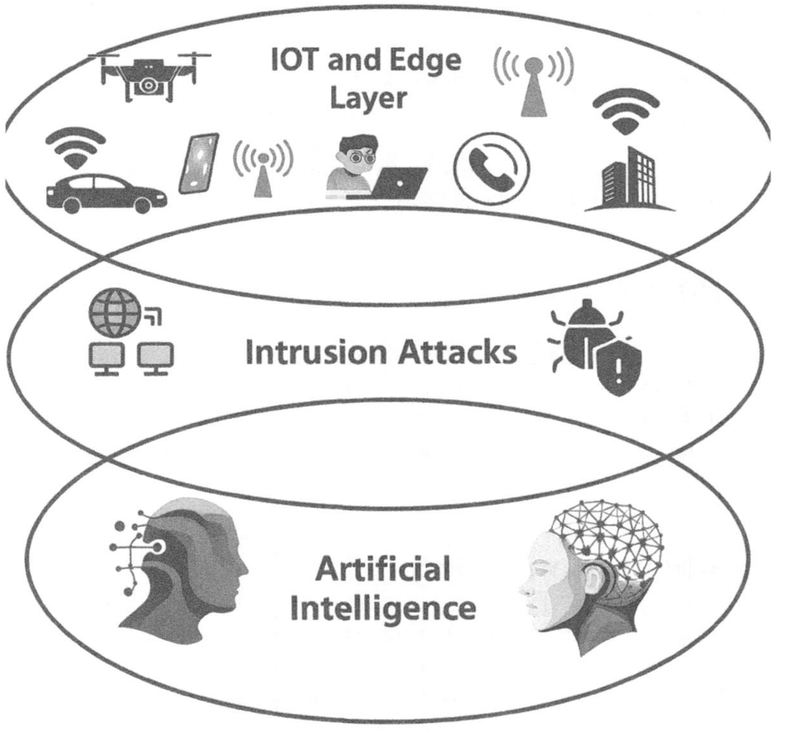

FIGURE 2.4 Threat detection with AI.

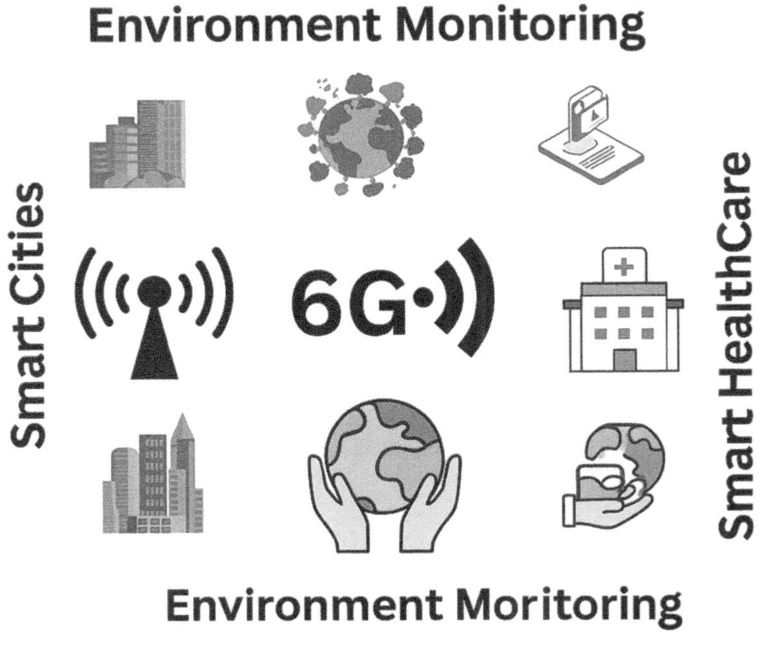

FIGURE 2.5 Applications of 6G in sustainable systems.

2.6.1 Smart Cities

The emerging electronic technologies have introduced the concept of smart cities. IoT is a key component in the next-generation wireless network communication for fast data transmission in smart cities [39]. 6G is a key component in smart grid, smart farming, smart education, smart environment, smart economy, smart healthcare, smart factory, smart transportation, and smart warehouse in smart cities. 6G has brought advancements in areas of sustainability, accuracy, and better performance [40].

2.6.2 Smart Healthcare

In today's era, there is a need to upgrade the healthcare sector, where an automated system can provide patient treatment away from hospitals. 6G makes IoT and AI capable of providing smart healthcare [41]. 6G technology, with its high capabilities, low latency, EE, and high data transfer rates, supports the hospital's connectivity. It promotes the lowest healthcare costs and better patient care through high connectivity among worldwide healthcare providers [42].

2.6.3 Environment Monitoring

Environment monitoring is crucial to watch out for the impact of human activities on our ecosystem. Remote sensing traditional techniques have provided good results; still, there is a need for advanced techniques to get better results. 6G-based sensing techniques are beneficial in environment monitoring, along with efficient resource management decision-making, with the characteristics of low latency, high data rates, and EE [43].

2.7 CHALLENGES AND FUTURE DIRECTIONS

6G technologies provide the benefits in various perspectives; still, some challenges are there as mentioned in Figure 2.6.

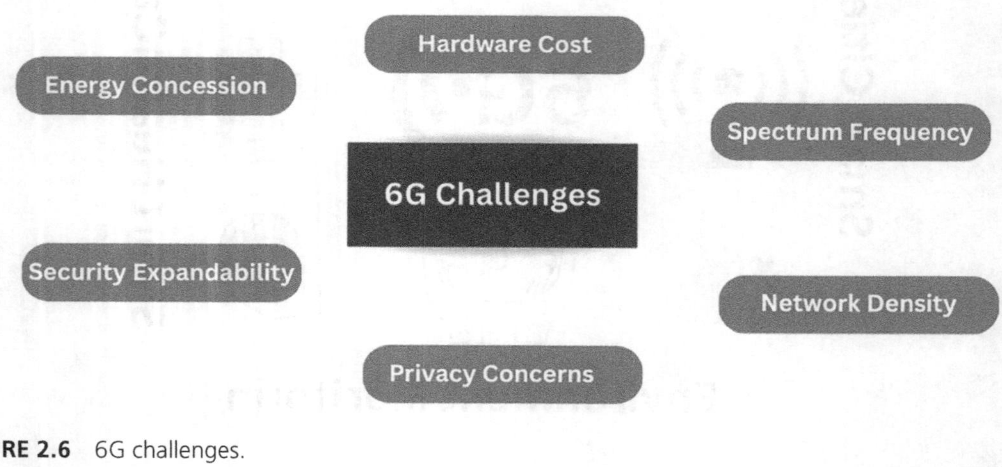

FIGURE 2.6 6G challenges.

2.7.1 Energy Concession

Maintaining low consumption of energy with a large bandwidth is a major challenge for 6G. PSO and Deep Reinforcement Learning have been developed to manage this trade-off. PSO optimization algorithms perform well to maintain this density. The integration of real-time AI, huge data rates, and dense networks consumes a lot of energy [44].

2.7.2 Security Expandability

Quantum-safe cryptography ensures the frequency and computational security in a vast number of IoT devices, which some solutions require. Post-quantum cryptography protocols safeguard the system from quantum attacks with scalability and balanced performance. As the network expands in terms of devices and connectivity, security measures also expand [45].

2.7.3 Spectrum Frequency

Managing the high-frequency spectrum in the terahertz is a challenge without interference and also offers large bandwidth. Dynamic allocation of resources in a wide spectrum is managed by advanced AI techniques without the waste of energy [46].

2.7.4 Network Density

Network density is a major challenge in 6G; network scalability must support the infrastructure without huge energy consumption, which is possible with network slicing and dynamic allocation of resources. Network expandability increases the network density. Managing this density efficiently is a key challenge for 6G [47].

2.7.5 Hardware Cost

Reconstructable, relocatable, and energy-saving hardware design is a key challenge to 6G networks. To maintain the sustainability of the network long time span, hardware components are required for 6G networks [48].

Future research should focus on integrating 6G (sixth-generation wireless networks) with renewable energy sources, advancing AI-driven optimization (the use of artificial intelligence to improve network performance) for ultra-low latency (extremely minimal delay), and developing ethical frameworks (guidelines to ensure responsible use) for secure, sustainable communication [49]. Hybrid classical-quantum optimization algorithms (methods combining traditional and quantum computing to solve complex problems more efficiently) could further enhance 6G's EE, paving the way for a resilient future [50].

2.8 CONCLUSION

6G demonstrates the framework of sustainable communication, AI usability, IoT, and quantum algorithms to develop energy-saving, safe, and resilient networks. 6G addresses the environmental and societal needs of recent connectivity by integrating low-power architectures, resource allocation, and safe and secure protocols.

6G applications lie in smart cities, smart healthcare, and the environmental future. 6G provides strong security mechanisms to make the network resilient and safe from quantum threats. The research goes into the energy concession and security concerns by keeping the ecosystem in mind. This will bring a drastic change in society since 2030 and make the cities smart enough.

REFERENCES

1. Lina Mohjazi, Bassant Selim, Mallik Tatipamula, and Muhammad Ali Imran. The journey toward 6G: A digital and societal revolution in the making. *IEEE Internet of Things Magazine*, 7(2):119–128, 2024.
2. Cheng-Xiang Wang, Xiaohu You, Xiqi Gao, Xiuming Zhu, Zixin Li, Chuan Zhang, Haiming Wang, Yongming Huang, Yunfei Chen, Harald Haas, John Thompson, Erik Larsson, Marco Di Renzo, Wen Tong, Peiying Zhu, Xuemin (Sherman) Shen, Vincent Poor, Lajos Hanz. On the road to 6G: Visions, requirements, key technologies, and testbeds. *IEEE Communications Surveys & Tutorials*, 25(2):905–974, 2023.
3. Rohit Singh, Aryan Kaushik, Wonjae Shin, Marco Di Renzo, Vincenzo Sciancalepore, Doohwan Lee, Hirofumi Sasaki, Arman Shojaeifard, and Octavia A Dobre. Towards 6G evolution: Three enhancements, three innovations, and three major challenges. *IEEE Network*, 2025. doi:10.48550/arXiv.2402.10781.
4. Shaista Farhat, Sumaiya Shaikh, M.A. Jabbar, and Shagufta Farhat. Unveiling challenges in migrating from 5g to 6G: Insights from wireless communication networks. In: *Proceedings of the 2023 International Conference on Wireless Communications, Networking and Applications (WCNA2023), Volume 1361 of Lecture Notes in Electrical Engineering*, Springer, Singapore, pp. 1–13, 2025.
5. Jagadeesha Rajendra Bhat and Salman Ali Alqahtani. 6G ecosystem: Current status and future perspective. *IEEE Access*, 9:43134–43167, 2021.
6. Tailai Song, David Lopez, Michela Meo, Nicola Piovesan, and Daniela Renga. High altitude platform stations: The new network energy efficiency enabler in the 6G era. In: *2024 IEEE Wireless Communications and Networking Conference (WCNC)*, Dubai, United Arab Emirates, pp. 1–6, 2024. IEEE.
7. Rohit Kumar, Saurav Kumar Gupta, Hwang-Cheng Wang, Chandani Shyamala Kumari, and Sai Srinivas Vara Prasad Korlam. From efficiency to sustainability: Exploring the potential of 6G for a greener future. *Sustainability*, 15(23):16387, 2023.
8. Lanlan Li. Research on future 6G green wireless networks. *Green Technologies and Sustainability*, 3(2):100156, 2025.
9. Onel Lopez, Ritesh Kumar Singh, Dinh Thuy Phan-Huy, Efstathios Katranaras, Nafiseh Mazloum, Kalle Ruttik, Riku Jantti, Hamza Khan, Osmel Rosabal, Pavlos Alexias, Prasoon Raghuwanshi, David Ruiz-Guirola, Bikramjit Singh, Andreas Hoglund, Dung Pham Van, Amirhossein Azarbahram, and Jeroen Famaey. Zero-energy devices for 6G: Technical enablers at a glance. *IEEE Internet of Things Magazine*, 8(3):14–22, 2025.
10. Anders Ellersgaard Kalør, Giuseppe Durisi, Sinem Coleri, Stefan Parkvall, Wei Yu, Andreas Mueller, and Petar Popovski. Wireless 6G connectivity for massive number of devices and critical services. In: *Proceedings of the IEEE*, Piscataway, NJ, USA, pp. 1–23, 2024.
11. Saeid Jahandar, Ibraheem Shayea, Emre Gures, Ayman Abdalla El-Saleh, Mustafa Ergen, and Mohammad Alnakhli. Handover decision with multi-access edge computing in 6G networks: A survey. *Results in Engineering*, 25:103934, 2025.
12. Vipul Vekariya and Arvind Kumar Pandey. Non-coherent and backscatter communications: Towards ultra-massive communication in 6G wireless connections. In: *2023 IEEE International Conference on Advanced Intelligent Systems and Communication (AISC)*, Greater Noida, Uttar Pradesh, India, pp. 1–6, 2023.
13. Robin Chataut, Mary Nankya, and Robert George Akl. 6G networks and the ai revolution—exploring technologies, applications, and emerging challenges. *Sensors*, 24(6):1888, 2024.
14. Bushra Sabir, Shuiqiao Yang, David Nguyen, Nan Wu, Alsharif Abuadbba, Hajime Suzuki, Shangqi Lai, Wei Ni, Ding Ming, and Surya Nepal. Systematic literature review of AI-enabled spectrum management in 6G and future networks. *arXiv* preprint, 2024. Available at arXiv:2407.10981.
15. Abiodun Gbenga-Ilori, Agbotiname Lucky Imoize, Kinzah Noor, and Paul Oluwadara Adebolu-Ololade. Artificial intelligence empowering dynamic spectrum access in advanced wireless communications: A comprehensive overview. *AI*, 6(6):126, 2025.
16. Weili Wang, Lun Tang, Tong Liu, Xiaoqiang He, Chengchao Liang, and Qianbin Chen. Toward reliability-enhanced, delay-guaranteed dynamic network slicing: A multi-agent DQN approach with an action space reduction strategy. *IEEE Internet of Things Journal*, 2023. doi:10.1109/JIOT.2023.3323817.

17. Kunaal Shirish Shindagi, Khushi Vikram Koppad, Pooja Rajendra Ekbote, Nagraj Krishna Bhandare, Darshan Digambar Revankar, Prakash Sonwalkar, Bharateesh Fadanis, and Jayashree Shreyas. Federated learning for enhancing cybersecurity in IoT-integrated 6G networks: Challenges, opportunities, and future directions. In: H. L. Gururaj, VinayaKumar Ravi, J. Shreyas, N. Pavithra, R. Sapna (Eds.), *Advanced Sciences and Technologies for Security Applications, volume Part F458 of Lecture Notes in Networks and Systems*, Springer, Singapore, pp. 249–262, 2025.
18. Prabhu Ramamoorthy, Sumaya Sanober, Luca Di Nunzio, and Gian Carlo Cardarilli. Sustainable power consumption for variance-based integration model in cellular 6G-IoT system. *Sustainability*, 15(17):12847, 2023.
19. Abdulrahman K. Alnaim and Ahmed M. Alwakeel. Zero trust strategies for cyber-physical systems in 6g networks. *Mathematics*, 13(7):1108, 2025.
20. Muhammad Ashfaq and Siti Nur. Iot sensor networks—orchestrating connectivity, efficiency, and intelligence across diverse domains. *International Journal of Innovative Research in Computer Science & Technology*, 12(3):154–161, 2024.
21. Martin Dratnal, Lukas Danys, and Radek Martinek. Bio-inspired optimization methods for visible light communication: A comprehensive review. *Artificial Intelligence Review*, 58(8), 2025. doi:10.1007/s10462-025-11251-5.
22. Jie Huang, Tao Yu, Xiaogang Zhu, Fan Yang, Xianzhi Lai, Osama Alfarraj, and Keping Yu. Energy efficiency maximization in uav-assisted intelligent autonomous transport system for 6G networks with energy harvesting. *IEEE Transactions on Intelligent Transportation Systems*, 2024. doi:10.1109/TITS.2024.3445088
23. Atul Agrawal, Pashupati Baniya, Shraiyash Pandey, Bharat Bhushan, and Nitin Rakesh. Particle swarm optimization versus simulated annealing for 5G and 6G wireless networks: A comparative study. In: *2024 27th International Symposium on Wireless Personal Multimedia Communications (WPMC)*, Greater Noida (NCR-Delhi), India, pp. 1–6, 2024. IEEE.
24. Sina Ebrahimi, Faouzi Bouali, and Olivier C. L. Haas. Resource management from single-domain 5G to end-to-end 6G network slicing: A survey. *IEEE Communications Surveys & Tutorials*, 26(4):2836–2866, 2024.
25. Hui Chen, Hadi Sarieddeen, Tarig Ballal, Henk Wymeersch, Mohamed-Slim Alouini, and Tareq Y. Al-Naffouri. A tutorial on terahertz-band localization for 6G communication systems. *IEEE Communications Surveys & Tutorials*, 24(3):1780–1815, 2022.
26. Tharindu Hewa, G¨urkan G¨ur, Anuradha Kalla, Mika Ylianttila, and Madhusanka Liyanage. The role of blockchain in 6G: Challenges, opportunities and research directions. In: *2020 2nd 6G Wireless Summit (6G SUMMIT)*, Levi, Finland, pp. 1–5, 2020. IEEE.
27. Faysal Marzuk, Andres Vejar, and Piotr Cho lda. Deep reinforcement learning for energy-efficient 6G v2x networks. *Electronics*, 14(6):1148, 2025.
28. HS Mohammed, Omer A. Abdulkareem, Amran Ahmad, and Captain Samuel Dowse. Energy-efficient wireless sensor networks using adaptive ant colony optimization and sixth generation (6G) technology. *International Journal of Engineering Trends and Technology*, 72(10):90–95, 2024.
29. Qihui Qi, Xiang Chen, Caijun Zhong, Chau Yuen, and Zhaoyang Zhang. Deep learning-based design of uplink integrated sensing and communication. *IEEE Transactions on Wireless Communications*, 2024. doi:10.1109/TWC.2024.3373797.
30. Abdul Hanan Khan, Nor Ulash Hassan, Chau Yuen, Jie Zhao, Ruicheng Zhang, Dusit Niyato, and Mohamed-Slim Alouini. Blockchain and 6G: The future of secure and ubiquitous communication. *IEEE Wireless Communications*, 28(6):130–137, 2021.
31. Hao Luo, Gang Sun, Jian Wang, Hao Yu, Dusit Niyato, Schahram Dustdar, and Zhu Han. Wireless blockchain meets 6G: The future trustworthy and ubiquitous connectivity. *IEEE Communications Surveys & Tutorials*, 2025. doi:10.1109/COMST.2025.3593918.
32. Yujie Liu, Shuang Peng, Meng Zhang, Shuo Shi, and Jie Fu. Towards secure and efficient integration of blockchain and 6G networks. *Plos ONE*, 19(2):e0266784, 2024.
33. Togu Novriansyah Turnip and Birger Andersen. Towards 6G authentication and key agreement protocol: A survey on hybrid post quantum cryptography. *IEEE Communications Surveys & Tutorials*, 2025. doi:10.1109/COMST.2025.3567439
34. Basma Kamil. Advanced 6G network protection using quantum key distribution: A systematic review. *Babylonian Journal of Networking*, 2025. doi:10.58496/BJN/2025/007.
35. Paul Scalise, Robert Garcia, Matthew Boeding, Michael Hempel, and Hamid Sharif. An applied analysis of securing 5G/6G core networks with post-quantum key encapsulation methods. *Electronics*, 13(5):856, 2024.
36. Enrico Paolini, Luca Valcarenghi, Lorenzo Maggiani, and Nicola Andriolli. Real-time clustering based on deep embeddings for threat detection in 6G networks. *IEEE Access*, 11:123456–123468, 2023.
37. Amir Yazdinejad and Ali Dehghantanha. Symbiotic federated learning for giant AI threat detection in 6G-IoT infrastructures. *IEEE Internet of Things Journal*, 99:1–10, 2025.

38. Bartosz Siniarski, Chandana Sandeepa, Shuang Wang, Madhusanka Liyanage, Cem Ayyildiz, Veli Can Yildirim, Hakan Alakoca, Fatma Güneş Kesik, Betül Güvenç Paltun, Giovanni Perin, Michele Rossi, Stefano Tomasin, Arsenia Chorti, Pietro Giardina, Alberto Pérez García, José María Jorquera Valero, Tommy Svensson, Nikolaos Pappas, and Marios Kountouris. Robust-6G: Smart, automated, and reliable security service platform for 6G. In: *Proceedings of the International Conference on Ubiquitous and Future Networks*, Budapest, Hungary, 2024. IEEE.
39. Sanjeev Sharma, Renu Popli, Sajjan Singh, Gunjan Chhabra, Gurpreet Singh Saini, Maninder Singh, Archana Sandhu, Ashutosh Sharma, and Rajeev Kumar. The role of 6G technologies in advancing smart city applications: Opportunities and challenges. *Sustainability*, 16(16):7039, 2024.
40. Ahmed Al Amin, Junho Hong, Van-Hai Bui, and Wencong Su. Emerging 6G/B6G wireless communication for the power infrastructure in smart cities: Innovations, challenges, and future perspectives. *Algorithms*, 16(10):474, 2023.
41. Edgar Batista, Pablo Lopez-Aguilar, and Agusti Solanas. Smart health in the 6G era: Bringing security to future smart health services. *IEEE Communications Magazine*, 62(6):74–80, 2023.
42. Arun Kumar, Mehedi Masud, Mohammed H Alsharif, Nishant Gaur, and Aziz Nanthaamornphong. Integrating 6G technology in smart hospitals: Challenges and opportunities for enhanced healthcare services. *Frontiers in Medicine*, 12:1534551, 2025.
43. G. Vinuja and Bharatha Devi. Applications of 6G-based remote sensing network in environmental monitoring. In: Narmadha, C., Mahmud, M., Albert, A. S., and Tripathi, S. L. (Eds.), *Development of 6G Networks and Technology*, pp. 145–167, 2024. Springer, New York.
44. Vivek Pathak, R. Chethan, Rahul Jashvantbhai Pandya, Sridhar Iyer, and Vimal Bhatia. Deep learning based energy, spectrum, and SINR-margin tradeoff enabled resource allocation strategies for 6G. *IEEE Access*, 12:74024–74044, 2024.
45. Shiang-Jiun Chen and Yi-Hsueh Tsai. Quantum-safe networks for 6G an integrated survey on PQC, QKD, and satellite QKD with future perspectives. *Computing & AI Connect*, 2(1):1–10, 2025.
46. David Cuellar, Muntadher Sallal, and Christopher Williams. BSM-6G: Blockchain-based dynamic spectrum management for 6G networks: Addressing interoperability and scalability. *IEEE Access*, 12:59643–59664, 2024.
47. Hewon Cho, Sudarshan Mukherjee, Dongsun Kim, Taegyun Noh, and Jemin Lee. Facing to wireless network densification in 6G: Challenges and opportunities. *ICT Express*, 9(3):517–524, 2023.
48. Ertugrul Basar, George C. Alexandropoulos, Yuanwei Liu, Qingqing Wu, Shi Jin, Chau Yuen, Octavia A. Dobre, and Robert Schober. Reconfigurable intelligent surfaces for 6G: Emerging hardware architectures, applications, and open challenges. *IEEE Vehicular Technology Magazine*, 19(3):27–47, 2024.
49. Vivek Pathak, R. Chethan, Rahul Jashvantbhai Pandya, Sridhar Iyer, and Vimal Bhatia. Deep learning based energy, spectrum, and sinr-margin tradeoff enabled resource allocation strategies for 6G. *IEEE Access*, 12:74024–74044, 2024.
50. Shiang-Jiun Chen and Yi-Hsueh Tsai. Quantum-safe networks for 6G an integrated survey on PQC, QKD, and satellite QKD with future perspectives. *Computing & AI Connect*, 2(1):1–10, 2025.

3 Towards Intelligent Connectivity

6G Network Architectures Empowering AIoT Applications

Yassir Moutia and Yassine Eljakani

3.1 INTRODUCTION

The convergence of artificial intelligence (AI) and the Internet of Things (IoT), commonly referred to as AIoT, is revolutionizing the conception and delivery of intelligent services. By interconnecting billions of context-aware devices with advanced AI algorithms, AIoT enables real-time analytics, autonomous decision-making, and personalized user experience. This paradigm shift has been accelerated by the vision of sixth-generation (6G) networks, which extend the capabilities of 5G networks by embedding native AI functions, sustainable edge intelligence, and resilient orchestration across high-density and heterogeneous environments.

Unlike conventional IoT systems, which rely heavily on centralized cloud infrastructure, the 6G-enabled AIoT architectures embrace distributed intelligence. Edge devices now perform data preprocessing, model inference, and collaborative learning, thereby reducing latency, preserving privacy, and optimizing bandwidth utilization. Federated learning (FL) has emerged as the cornerstone of this transformation, enabling distributed AI training without sharing raw data or addressing privacy and security concerns [1]. Despite rapid advances, several challenges have hindered the realization of AIoT in the 6G era. Existing frameworks often lack unified mechanisms for managing the end-to-end quality of service (QoS) when AI workloads coexist with traditional traffic. Furthermore, the orchestration of heterogeneous resources from constrained sensors to powerful edge servers remains fragmented, and end-to-end security across multiple administrative domains remains an open problem. This chapter provides a comprehensive survey of the architectural principles, AI enablers, and network-centric innovations driving AIoT in 6G. Specifically, we:

- The convergence of AI and IoT and the transformative role of 6G networks in enabling advanced AIoT services.
- The vision, requirements, and challenges for deploying scalable and intelligent AIoT solutions over 6G infrastructures.

DOI: 10.1201/9781003683599-3

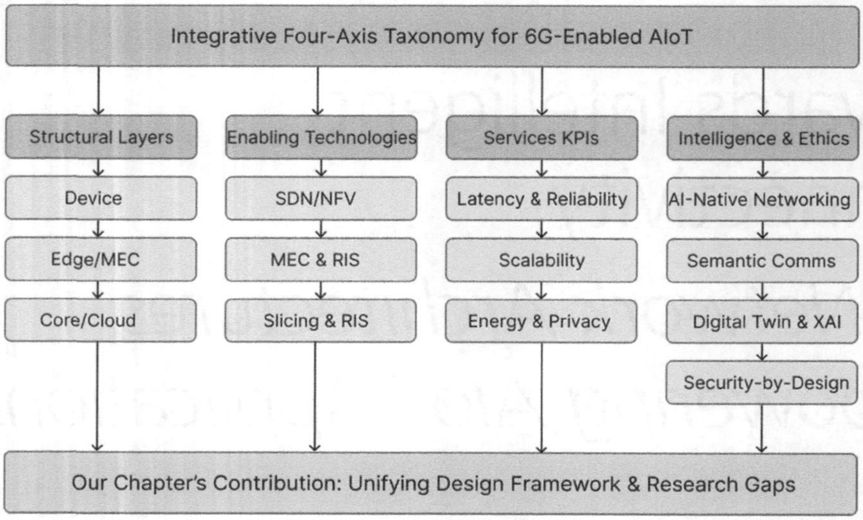

FIGURE 3.1 Integrative four-axis taxonomy for 6G-enabled AIoT.

- The architectural foundations of 6G for AIoT, including distributed hierarchical designs, intelligence-native networking, and enabling technologies.
- The integration of QoS frameworks tailored for dynamic AIoT environments, with a focus on network slicing, adaptive resource management, and multidimensional QoS metrics.
- Emerging open challenges and future research directions shaping the path toward next-generation cognitive AIoT systems.

This survey synthesizes recent advances and identifies key gaps to guide researchers and practitioners in developing scalable, secure, and sustainable AIoT systems for the upcoming 6G era. Figure 3.1 illustrates the four-axis taxonomy introduced in this chapter, with our contribution connecting structural architecture, enabling technologies, multidimensional service Key Performance Indicators (KPIs), and intelligence-ethical considerations.

3.2 AIoT IN THE ERA OF 6G: VISION AND REQUIREMENTS

3.2.1 The Convergence of AI and IoT

The Internet of Things has expanded significantly over the past decade, evolving from isolated sensor-actuator architectures to interconnected ecosystems. However, billions of heterogeneous devices that generate continuous data streams pose challenges to the traditional cloud-centric processing paradigms. These challenges include limitations of real-time responsiveness, network scalability, context awareness, and energy efficiency, which are critical for dynamic applications. AIoT represents the next stage of IoT, enabling on-device learning, autonomous decision-making, and intelligent actuation by integrating AI with IoT systems. This convergence enables systems to sense, transmit, understand, predict, and adapt in real-time. AIoT systems reshape various sectors through context-aware intelligence and adaptive functionality, from predictive maintenance in smart factories to intelligent transportation, precision agriculture, and personalized healthcare.

3.2.2 Why 6G for AIoT?

Although 5G networks have introduced key innovations, including enhanced mobile broadband (eMBB), ultra-reliable low-latency communication (URLLC), and massive machine-type communication (mMTC), they are increasingly inadequate for meeting the expanding demands of AIoT systems. The challenges posed by distributed AI workloads, heterogeneous data modalities, and time-sensitive decision loops require a fundamental rethinking of the operation of future networks.

Beyond the enhanced mobile broadband, URLLC, and mMTC services of 5G, 6G will support new types such as Computation-Oriented Communications (COC). The COC flexibly adjusts operations based on communication resources to achieve a specific computational accuracy, rather than just the classical QoS [2].

6G technology is being developed to address these limitations by not only relying on communication but also integrating intelligence into the network, providing a new intelligent method for managing communication and control, surpassing the previous generation. Additionally, 6G allows the building of a scalable and responsive AIoT infrastructure owing to its ability to emphasize cognitive networking, edge intelligence, and AI-driven orchestration.

3.2.3 Vision for AIoT in 6G

Emerging applications of AIoT, such as autonomous vehicles and real-time diagnostics, require capabilities that surpass those offered by 5G networks. These applications require substantial computational intelligence, ultra-low latency, high reliability, and scalable connectivity. Although 5G has introduced significant advancements, it remains inadequate for addressing the dynamic and multidimensional requirements of next-generation intelligent systems.

In contrast, 6G is conceptualized as a transformative advancement intended to address these limitations by inherently incorporating intelligence into network architectures and closely integrating communication, computing, and control functionalities. Within the 6G framework, AIoT systems are expected to surpass the constraints of current architectures through the implementation of several innovative and synergistic technologies.

- **Edge-Distributed Intelligence and Federated Learning:** 6G networks decentralize AI processing across IoT devices, edge nodes, and core elements, allowing systems to make faster, smarter, and more secure decisions in real-time. FL plays a key role by allowing collaborative model training without sharing raw data, thereby improving the scalability and security of AIoT systems [3].
- **AI-Native and Self-Optimizing Network Architectures:** 6G networks are expected to transition from conventional software-defined paradigms to inherently intelligent infrastructure. These AI-native systems incorporate real-time feedback loops to support autonomous fault detection, control-plane automation, and dynamic optimization with minimal human oversight [3].
- **Predictive and Proactive Network Behavior:** By leveraging continuous data-driven insights, 6G networks can anticipate traffic dynamics, user mobility, and service demand. This supports proactive resource allocation, seamless orchestration, and consistent Quality of Experience (QoE) in dynamic environments.
- **Goal-Oriented and Semantic-Aware Communication:** The 6G network prioritizes transmission based on data relevance and intent rather than raw throughput. Paradigms such as semantic communication, learn-to-optimize, and generative AI reduce the computational overhead while enabling the adaptive control of complex technologies such as ultra-massive multiple-input multiple-output (UM-MIMO) and Reconfigurable Intelligent Surfaces (RIS) [4].

- **Ubiquitous Intelligence Across the Network Stack:** AI capabilities are embedded throughout the 6G architecture, from the physical and MAC layers to service orchestration and user interfaces, ensuring pervasive intelligence that supports autonomous, context-aware AIoT services [3].
- **Digital Twin and Network Slicing:** Digital twin technology facilitates real-time modeling of network and device states, thereby enabling predictive management and efficient resource optimization. Furthermore, network slicing provides the ability to partition network resources dynamically, fulfilling the specific requirements of various AIoT use cases [5].

In conclusion, these innovative paradigms constitute the foundation of the vision of the 6G-enabled AIoT, achieving intelligent, adaptive, and resilient digital systems. This next-generation system is ready to support the implementation of mission-critical and sensitive systems, such as cooperative robotics, autonomous transport systems, and digital twins, by providing unprecedented levels of performance and autonomy.

3.2.4 Key Technical Requirements for AIoT over 6G

As discussed in the previous section, 6G is a promising technology that provides a good infrastructure for implementing AIoT systems; however, the latter has various requirements. Unlike previous generations, which focused on throughput and connectivity, 6G aims to incorporate intelligence across all network layers, from the core infrastructure to the edge devices. This evolution is driven by AIoT systems, which require high-speed data exchange and the ability to sense, analyze, and act autonomously in real-time. 6G must support diverse use cases, with unique demands for responsiveness, scalability, and contextual awareness. Addressing these needs requires the development of new network capabilities, including URLLC, massive-scale connectivity for IoT environments, and distributed intelligence for localized learning. Energy efficiency, data privacy, and real-time adaptability are essential to ensure the sustainability and security of AI-driven ecosystems. Table 3.1 delineates the technical criteria required to realize AIoT in a 6G-enabled landscape.

Meeting these requirements requires cross-layer innovation in communication protocols, edge computing infrastructure, AI models, and security frameworks. Ultimately, the success of AIoT in the 6G era hinges on the co-design of intelligent networks and decentralized AI systems underpinned by an architecture that is both adaptive and purpose-driven.

TABLE 3.1 Key technical characteristics for AIoT over 6G Networks

REQUIREMENT	*DESCRIPTION*
Latency	Sub-millisecond end-to-end delay to support closed-loop control systems such as robotics and telesurgery.
Reliability	Ultra-reliable communication with availability exceeding 99.999% for safety-critical operations.
Scalability	Seamless support for up to 10^7 devices per km^2 with diverse capabilities, data rates, and energy profiles.
Edge intelligence	Real-time AI inference and federated learning at the edge to reduce central processing bottlenecks.
Energy efficiency	Power-optimized protocols and computations to support battery-constrained or energy-harvesting devices.
Privacy & security	Integrated mechanisms for secure data aggregation, federated learning, and privacy-preserving AI.
Context-awareness	Dynamic adaptation of behavior, QoS, and resource allocation based on environmental and operational context.

3.3 ARCHITECTURAL FOUNDATIONS OF 6G FOR AIoT

The previous sections discussed the integration of AIoT into 6G networks and the requirements for application use cases. However, the ambitious vision of the AIoT in the 6G era necessitates the fundamental architectural transformation of next-generation communication networks. Rather than relying on incremental advancements over 5G, 6G introduced a paradigm shift that tightly integrated AI capabilities into every layer of the network stack. This integration aims to support large-scale intelligent IoT ecosystems capable of autonomous decision-making, adaptive behavior, and real-time responsiveness across diverse and dynamic environments.

3.3.1 Distributed, Hierarchical Architecture

To fully harness the potential of AIoT in the 6G era, it is imperative for the network to adopt a distributed hierarchical architecture that enables scalable intelligent deployment and efficient resource utilization. This multilayered design decentralizes the processing and learning tasks across three primary tiers:

- **Device Layer:** This layer consists of end devices, such as sensors, actuators, wearables, and smart appliances, equipped with embedded AI capabilities and lightweight inference engines. Although constrained by power, memory, and computational capacity, these devices can perform local data filtering, anomaly detection, and context-aware preprocessing, thereby reducing the burden on the upstream layers and enhancing privacy.
- **Edge Layer (Multi-Access Edge Computing, MEC):** Located near data sources, such as base stations, gateways, or access points, the edge layer serves as a critical intermediary for low-latency services. It enables localized model training via FL, performs real-time analytics, and coordinates collaborative intelligence between edge nodes and devices. This proximity allows for faster responses and alleviates the load on central servers, making it ideal for mission-critical and delay-sensitive AIoT applications.
- **Core/Cloud Layer:** Core networks and cloud platforms form the backbone of global intelligence orchestration. These centralized resources support highly complex tasks, such as large-scale model training, cross-domain knowledge fusion, long-term storage, and system-wide optimization. Although latency is less critical at this level, the cloud plays a vital role in aggregating insights, disseminating updated AI models, and ensuring coherence across distributed AI systems.

This hierarchical AI-native architecture offers several benefits, such as minimizing end-to-end latency, improving system scalability, enhancing resilience through local autonomy, and supporting privacy-preserving mechanisms by limiting exposure to raw data. More importantly, it enables adaptive and context-aware service provisioning, where decisions can be made locally or globally, depending on the application's urgency, sensitivity, and complexity. This design is essential to meet the stringent performance, security, and intelligence requirements of next-generation AIoT systems.

3.3.2 Intelligence-Native Networking

AI has shown good results for many applications, such as precision agriculture and autonomous vehicles. However, the application of intelligence to the network itself has not yet been developed, and 5G has not yet been customized to support integrated intelligence in its infrastructure. Thus, the arrival of the new sixth-generation (6G) brings new capabilities to proceed with an intelligence-native approach. The application of AI to 6G networks may yield good results for needs such as the optimization of network resources, dynamic spectrum allocation, and predictive maintenance. Moreover, because AIoT applications are dynamic and evolving, the native integration of AI allows the network to learn and evolve over time. Key components include:

- **AI for Network Optimization:** AI models are integrated into the core network for different goals, such as traffic prediction, congestion control, and dynamic spectrum allocation. Thus, the network can adapt to the dynamic and heavy loads of the AIoT systems.
- **Physical Layer Intelligence:** The 6G network may benefit from the integration of AI, especially in physical technologies such as sub-terahertz frequencies and advanced radio sensing. This results in ultra-low latency, high reliability, and precise synchronization, which are essential for real-time AIoT interactions [6].
- **AI-Driven Resource Allocation:** Computing, bandwidth, and storage resources in real-time can be orchestrated via AI networking and slicing. Using this technology, AI brings the edge closer to distributed learning and decision-making at the edge of a network. This implies that the network benefits from latency, scalability, efficiency, and privacy [7].
- **Semantic Communication:** Only meaningful and contextually relevant information is transmitted, which significantly reduces data redundancy and latency. This is particularly beneficial for intent-aware communication between AIoT devices and services [8].
- **Digital Twin Integration:** Digital Twin Integration combines virtual models of physical systems and network components intended to monitor and optimize resources in a predictive manner. With intelligent networks, digital twin capabilities will introduce customized and optimal handling of various AIoT use cases [9].

These intelligence-native networks are designed to be self-learning, self-healing, and context-aware, thereby forming the foundation for scalable and autonomous AIoT deployment. They rely on cloud-native infrastructure, adaptive architecture, and secure AI-enhanced communication protocols to create a hyperconnected and intelligent IoT ecosystem. As illustrated in Figure 3.2, the proposed 6G architecture adopts a hierarchical structure comprising the device, edge, and core layers to enable efficient AIoT deployment.

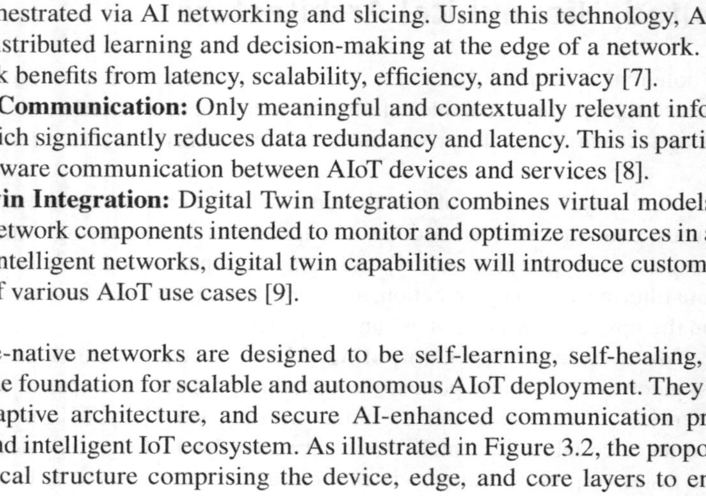

FIGURE 3.2 Distributed, hierarchical architecture for 6G-AIoT.

3.3.3 Enabling Technologies at the Network and Edge Layers

To realize the vision of a 6G-powered AIoT, a comprehensive suite of enabling technologies is essential, particularly at the network and edge layers of the AIoT architecture. These technologies are crucial for achieving the low-latency, high-throughput, and context-aware capabilities required for intelligent, distributed IoT applications. They enhance the flexibility, scalability, and responsiveness of the AIoT infrastructure while facilitating autonomous network operations.

At the network level, the concept of Software-Defined Networking (SDN) introduces centralized programmability that allows dynamic configuration and effective orchestration of heterogeneous AIoT traffic. By separating the control and data planes, SDN renders networks adept at responding to ever-changing operating conditions in real-time, catering to various QoS requirements. In this regard, network functions virtualization (NFV) complements SDN by decoupling network services from proprietary hardware, thus enabling rapid deployment, cost savings, and scaling-up of operations in AIoT settings [10].

MEC distributes all cloud potentialities to the edge so that activities can be processed and determined locally with minimal delay. Applications that fall under this category include autonomous vehicles, augmented reality, and industrial automation, all of which are time-sensitive [11]. FL is another enabling technology that allows AI models to be trained in a decentralized manner on devices or edge nodes with the benefits of conserving data privacy and lessening communication overhead.

Moreover, Network Slicing offers opportunities for creating many virtualized network segments according to the needs of different AIoT services from mission-critical control to large sensor deployments. This flexibility in architecture guarantees that the optimization of performance and resource allocation can be achieved within disparate service environments.

Contrary to the reactive scheme of 5G URLLC, extended ultra-reliable low-latency communications (xURLLC) relies on a predictive scheme backed by recent developments in machine learning (ML). Their prediction is related to major parameters in the network, such as channel states, traffic dynamics, and system performance indicators [12]. Although 5G URLLC is extremely reliable, it sacrifices spectral efficiency and is based on static channel conditions, making it incompletely reliable in dynamic, non-stationary environments. In contrast, xURLLC takes a proactive approach, allowing more scalable and adaptable communications to fit the evolution required by the ever-changing AIoT.

Simultaneously, MEC brings computation, storage, and networking capabilities closer to the user by embedding IT functionalities within the radio access network. The effective orchestration of these edge resources requires innovative mechanisms for mobility management, dynamic service placement, and seamless API exposure to third-party applications [13]. Complementing this, Self-Organizing Networks leverage AI-driven algorithms to automate configuration, fault detection, and performance optimization, ensuring resilient and context-aware network operations in real-time.

3.3.4 Design Principles for AIoT-Aware 6G Architecture

To design a 6G network that effectively addresses the requirements of AIoT, it is imperative to transcend traditional paradigms and develop architectures that are intelligent, adaptable, and ethically conscious. These systems must concurrently manage the limitations of edge devices, ensure real-time responsiveness, and satisfy the increasing demands for reliable and scalable operations. In this section, we introduce the principles of the architecture.

The AIoT-aware 6G framework binds heterogeneous sensing, ubiquitous high-bandwidth/low-latency communication, and distributed AI computing solutions. It optimizes communication, computing, and storage across devices, edges, and cloud data centers [2]. This development gives rise to several service types, such as COC, which prioritize the accuracy of computation in consideration of the available communication resources rather than adhering to the conventional QoS provision [2].

Unlike the previous generations of networks, which expected 6G networks to take a hands-on, comprehensive approach, everything from network management and operations to the provision of intelligent vertical services will be open to AI adoption [3]. The designs must be guided by the "AI native" philosophy to enable the network to be intelligent, agile, learning, and evolving concurrently with rapid changes to the dynamics of the system. AI is considered essential for intelligent learning, reasoning, and decision-making in 6G wireless networks. This transformation aims to achieve the notion of "connected intelligence," going beyond just "connected things" [2].

As new radio access technologies, such as millimeter-wave and terahertz (THz) bands and resource-constrained IoT devices, have become more ubiquitous, hardware limitations will accelerate further in the design of 6G networks [3]. This compels emerging technologies to adopt a "hardware-aware communication" paradigm involving hardware-algorithm co-design in performance-critical scenarios and application-aware communications [2,14].

Scalability and interoperability are prime factors for consideration in the 6G world, which is meant to connect a multitude of intelligent devices called the Internet of Everything (IoE), with a high connection density exceeding 10^7 devices per square kilometer [3]. Furthermore, to implement advanced technologies, such as ultra-massive multiple-input multiple-output (UM-MIMO) and RIS, many antenna elements must be employed, thus presenting an additional challenge to the scalability of AI-based algorithms [4]. Moreover, there is a problem of interoperability across different protocols, which is another significant challenge for the interlinking of 6G with different network systems [15].

Taken together, these principles outline a 6G architecture for AIoT that is agile, intelligent, adaptive, and sustainable, characterized by the autonomous orchestration of resources, establishment of trust, and generation of real-time intelligence across devices, edge nodes, and core networks.

3.4 QUALITY OF SERVICE IN AIoT: A NEW GENERATION OF DEMANDS

As AIoT systems become increasingly integrated into sectors such as healthcare, industry, and mobility, the demand for dynamic and intelligent QoS has become more critical than ever. In contrast to traditional networks that rely on static parameters, 6G technology must enable adaptive and predictive QoS that corresponds to the dynamic behaviors of intelligent applications and services. This section examines how 6G redefines QoS in the AIoT era.

3.4.1 Dynamic QoS Requirements in AIoT Systems

AIoT applications span various domains, including smart cities, precision agriculture, industrial automation, autonomous vehicles, and remote healthcare. Each domain has unique QoS requirements that change over time, with operational context, impact, user behavior, and device mobility. Such built-in dynamism calls for a flexible QoS framework that monitors, predicts, and adapts in almost real-time.

As listed in Table 3.2, each application area imposes specific QoS requirements. In such cases, the QoS must be context-aware and time-adaptive. For instance, with normal readings on a wearable device, the system may accept packet loss, whereas the actual transmission of information upon the detection of abnormal readings must be given immediate priority. Capabilities such as edge computing, FL, and semantic communication in 6G can satisfy finer and situationally dependent QoS expectations.

TABLE 3.2 QoS priorities for representative AIoT use cases

USE CASE	*QoS PRIORITY*
Autonomous vehicles	Ultra-low latency (<1 ms) and high reliability.
Industrial automation	Deterministic delay, minimal jitter, and time-synchronized operations.
Remote healthcare (e.g., telesurgery)	Ultra-high reliability, guaranteed bandwidth, and low packet loss.
Smart homes & wearables	Low energy consumption with periodic connectivity.
Drone swarms/robotics	Real-time feedback, low distributed-control latency, and coordinated mission execution.

3.4.2 Multidimensional QoS Metrics (Latency, Energy, Reliability,...)

6G networks are being transformed significantly toward totally intelligent networks and accommodating various smart services and new solutions using AI and ML [3]. As the advancement of AI-powered IoT devices proliferates alongside other well-connected devices continually seeking the prospect of real-time, data-hungry applications, QoS must be implemented across metrics such as latency, dependability, energy efficiency, throughput, scalability, privacy, security, and resource management. Hence, it is transitioning from a reactive threshold-based model toward a predictive goal-oriented model, which requires not only a complete architectural overhaul but also algorithmic structures, all of which are only possible with intelligent 6G [16].

In the aforementioned evolution, which will be introduced by 6G networks, this latter permits the realization of ultra-low end-to-end latency, targeting values as low as 0.1 ms, with processing delays reduced to under 10 ns. This represents a substantial leap from the 1ms target of 5G and is essential for latency-sensitive AIoT applications such as autonomous vehicles, tactile Internet, and real-time industrial automation [3]. It concentrates on edge intelligence, which reduces dependence on the centralized cloud infrastructure by performing computational tasks near data sources and thus considerably reduces the response time [17].

In terms of reliability, 6G aspires to achieve "seven nines" (99.99999%) or even "nine nines" (99.9999999%) availability, with proposals extending reliability guarantees to probabilities as low as 10^{-5} to 10^{-7} [12]. The International Telecommunication Union (ITU) defines reliability as the probability of successfully transmitting a predefined amount of data within a given time frame. To satisfy such stringent reliability benchmarks, 6G leverages advanced techniques, including network slicing for extremely ultra-reliable low-latency communications and redundancy mechanisms, to maintain service continuity even in the face of device or link failures [18].

Energy efficiency remains a critical concern, particularly given the increase in high-frequency bands (e.g., mmWave and THz) and the training demands of large ML models. AIoT systems must minimize the energy consumption per data packet processed, transmitted, or received across both device and edge layers [2]. However, resource-constrained edge devices face a substantial energy burden, particularly during the training of the FL models. In response, techniques such as TinyML, intelligent task offloading, and dynamic load balancing have been adopted to reduce computational overhead while maintaining QoS standards.

However, security in AIoT systems encompasses the confidentiality, integrity, and availability of data and control flows. Although centralized learning frameworks heighten privacy risks by requiring raw data transmission, even decentralized paradigms such as FL are not immune to threats. Frequent exchanges of updated model parameters can expose systems to privacy vulnerabilities. Attacks may target the data itself through tampering or poisoning, or they may target learning models through extraction, inversion, or adversarial manipulation [3]. To address these multifaceted risks, a combination of edge-based security mechanisms, privacy-preserving FL, blockchain-enabled authentication, and data integrity frameworks has been proposed. These measures collectively reinforce the resilience of AI-driven 6G networks to emerging security and privacy challenges [19].

Nonetheless, achieving these multidimensional QoS objectives simultaneously is challenging due to trade-offs among competing requirements. For instance, increasing the throughput via higher-frequency bands may limit coverage and affect reliability, owing to the higher susceptibility to signal blockage. Similarly, reducing latency through edge computing may hinder scalability and resource optimization in ultra-reliable applications [4]. Overcoming these tensions requires the integration of enabling technologies such as UM-MIMO, RIS, Non-Terrestrial Networks (NTN), AI-based orchestration, quantum computing, and Network Digital Twins (NDT) [4].

3.4.3 QoS-aware Network Slicing for Contextual Adaptation

6G networks are expected to transform wireless communication by adopting a comprehensive approach that intelligently supports a broad spectrum of smart services and innovative applications in the near future. This vision is dependent on advanced capabilities such as QoS-aware network slicing for contextual adaptation in view of the unprecedented complexity and heterogeneity of future networks [20]. Network slicing has come to be viewed as a basic feature of 6G architectures, which allow partitioning of shared physical infrastructures into multiple, virtually and independently managed logical networks. Each slice is specifically customized for the salient requirements of an AIoT service class, so that differentiated service delivery can be entertained across a diversely demanding application.

Network slicing, which is a fundamental capability of 5G and subsequent technologies, involves the creation of multiple virtual networks or slices on a single physical infrastructure. Each slice consists of a tailored set of resources, including memory, computation, and network capacities, along with others, such as virtual network functions, to be specifically differentiated to support particular services. These applications will also bring new challenges for the management and orchestration of these slices in 6G with respect to new use cases, making it necessary to adopt mechanisms that can manage new trade-offs for throughput, reliability, latency, and possibly privacy. The aim is to achieve automated and flexible management of resource allocation among various operational slices [3].

To address these challenges, researchers have also proposed resource allocation algorithms for slice management, such as Algorithm Selection (AS), Deep Reinforcement Learning (DRL), and Squatting and Kicking Strategies (SKM) [21], so that scarce resources such as bandwidth, energy, and storage can be utilized efficiently without sacrificing the various QoS requirements needed in applications, while also balancing the cost, latency, and reliability. According to recent studies in references [22,23], more emphasis has been placed on the combination of AI-based orchestration and network slicing for future 6G networks, along with their expected contributions to ultra-reliable and context-aware connectivity in highly dense ecosystems of AIoT. Hence, QoS-aware slicing is not merely a feature of the technology; rather, it is a strategic enabler to reach intelligent, context-sensitive, and mission-critical AIoT services at scale. This combination of AI-driven orchestration, slicing, and context-aware resource allocation shapes the evolution of 6G networks, preparing them to meet the diverse and complex requirements of future AIoT applications.

3.5 OPEN CHALLENGES AND FUTURE DIRECTIONS

The more 6G networks incorporate AIoT technologies, the greater the opportunities and challenges for future research. Advanced FL techniques designed to handle heterogeneous data and unreliable networks are vital enablers for collaborative artificial intelligence, AI model training, and distributed IoT devices, while remaining closely aligned with the principles of data privacy and network resilience. This approach allows improved use of the common intelligence of ceiling devices without transferring sensitive data; therefore, it has a significant impact on the design of algorithms to handle conditions created by statistical heterogeneity and communication constraints.

Optimizing AI for hardware-constrained IoT devices is a significant research trajectory. As IoT deployments proliferate, there is an increasing demand for lightweight AI models and efficient inference methodologies that can function within the restricted computational, memory, and energy capacities of the edge devices. Furthermore, AI-driven autonomous network orchestration and self-optimization are imperative for managing the complexities of 6G AIoT ecosystems, facilitating dynamic resource allocation, intelligent routing, and adaptive network configurations to enhance performance and efficiency. As AI becomes deeply embedded in network operations, it is also important to build transparency and trustworthiness in decision-making. Indeed, Deep Learning (DL)-based AI models have frequently been criticized for their limited explainability and inherent black-box nature, which undermines trust and limits their adoption by stakeholders. Therefore, there is a strong need to implement eXplainable AI (XAI) [3]. In addition to AI-based orchestration, the concept of Network Digital Twins provides common opportunities for improving network management and optimization [4].

Nonetheless, the full realization of its vision of universally distributed intelligence depends on overcoming considerable technical challenges. Interdisciplinary collaboration among AI, networking, security, and domain experts is crucial. As research progresses, we can expect AIoT capabilities to gradually expand, enabling increasingly autonomous and intelligent systems that seamlessly blend physical and digital worlds. However, careful consideration of the ethical and societal implications should guide technological evolution.

3.6 CONCLUSION

This chapter explored the convergence of AI and IoT within the context of the emerging 6G landscape, highlighting the architectural, technical, and operational transformations required to enable intelligent, autonomous, and scalable AIoT services. By incorporating distributed intelligence, edge computing, semantic communication, and FL, 6G represents a significant advancement toward fully cognitive networks. As 6G evolves, its success will hinge on the development of systems that are not only high-performing and adaptive but also secure, trustworthy, and resource-aware, ultimately facilitating the seamless integration of intelligence into every layer of the digital ecosystem.

REFERENCES

1. F. R. Mughal, J. He, B. Das, F. A. Dharejo, N. Zhu, S. B. Khan, and S. Alzahrani, "Adaptive Federated Learning for Resource-Constrained IoT Devices through Edge Intelligence and Multi-Edge Clustering," *Sci. Rep.*, vol. 14, no. 1, 2024. doi:10.1038/s41598-024-78239-z.
2. K. B. Letaief, W. Chen, Y. Shi, J. Zhang, and Y.-J. A. Zhang, "The Roadmap to 6G: AI Empowered Wireless Networks," *IEEE Commun. Mag.*, vol. 57, no. 8, pp. 84–90, 2019. doi:10.1109/MCOM.2019.1900271.
3. H. Hafi, B. Brik, P. A. Frangoudis, A. Ksentini, and M. Bagaa, "Split Federated Learning for 6G Enabled-Networks: Requirements, Challenges, and Future Directions," *IEEE Access*, vol. 12, pp. 9890–9930, 2024. doi:10.1109/ACCESS.2024.3351600.
4. A. A. Shamsabadi, A. Yadav, Y. Gadallah, and H. Yanikomeroglu, "Exploring the 6G Potentials: Immersive, Hyperreliable, and Low-Latency Communication," *IEEE Veh. Technol. Mag.*, vol. 20, no. 1, pp. 74–82, 2025. doi:10.1109/MVT.2025.3531346.

5. G. Liu, J. Sun, J. Deng, N. Li, Y. Huang, and Y. Wang, "The SOLIDS 6G Mobile Network Architecture: Driving Forces, Features, and Functional Topology," *Engineering*, vol. 8, pp. 42–59, 2022. doi:10.1016/j.eng.2021.07.013.
6. V. Ziegler, V. Raisanen, H. Flinck, K. Hatonen, M. Hoffmann, and H. Viswanathan, "6G Architecture to Connect the Worlds," *IEEE Access*, vol. 8, pp. 173508–173520, 2020. doi:10.1109/access.2020.3025032.
7. H. Yang, J. Zhao, Z. Xiong, D. Niyato, A. Alphones, and K. Wu, "Artificial-Intelligence-Enabled Intelligent 6G Networks," *IEEE Netw.*, vol. 34, no. 6, pp. 272–280, 2019. doi:10.1109/mnet.011.2000195.
8. W. Yang, H. Du, Z. Qin Liew, W.Y. B. Lim, Z. Xiong, D. Niyato, X. Chi, X. S. Shen, and C. Miao, "Semantic Communications for Future Internet: Fundamentals, Applications, and Challenges," *IEEE Commun. Surv. Amp Tutor.*, vol. 25, no. 1, pp. 213–250, 2023. doi:10.1109/comst.2022.3223224.
9. X. Shen, J. Gao, W. Wu, C. Zhou, M. Li, and W. Zhuang, "Holistic Network Virtualization and Pervasive Network Intelligence for 6G," *IEEE Commun. Surv. Amp Tutor.*, vol. 24, no. 1, pp. 1–30, 2022. doi:10.1109/comst.2021.3135829.
10. F. Paolucci, T. Osinski, F. Cugini, and P. Castoldi, "Enhancing 5G SDN/NFV Edge with P4 Data Plane Programmability," *IEEE Netw.*, vol. 35, no. 3, pp. 154–160, 2021. doi:10.1109/mnet.021.1900599.
11. S. Abimannan, E.-S.M. El-Alfy, S. Hussain, Y.-S. Chang, S. Shukla, D. Satheesh, and J. G. Breslin, "Towards Federated Learning and Multi-Access Edge Computing for Air Quality Monitoring: Literature Review and Assessment," *Sustainability*, vol. 15, no. 18, p. 13951, 2023. doi:10.3390/su151813951.
12. J. Park, S. Samarakoon, H. Shiri, M. K. Abdel-Aziz, T. Nishio, A. Elgabli, and M. Bennis, "Extreme URLLC: Vision, Challenges, and Key Enablers," *arXiv preprint* arXiv:2001.09683, 2020. doi:10.48550/ARXIV.2001.09683.
13. H. Tabatabaee Malazi, S. Rasool Chaudhry, A. Kazmi, A. Palade, C. Cabrera, G. White, and S. Clarke, "Dynamic Service Placement in Multi-Access Edge Computing: A Systematic Literature Review," *IEEE Access*, vol. 10, pp. 32639–32688, 2022. doi:10.1109/access.2022.3160738.
14. L. Tan, X. Cheng, G. Srivastava, F. Ming, and K. Yu, "Secure and Resilient Artificial Intelligence of Things: A HoneyNet Approach for Threat Detection and Situational Awareness," *IEEE Consum. Electron. Mag.*, vol. 11, no. 3, pp. 69–78, 2021. doi:10.1109/mce.2021.3081874.
15. A. Mukherjee, P. Mukherjee, D. De, and N. Dey, "QoS-aware 6G-enabled ultra low latency edge-assisted Internet of Drone Things for real-time stride analysis," *Comput. Electr. Eng.*, vol. 95, p. 107438, 2021. doi:10.1016/j.compeleceng.2021.107438.
16. C. Wang, X. Yu, W. Wang, X. Cheng, L. Xu, and F. Jiang, "QoS-Aware Offloading based on Communication-Computation Resource Coordination for 6G Edge Intelligence," *China Commun.*, vol. 20, no. 3, pp. 236–251, 2023. doi:10.23919/jcc.2023.03.017.
17. K. B. Letaief, Y. Shi, J. Lu, and J. Lu, "Edge Artificial Intelligence for 6G: Vision, Enabling Technologies, and Applications," *IEEE J. Sel. Areas Commun.*, vol. 40, no. 1, pp. 5–36, 2022. doi:10.1109/JSAC.2021.3126076.
18. F. Rezazadeh, C. Verikoukis, L. Alonso, and H. Chergui, "SliceOps: Explainable MLOps for Streamlined Automation-Native 6G Networks," *IEEE Wirel. Commun.*, vol. 31, no. 5, pp. 224–230, 2024. doi:10.1109/mwc.007.2300144.
19. S. W. Turner, E. Guler, S. Uludag, and M. Karakus, "A Promising Integration of SDN and Blockchain for IoT Networks: A Survey," *IEEE Access*, vol. 11, pp. 29800–29822, 2023. doi:10.1109/access.2023.3260777.
20. C. Sergiou, M. Lestas, P. Antoniou, C. Liaskos, and A. Pitsillides, "Complex Systems: A Communication Networks Perspective towards 6G," *IEEE Access*, vol. 8, pp. 89007–89030, 2020. doi:10.1109/ACCESS.2020.2993527.
21. V. R. K, R. B, R. Changala, A. S. S. T, P. K. Kalangi, and B. M, "Optimizing 6G Network Slicing with the EvoNetSlice Model for Dynamic Resource Allocation and Real-Time QoS Management," *Int. Res. J. Multidiscip. Technovation*, pp. 325–340, 2024. doi:10.54392/irjmt24324.
22. W. Wu, C. Zhou, M. Li, H. Wu, H. Zhou, N. Zhang, X. Shen, and W. Zhuang, "AI-Native Network Slicing for 6G Networks," *IEEE Wirel. Commun.*, vol. 29, no. 1, pp. 96–103. 2022. doi:10.1109/mwc.001.2100338.
23. S.-P. Yeh, S. Bhattacharya, H. Moustafa, and R. Sharma, "Deep Learning for Intelligent and Automated Network Slicing in 5G Open RAN (ORAN) Deployment," *IEEE Open J. Commun. Soc.*, vol. 5, pp. 64–70. 2024. doi:10.1109/ojcoms.2023.3337854.

Regulating the Speed of Innovation

A Legal and Ethical Framework for 6G Deployment in Smart Societies

4

Shashank Solanki, Vijeta Verma, and Vijaishree Dubey Pandey

4.1 INTRODUCTION

4.1.1 Background

The sixth generation of wireless communication (6G) is a revolution in the development of digital infrastructure, which is expected to be implemented in the whole world by 2030. This time, however, unlike in the case of its predecessors, 6G is not only going to be simply an acceleration of 5G or a decrease in latency rates; this is going to be a fundamental rearrangement of how the human being interacts with its environs, the devices, and even its own cognitive cities. The main 6G promises are holographic telepresence, immersive experience powered by AI (AI-immersivity), and nano-level machine-to-machine (M2M) networks, which all will work with near-reality. All this is likely to support the smart society architecture, i.e., the architecture of smart cities, digital healthcare, and intelligent transport systems, and of real-time governance [1].

But as hyperconnected as people are, so too is the increase of legal and ethical issues. With the ever-growing infiltration of the 6G technologies into the fields where human autonomy, body integrity, and information self-determination are at stake, the requirement for a legal framework that can withstand the weight of the technological change becomes imminent and absolutely essential. This issue is added to

DOI: 10.1201/9781003683599-4

the fact that history continuously demonstrates the trend of technological innovation continually outpacing legal control. As an example, the proliferation of cloud computing, social media websites, and data analytics in the early 2010s ultimately led to regulatory reactions such as the European Union's General Data Protection Regulation in 2018, over 9 years after such technologies had already become well-established socio-economic entities [2]. The same delay in the regulation of 6G may mean irreversible losses such as widespread surveillance, algorithmic biases, and uncontrolled use of personal data by states or companies.

Against this backdrop, the chapter is, thus, able to interact with the central question that becomes: How should the speed of innovation be pre-emptively governed by legal systems so that the development of 6G technologies advances within the spectrum governed by ethics and rights?

4.1.2 Scope and Objectives

This chapter will seek to conduct a cross-legal inquiry into the regulatory framework needed to regulate 6G-enabled ecosystems, especially protecting human rights, data protection, and accountability and regulatory control of algorithms in smart societies. Their legal area of activity, as covered in this chapter, encompasses the spheres of the public and by now the private law of so-called emerging technologies with the focus on:

- What regulatory frameworks their AI-incorporated XR/VR applications might be subject to, both in existing and upcoming regulatory regimes;
- The privacy and data management problems bred with real-time biometric tracking and nanonetworks;
- The ethics of artificial intelligence and digital profiling in communication environments of augmented reality;
- The contribution of sovereignty and cross-border legal enforcement regarding 6G networks regulation.

The main aim is to suggest a legal and ethical framework that is not an inhibition of innovation but rather uses it in a responsible manner, where technological advances have distinct standards of law, procedural protection, and rights enforcers. The chapter does not support hard anti-innovation regulation but a precautionary, proportional, and rights-based approach to governance that meets the ultra-fast, transnational, and automated reality of the 6G technologies [3].

4.1.3 Methodology

In dealing with the foregoing purposes, the chapter employs the approach of the comparative legal method, carrying out in-depth analysis of the existing and proposed regulatory frameworks in three major jurisdictions, i.e., in the United Kingdom, Canada, and Australia.

- The European Union, whose risk-based platform is reflected in the Digital Services Act (DSA) and the initiated Artificial Intelligence Act (AI Act);
- The United States, where soft-law measures, such as the AI Bill of Rights and the regulatory activity of the Federal Communications Commission (FCC), demonstrate an approach of market orientation and rights sensitivity at the same time;
- India, having recently adopted the Digital Personal Data Protection Act (DPDPA, 2023), is now engaged in discussions about digital sovereignty and the automation of decision-making.

On a more broadly legal basis, this chapter uses the method of legal reasoning grounded in cases to consider the practical implications of 6G technologies, in the context of two opposite national evidentials [4]:

- Estonia, considered by many as the country of democratic digital government and, specifically, a user of AI and XR technologies in government;
- China, in which the infrastructure has been installed in the context of the surveillance-based governance system, and there are worsening concerns regarding algorithmic discrimination, mass data gathering, and adherence to human rights.

Among others, the chapter also makes use of the analysis of international legal materials, among which:

- Budapest Convention on Cybercrime, covering the transnational cybercrimes;
- The UN Human Rights Committee General Comment No. 25 (2021), which offers an authoritative interpretation of the right to privacy in the digital realm;
- OECD and ITU policy papers and recommendations on the topic of emerging technologies and management of these across borders.

This triangulated approach that runs through the doctrinal, comparative, and case-based approach allows for a theoretical sufficiency and practical relevance of the analysis [5].

4.1.4 Relevance to Book Theme

The theme of this chapter, called out by the edited volume, Landscaping 6G: Unlocking the Power of Ultra-Fast Communication, is intrinsically that of the act of unlocking, which is here considered in a critical legal-ethical sense. Although the book is probably going to contain a myriad of contributions focusing on the technical, engineering, and design of 6G, this chapter should be understood as a polemic against it, as 6G needs to be unlocked in the context of civil liberties, procedural fairness, and transnational justice being jeopardized.

By so doing, the chapter highlights the regulatory input of states, international organizations, and private companies when it comes to having a future that encompasses 6G technologies, besides being fast and efficient; they must be legal, ethically responsible, and socially fair. Locating legal governance as a co-equal actor to the discourse of 6G—not necessarily a corrective one—the chapter will shift the legal community narrative from a focus on reactive litigation to proactive norm setting [6].

4.2 LEGAL-ETHICAL CHALLENGES IN THE 6G ECOSYSTEM

The advent of the deployment of 6G communication systems to the world forms a turning point in the governance of digital societies. In contrast to the previous generations of mobile connectivity, such as 5G, 6G is likely to support nearly immediate delivery of enormous volumes of data, a mix of real-time AI, and smooth functionality among virtual, physical, and organic spaces. Although seemingly progressive in an industrial sense, these changes raise exigent and unprecedented legal and ethical considerations on issues of preservation of individual rights, the transparency of autonomous systems, and the extent of surveillance technologies that are acceptable. This section discusses the main legal-ethical dilemmas 6G infrastructure poses with regard to three interrelated aspects: privacy and AI decision-making, the empirical spaces and physio-emotional integrity, and growing menace of surveillance capitalism and totalitarian technologies.

4.2.1 Data Privacy and AI Decision-Making in 6G Environments

Real-time data generation omnipresent in the 6G ecosystem represents a major complication to current data privacy frameworks. 6G will allow a continuous feedback loop of biometric, geospatial, behavioral, and cognitive data via advanced XR wearables, smart glasses, nano-scale biosensors, and devices embedded in or on the human body. This data ecology will not only blur the distinction between online and offline life, but also transform the very idea of personal data beyond the common identifiers into areas that concern thought patterns, affective states, and neurobiological responses [7].

One of the most serious legal issues occurs concerning AI-assisted profiling and decision. According to the existing international and national data protection framework, including, but not limited to, the General Data Protection Regulation (GDPR) in use in the European Union, when such automated profiling has a major impact on individuals, strict controls are involved, including the right not to be affected by purely automatic decisions and the existence of relevant human oversight. At 6G, all of these protections are becoming more challenging to operationalize, however. The intention of AI systems within 6G networks is that they can act independently, without requiring command of any human input, and process their results and forecasts in real-time with constant data flow. The scope of this process is large and fast, questioning the possibility of informed consent and substantial control.

Also, the envisaged application of nanorobotics and bio-implanted devices to monitor health or augment mental capabilities brings out a special type of data that would be obtained inside the human body itself. This makes a new point of contact between the data protection law and the right to the bodily integrity [8]. The legal regimes have failed to identify whether and how any consent to internal, non-homeostatic data collection can be obtained and whether users can reasonably revoke such consent after embedding systems have been obtained. This has led to calls for the need for an ex ante legal measure—one of which is Technology Impact Assessment (TIA), that would attempt to judge the legality, proportionality, and ethical considerations of data practices enabled by AI before their implementation and not after.

4.2.2 Immersive Environments, Consent, and Bodily Integrity

The 6G infrastructure will support the propagation of immersive environments fusing AI, haptic technologies, augmented reality (AR), and virtual reality (VR), and resultantly create alternate realities that cannot be distinguished in relation to the physical world. The application of such technologies will cut across education, mental healthcare, industrial training and tourism among other fields. Such experiences have the potential to be revolutionary, but they implicate the very constitutional rights from which they are offered: the right to privacy, the right to autonomy, and the right to bodily integrity and psychological integrity. The principle of informed consent, which is the foundation of privacy law and medical ethics, is susceptible to substantial undermining when immersive experiences are involved. At its conventional manifestations in the digital realm, the interacting users communicate via discrete interfaces with choices they can make with clarity. In contrast, immersive 6G environments are continuous and perceptively responsive, and without a point of entry or exit. The users might also not know what they are doing, that they might be observed, and that the posture they are taking or the tone of their voice is being recorded and analyzed in order to make the experience sharper. The question remains of finding legal formats that could capture consent meaningfully as it occurs, and not turn it into an empty, automated, box-ticking exercise in a changing simulation [9].

Furthermore, the introduction of AI-powered emotional computing, i.e., when systems recognize emotions of users and react in real-time, introduces the specter of manipulation and behavioral nudging. Despite the fact that these practices are common in commercial promotion and political slogans, the use of such strategies in the context of the 6G-enabled environment provides a whole new dimension of involvement and engagement [10]. This can violate the autonomy of users, and this effect could be more pronounced when it is used secretly or massively. There is concern by legal scholars that such practice may constitute emotional manipulation, particularly when used on vulnerable groups, e.g. children, old age people, and cognitively impaired persons.

Legal protection must also extend to the legalization, as a protection of bodily integrity, of digitally mirrored bodies, e.g., in the form of avatars or neural connections. Global law on human rights, such as Article 8 of the European Convention on Human Rights and Article 17 of the International Covenant on Civil and Political Rights, exists to guarantee against arbitrary interference with private and family life, home, and correspondence of people. Such tools might require reassessment to cover the virtual invasions such as the violation and abuse of the digital bodies in immersive virtual realities. Noticeable legislative innovations, such as the amendments to the so-called "neuro-rights" in Chile, are valuable initial steps toward a paradigm of cognitive liberty and cognitive privacy when applied to the digital world [11].

4.2.3 Surveillance Capitalism and Authoritarian Technology Governance

Arguably, the most politically acute scenario posed by 6G is that it will further feed into surveillance capitalism and consolidate authoritarian patterns of government. Surveillance capitalism is the economic system when information about the user is collected, turned into commodities and sold to determine and influence behavior. Under 6G conditions, such a model becomes many times more intrusive because companies will have access not only to user preferences and habits but also to their physiological and emotional states due to biometric sensing and XR technologies.

Its legal consequences are grim. In the absence of highly regulated responses, it is possible that 6G infrastructure power ends up being exploited by the private sector to build opaque systems of behavioral prediction and control, enforcing power asymmetries and restricting democratic accountability [12]. A legal rights-based response must, as such, take into account not only personal harms but also systemic risks to civic participation, information autonomy, and digital self-determination. At the same time, 6G can enable authoritarian governments to establish the politically controlled digital surveillance systems in the name of the safety of the people or social stability. One bright example is the Chinese social credit system, whereby data gathered by various sources, including financial, behavioral, and political aspects, are used to rank the citizens with an actual impact on real life. The combination of such programs with real-time AI analytics and ubiquitous sensors is alarming as far as international human rights law is concerned, especially in terms of proportionality, due process, and non-discrimination.

The normative baseline offered by the Budapest Convention on Cybercrime and the General Comment No. 25 (2021) on the right to privacy in the digital age by the United Nations is relevant, but the enforcement of these standards is not uniform throughout the juridical systems. In addition, the instruments possess low applicability in national security exception or in cases where there is transnational activity by a private actor that any regulator lacks jurisdiction to control [13]. To implement such a change, it is not enough to provide technical measures of protection, structural legal changes are needed: algorithmic audits should become compulsory, court oversight of AI-maintained law enforcement tools/AI-based governance systems like an automated drug trial process (as well as their source code), and recognition of digital rights as having standing in the domestic state constitution or international legal treaties should be founded. The hasty introduction of 6G without such legal architecture could end up enhancing the deterioration of liberal democratic values in both the public and the privatized space [14]. The described stratified interaction between developing 6G technologies and the related legal-ethical control can be illustrated as in Figure 4.1.

Figure 4.1 demonstrates the stratified interaction between the developing 6G technologies and law-ethical control. The technological stack allows capturing data in real-time, autonomous processing of AI, and an engaging experience of users. Concerning this, legal systems have developed in response in order to protect privacy, control AI, and hold systems accountable in the process. Below, the two layers are underpinning principles of ethics, that is, human dignity, bodily integrity, and justice, which can drive the designing of regulations and make responsible innovation possible.

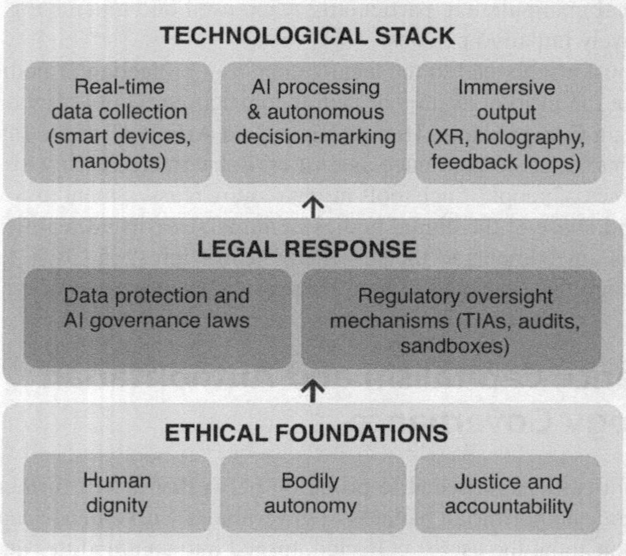

FIGURE 4.1 The Legal-Technology interface in 6G.

4.3 COMPARATIVE LEGAL REVIEW: REGULATORY FOUNDATIONS ACROSS JURISDICTIONS

With the appearance of 6G technology, emerging legal and ethical issues come to the fore that require active regulatory interaction. That combination of super-speedy connections, real-time information exchange, AI, and extended reality (XR) spaces requires the rethinking of the current regulatory paradigm. This section presents the comparative legal study of the readiness of the legal systems of three leading jurisdictions—the European Union, the United States, and India—to deal with the intricacies related to 6G deployment. Using some of the foundational instruments, including data protection laws, laws concerning AI specifically, and digital governmental administration policies, it is possible to derive strengths and regulatory gaps that might affect the governance of 6G technologies over the next decade in this section of this chapter.

4.3.1 The European Union: A Risk-Based Model Anchored in Fundamental Rights

The European Union (EU) has become a world leader in shaping a normative legal framework that guides the regulation of new technologies. At the heart of its strategy is the inculcation of human rights protections on the technological environment with a risk-based legal framework. The current proposals of the DSA and the Artificial Intelligence Act (AI Act), combined with the already in place GDPR, are the central legal support of the 6G infrastructure and applications [15].

The Digital Services Act (2022) does imposes a horizontal regulatory framework upon digital services, especially very large online platforms (VLOPs) and systemic risks arising through algorithmic content distribution, disinformation, and non-transparent recommenders. DSA mandates that platforms to be assessed on the risk on a yearly basis and mitigation against the spread of illegal content and any negative impact on the basic rights. The focus of algorithmic transparency, independent audit, and control systems that the DSA outlines will be of particular relevance in a 6G environment where live data processing and XR interface can potentially dominate social and economic interactions [16]. It creates a precedent on how to control a 6G service provider and platform operator including the ones that incorporate AI decision-making in immersive and customizable spaces.

In addition to this, there is the proposed AI Act that takes a tiered risk-classification approach to AI systems which involve unacceptable risk (prohibited), minimal risk (self-regulated), among others. Such systems are characterized by high risks, i.e. biometric identification, AI in critical infrastructure, and emotion recognition technologies: they should meet the high requirements including data quality requirements, human surveillance, and post-market surveillance. The high-risk category of applications is likely to include in a 6G world many drones powered by AI, XR-based learning tools, or healthcare involving nanobots, among others, which will need conformity assessments and enforced documentation. Thus, the AI Act would expect a form of regulation that would correspond to the complexity and invasiveness of 6G technologies [17].

Although the GDPR was adopted and established prior to the 6G era, it is a technology-neutral document, and the legislation still offers potent data protection principles that can be applied to 6G settings. The most important of them are data minimization, purpose limitation and lawfulness of processing and will be challenged in 6G ecosystems that are characterized by pervading surveillance, and real-time collection of behavioral data and cross-border data flows. As an example, the wearable devices powered by 6G networks can collect sensitive personal data during persistent use through biometric data, location data, etc. This aspect needs a clear consent mechanism and stricter data protection by design provided by Article 25 GDPR. Besides, data subjects rights need easy and non-discriminatory operationalization in super-connected spaces, including the right to access, rectify, erasure and portability.

Overall, the legal framework of the EU can be described as highly normative in nature, being founded on the essential rights focus, risk-based regulation, and institutional responsibility [18]. This makes the EU an example in terms of the regulation of the ethical application of 6G technologies. Nevertheless, with advancing technology, there is still a necessity to adjust such instruments to concentrate on the peculiarities of ultra-fast, real-time, and immersive communication networks.

4.3.2 The United States: Sectoral Fragmentation and Soft Law Challenges

The United States has a divided regulatory environment as compared to the EU, which utilizes a more coordinated strategy of regulatory policy. Although the U.S. has excelled in terms of AI innovations and telecommunications infrastructure, research in terms of preparing its regulation in terms of 6G is fairly undeveloped in regard to a rights-based governance approach.

Smartphone blueprints. The Blueprint of an AI Bill of Rights issued by the White House Office of Science and Technology Policy in 2022 sets out five core principles, namely safe and effective systems, protection against algorithmic bias, data privacy, explanations and transparency, and human alternatives to executed automation. Whereas it comes in handy as a moral guide, the document is not binding, and it lacks the force of law. Consequently, there is no legal requirement that binds the private players to comply, and the enforcement agencies do not have the authority to punish non-compliance. Since the 6G environment promises to be a world full of autonomous systems and algorithmic decision-making that might become practically integrated in everyday services such as predictive policing or automated healthcare diagnostics, such a lack of enforceable legal regulations can make people susceptible to exploitation and systemic prejudice [19].

The FCC will have the responsibility of administering the electromagnetic spectrum and spreading broadband services at the federal regulatory tier. Since 6G technology will be in need of spectrum greater in frequencies than those previously used such as sub-terahertz and visible light, the involvement of FCC in permitting and governing spectrum becomes extremely vital. Nevertheless, the FCC jurisdiction is mostly confined to technical requirements and infrastructure roll out and does not concern itself with the privacy, algorithm accountability, and human rights consequences of top-level services using the 6G network infrastructure. To make matters worse, there is a dearth in the full-scale federal law dealing with data protection. There are also state-level measures provided by the existing laws, like the California Consumer Privacy Act (CCPA) and the California Privacy Rights Act (CPRA), which have little to no coverage, and their application varies between jurisdictions. This decentralized solution poses the risk of compromising coherence and predictability needed to comply with the laws within cross-border jurisdictional areas in a 6G world that is characterized by universally enabled connectivity and data flows. Moreover, it imposes too much burden on users and organizations to traverse the spectrum of data protection, and thus causes ambiguity in the implementation and realization of the rights [20].

Thus, including regarding the occurrence and forms of innovation in wireless technologies, the United States could be considered as a country that still does not quite provide a coherent system of laws that are responsive to the ethical issues of 6G ecosystems. The presence of a federal AI statute or consolidated statute of data protection can assist the U.S. in ensuring that the rights of individuals and safety of the citizens are acceptable to a majority of people, despite the steady rise in collision and compromise involving next-generation communication infrastructures.

4.3.3 India: Asserting Digital Sovereignty through Emerging Legal Instruments

Regulatory approach to digital technologies has gained substantial traction in India in recent years, not only because of the sheer size of the digital population but also due to the ambitions that India had to turn into a global technology hub. The issuance of Digital Personal Data Protection Act (DPDPA), 2023 marks a turning point in the nationwide move to control the data ecology by balancing it on the point of the worldwide standards of privacy [21].

DPDPA provides a consent-driven regime of lawful processing of personal information, which is bolstered by the access, correction, and redressal of grievances. It provides the establishment of the Data Protection Board of India, a quasi-judicial body in charge of supervising compliance, adjudication of disputes. Nevertheless, the Act too has broad exemptions given to the central government to the extent of Clause 17(2) which grants the state a right to process personal data without consent in relation to matters of national security, law enforcement and maintenance of public order. Such exemptions pose constitutional questions in a setting of 6G implementation when a great deal more monitoring capability is available in real time. Lack of substantial procedural protections and due process of law could lead to arbitrary interference with the right to privacy, which has been listed as a fundamental right by the Supreme Court in Justice K.S. Puttaswamy v. Union of India (2017).

It is India, shifting its position on digital sovereignty, that adds to the content of the regulatory methodology. The proposed Digital India Act and the National Data Governance Framework Policy, as well as the prohibition of the cross-border transfer of data, indicate a transition to a centralized manner of operation of digital infrastructure and data flows. Although this can provide strategic benefits of cybersecurity and national scope of innovation, it can potentially divide the global internet and threaten interoperability—one of the essential characteristics of the 6G systems across borders and infrastructures. India is not a country that has come forward with a specific framework of regulation of AI. Although the policy paper on responsible AI by the NITI Aayog is laudable in stating the principles of ethics, this policy paper does not oblige anyone. Without the possibility of legal rules on algorithmic transparency, bias prevention, and accountability, one can expect that AI-enhanced 6G applications, including predictive diagnostics, surveillance, or automated education systems, will be implemented without the establishment of appropriate counterbalancing and controls [22].

TABLE 4.1 Comparative table—6G-relevant legal provisions (EU Vs US Vs India)

AREA OF CONCERN	*EU (DSA, AI ACT)*	*US (AI BoR, FCC)*	*INDIA (DPDPA, 2023)*
AI profiling	Risk-based+DPIA	Ethical guidelines, non-binding	Consent-based, limited TIA
Data sovereignty	Data localization debate	Cross-border transfers allowed	Stronger localization proposed
Algorithmic opacity	Transparency by design	Algorithmic impact statements	No clear framework yet

Wrapping up, the Indian legislative trend, as discussed, is rather a juggling act of shoving in a lot of innovation, inviolability of personal data, and barrel-shoving its way in regulatory control. Although the DPDPA forms the basis of a more rights-oriented approach to data protection, it requires additional legal advancement due to the absence of inclusive legislation regarding AI, as well as the permissive attitude to state surveillance. As India considers itself a central actor in the eyes of the global 6G network, it is likely that any regulation option in the future will greatly affect not only the rights protection of the local citizenry but also collaboration abroad.

A comparative overview of these three jurisdictions is summarized in Table 4.1.

4.4 INTERNATIONAL FRAMEWORKS AND CROSS-BORDER DATA LAW

The implementation of 6G infrastructure in the smart societies will demand not only national control but also the harmonized international legal response. The legal imperative to regulate sovereignty, jurisdiction, interoperability, and protection of human rights at the global level emerges, as the use of ultra-fast communications systems connecting data flows, immersive technologies, and AI systems in real-time defies national borders. This section reviews some of the major international legal regimes and collaborative networks that 6G governance may possibly access with reference, by way of example, to privacy, cybercrime, data mobility, and institutional harmonization.

4.4.1 UN & International Human Rights Law

The discussion on human rights law at international law also plays a pivotal situation in the digital landscape to be created by 6G. Among the more foundational tools, General Comment No. 25 (2021) of the UN Human Rights Committee provides an official interpretative treatise to Article 17 of the International Covenant on Civil and Political Rights (ICCPR), namely in the context of the digital age. The Comment reaffirms that the right to privacy covers all the scope of digital communications and new technologies, as well as new possibilities such as AI, identification of the human biometric system, and immersive spaces. It warns the states that any surveillance, gathering of data or use of AI should not violate the principles of legality, necessity and proportionality [23].

The importance of General Comment No. 25 to the 6G is the fact that it demands that states adopt proper legislative frameworks to ensure that players in the private sector, which either store or commercialize personal data, are regulated. The demand of transparency and redress mechanisms, as urged in the Comment, will thus offer cross-border governance a normative basis given that 6G networks will be more privately sector-oriented, and AI systems will be black-box tasks, more often than not.

As regards cybercrime and transborder enforcement, the most extensive treaty dealing with crimes committed through the internet and other computer networks is Budapest Convention on Cybercrime (2001), which is a treaty within the Council of Europe. It will provide compatible criminal policy on unauthorized access offenses, data interference offenses, and offenses involving the misuse of devices, which will increase with the growth of immersive and autonomous systems that will trade on the 6G network. Notably, the Convention is also the first to include provisions on international cooperation mechanisms, the mutual legal assistance, and those on transnational investigatory powers, which are relevant when it comes to the enforcement of law in a hyperconnected 6G global environment in which the borders may not be technically obsolete; however, they are legally relevant. Nevertheless, it is notable that a number of key jurisdictions, such as India and most of the African states, do not participate in the Budapest Convention due to limitations regarding sovereignty issues and insularity in the drafting thereof. This is an issue for a genuine worldwide cyber-legal regime, and one of the reasons why there is greater urgency to either universalize accession or a more representative successor treaty that regulates 6G specific risks [24].

4.4.2 G7/G20 and ITU Recommendations on 6G Governance

On the policy level, multilateral forums, such as G7, G20, and the International Telecommunication Union (ITU) have made the first steps in developing the global consensus over next-generation communication technologies. The G7 Digital and Technology Ministers Declaration (2023) reinforced the vision of the promotion of human-centric AI and digital infrastructure and urged frameworks that support security, openness, and interoperability in new technologies, such as 6G. The ITU, a specialized agency of the United Nations has been engaged in technical and policy design of 6G, in particular through the ITU-T Focus Group on IMT-2030. The group suggests that there should be the formulation of international standards and implementation of privacy-by-design, trustworthy AI practices, and spectrum harmonization. The recommendations also present a significant degree of normative power since they serve to persuade member states to embrace the pre-regulatory impact assessment and coordinate at broader levels internationally before 6G implantation becomes too serious [25].

Nonetheless, as good as these recommendations are they are not enforceable since the document is non-binding. They heavily rely on free will and political commitment of nations, at the same time, and this may easily be subjected to strong fluctuations among jurisdictions. There are weak binding commitments at multilateral level that could be concerning for fragmented execution and establishment of regulatory arbitrage regions in the world in digital economy.

4.4.3 The Need for Harmonization

Considering the transnational nature of 6G ecosystems where AI, cloud-edge hybrid computing, real-time XR environments, and autonomous data exchange are present, the need to harmonize the law ceases to be an option and becomes a necessity. The non-harmonized land mine of data protection regimes, surveillance frameworks, and AI regulatory frameworks across jurisdictions impose substantial compliance, jurisdictional uncertainties, and rights gaps on users who perform their actions within 6G networks globally. The cross-border attribution of legal liability to the immersive interactions is one of the most significant issues. Consider, for example, that a person is a victim of a rights-violating XR simulation that was launched by a server in another jurisdiction; under which law should we catalog the activity, who is responsible—the platform, developer, or the network provider—and what court could provide relief, jurisdictionally? Any existing jurisdictional frameworks within the framework of the private international law and mutual legal assistance treaties are not well suited to respond to such cross-jurisdictional, rapid, and AI-mediated interactions.

On the same note, transnational data flows that are part and parcel of 6G operation continue to be subjected to varying regimes. Each of the EU GDPR, the United States industry-specific approach, and new provisions in India in the sphere of data localization would provide the organizations with diverse compliance requirements [26]. Such variance threatens to split asunder the worldwide digital ecosystem unless complemented by cross-boundary legitimate tools, including customary contractual conditions, adequacy arrangements, or multilateral conventions providing the execution of enforceable commitments. Work is

TABLE 4.2 Key international legal instruments relevant to 6G deployment

INSTRUMENT/FRAMEWORK	*DOMAIN OF REGULATION*	*RELEVANCE TO 6G DEPLOYMENT*
UN General Comment No. 25 (2021)	Right to privacy in digital age (ICCPR)	Provides normative standards on privacy, AI, surveillance in tech.
Budapest Convention on Cybercrime (2001)	Cybercrime and transnational enforcement	Facilitates international cooperation on cyber offenses.
G7 Digital & Technology Declaration (2023)	Human-centric digital governance	Encourages democratic digital governance and AI responsibility.
ITU-T Focus Group on IMT-2030	Technical and policy recommendations for 6G	Promotes privacy-by-design, spectrum harmonization, and safety.
OECD Privacy Guidelines	Cross-border data flows	Basis for interoperable data governance models.

thus needed that coalesces around designing a truly global normative framework around 6G deployment, maybe through the United Nations or an emerging multilateral treaty organization. It must establish baseline legal protection of users, clear jurisdiction over cross-border harms, impact assessments, and hold real-time AI and XR systems to account operating with 6G networks.

The key international legal instruments that guide 6G deployment are presented in Table 4.2.

4.5 CASE STUDIES: THE GOVERNANCE CHALLENGE IN ACTION

As the 6G technologies become the center of the next smart societies, the type of governance that country-states will have implemented will have a significant impact on legal, ethical, and societal implications of hyperconnected infrastructures. This part discusses two very different country experiences Estonia and China, which have both embraced progressive digital systems but whose legal philosophies, rights-based approaches and governance systems differ greatly. These case studies can be used to illustrate why a principled international governance framework that has the potential to ensure 6G is operationalized in a rights-respecting, rule of law, and democratic manner is required forthwith [27].

4.5.1 Case Study: Estonia—Rule-Based e-Governance and Transparent Automation

Estonia is currently considered to be a worldwide leader in terms of the digitally-enabled rule-of-law-based governance. Estonia has, since the early 2000s, invested in highly integrated e-governance, with secure digital identity (e-ID) solution, interoperable databases, and algorithm-based delivery of government services. The government has also implemented an advanced use of new technologies like AI, XR, and blockchain into its administrative system to the point where over 99% of services offered by the government can be accessed online as of 2025. The likely defining element of the Estonian model is that it has the legal obligation to provide transparency and auditability of algorithms. The Public Sector Information Act and other administrative legislation in the country require human oversight, explainability, and traceability of automated decision-making tools utilized in the provision of services to the nation. In 2021, the Estonian Parliament adopted the Algorithmic Transparency Standard that is a soft law instrument requiring state agencies to have publicly accessible registries of AI systems used, with their legal basis, description of their functioning, and data inputs. It is especially applicable to the situations of the upcoming 6G settings, where artificially intelligent based choices, such as this of social good allotment, and even with the aid of XR in the context of policing, can have a profound and deeply personal impact on the rights of individuals [28].

In addition, the X-Road infrastructure used in Estonia, which enables secure transmission of data between databases of the state and the private sector, is a legal, encrypted and restricted environment capable of supporting communications requirements in the future 6G. The features of data minimization, access based on consent, and digital sovereignty proposed in this framework correspond with the EU legal framework overall, such as the GDPR and the proposed AI Act. Estonia therefore proves that rapid technological advancement can not necessarily come at the expense of legal protection or democratic oversight. Yet even in such advanced environment, there are dilemmas. The possible entry of immersive technologies like XR to public administration, where it will be most likely to be used in healthcare, education, and law enforcement gives normative considerations of emotional manipulation, informed decision-making, and ambient data-gathering legality [29]. The Estonian government is working on these problems via its AI Risk Mitigation Program (ARMP), which involves TIAs as a prerequisite to deploying the systems. This pre-emptive regulatory approach provides a valuable example to other countries considering readiness to 6G.

4.5.2 Case-Based Analysis: China—Centralized Surveillance and the Erosion of Personal Autonomy

By stark contrast, the People's Republic of China is an authoritarian type of governance through which highly technologically developed digital infrastructures are closely intersected with the same world of centralized control, mass surveillance, and profiling via algorithms. The Chinese government strategy regarding smart governance combines the combination of AI, the use of biometric databases, and real-time video analytics in a centralized 6G-ready surveillance system in a single ecosystem. Important examples of this are Sharp Eyes Program and the Skynet Project, which implement facial recognition, behavior prediction, and citizen scoring by using the interlocking surveillance networks [30]. These aspects have severe legal and ethical implications. Currently, China does not have a holistic rights-based framework such as GDPR or EU AI Act. Though Personal Information Protection Law (PIPL) and Data Security Law (DSL) have been passed in the recent years, their application is likely to be placed under the umbrella priorities of the state security. Furthermore, it lacks the judicial system to review judicial decisions and transparency in algorithmic decision-making that indicate the shortcoming of rights protection in the Chinese legal system.

Such actions are arguably not in line with the ICCPR, i.e. Articles 17 (protection of privacy), 19 (freedom of expression) and 21 (freedom of peaceful assembly) in the normative international law respect. Though China is a non-ratifying signatory of the ICCPR, surveillance arrangements, including mass profiling using AI-powered systems, without authorization as well as without judicial approval, perhaps need to be perceived as contravening the global norm on digital rights as promulgated in the General Comment No. 25 (2021) on the right to privacy of the UN Human Rights Committee. The Comment will highlight that such interference with digital privacy is subject to requirements of legality, necessity, and proportionality- principles that are largely lacking in the Chinese current system of surveillance. In addition, the use of AI-based programs to score individuals digitally based on behavioral risks and predictive policing creates another high-stakes form of algorithmic governance without effective transparency or control. Realizing such models in combination with 6G-compatible real-time communication networks and immersive surveillance capabilities would expose the extremely high potential risk of an entirely digitally enforced panopticon being created-wherein the independence of personhood is persistently interfered with, and where legal recourse either does not exist or proves ineffective. Against these trends, the Chinese paradigm has thus served as a warning model of how absence of control or mismanagement of technological advance may undermine basic freedoms and constitutionalism. It highlights the need

TABLE 4.3 Governance models of smart societies — Estonia vs. China

CRITERIA	*ESTONIA*	*CHINA*
Governance approach	Democratic, rule-based	Centralized, authoritarian
Transparency standards	Algorithmic registries, explainability	Opaque, state-controlled AI systems
Public participation	Citizen involvement in tech governance	Limited civic engagement
Data governance	Consent-based, GDPR-compliant	Broad state access with minimal rights
Legal protections	Judicial oversight, human-in-the-loop	Minimal due process, no independent oversight
International alignment	EU-aligned legal instruments	Divergence from global human rights norms

to create binding legal frameworks or governance charters that tie state and non-state players that use high-risk technologies to the ultra-fast communication world [31]. Estonia and China, which have different governance systems in relation to smart societies are illustrated in Table 4.3.

Table 4.3 provides a comparison between the governance architectures of smart societies in Estonia and China along dimensions of six major facets. Estonia is an example of the rights based and democratic approach, which is based on transparency, citizen inclusion, and straightforward laws coinciding with the EU. It suggests the pre-eminence of algorithm responsibility, data practices well-aligned with GDPR laws, and judicial oversight. By contrast, China has a centralized model where there is little civic influence, a lack of transparency in AI governance, and broad control of personal data by the state with little procedural safeguards and non-conformance to international human rights standards. This comparative structure highlights the normative conflict between the techno-liberal governance of the law in technology with the authoritarian governance of the digital realm.

4.6 THE NEED FOR PRECAUTION AND LEGAL TOOLS FOR 6G GOVERNANCE

4.6.1 Precautionary Principle in Tech Law

The implementation of 6G technologies allows an unprecedented level of complexity, autonomy, and pervasiveness of digital systems to be provided. Such high stakes and perpetually uncertain conditions necessitate the use of the precautionary principle which is widely linked to the acts of environmental and biotechnological regulation as a crucial normative instrument of technology governance. The precautionary principle holds that, when there are threats of serious or irreversible environmental damage, the lack of scientific certainty should not be used as a reason to postpone cost-effective prevention measures [32]. The precautionary principle is an international instrument as specified in the Rio Declaration on Environment and Development (1992). The methodology has crept slowly into the sphere of emergent technologies, especially its control over AI, biotechnology, and nanotechnology.

The precautionary reasoning approach is also increasingly formalized in the legal side regarding artificial intelligence in documents like the proposed Artificial Intelligence Act of the EU, taking a risk-based approach that bans or heavily constrains high-risk AI systems. It is not only under the model of reactive liability but in strict ex ante obligations, which are focused on the aspects of safety, transparency, and human supervision before deployment. By paralleling with the 6G, which will feature immersive technologies like XR or will provide real-time AI calculations along with human-machine symbiosis, it is recommended that the precautionary principle be accepted so that harms to societies and human rights can be identified prior and addressed. Considering that 6G applications can lead to the violation of neuro-rights, emotional surveillance, and biometric behavioral forecasting, assumptions of utilizing standard post-hoc regulatory processes are legally insufficient. In place, there must be a requirement by the law to show that the developers and implementing bodies are satisfied that the technology is not merely innovative, but it also conforms to human dignity, proportionality, as well as fundamental rights. Here, the precautionary principle moves out of the realm of scientific principle into one of law, with an expectation of regulatory vigilance, advancement governance and normative responsibility to respond to technological uncertainty [33].

4.6.2 Technology Impact Assessments

TIAs, as the practical instrument of the precautionary norm, will have to be formalized as a binding legal obligation regarding systems involved in the 6G, including systems including systems involving high-risk applications of AI, XR, and biometric data analytics. A TIA is a guided legal-technical document that mediates potential harms associated with a new or substantially upgraded technology by reviewing the potential consequences and mitigating said harms prior to implementation. Unlike the looser (and generic) Environmental or Social Impact Assessments, a TIA is specific to technology systems and sociotechnical complexities [34]. Such assessments are already required by the European GDPR, in Article 35, particularly in those cases where the operations risk posing significant threats to the rights and freedoms of individuals. In the same way, the providers of high-risk AI systems under the EU AI Act are expected to carry out conformity assessments to determine correspondence with the stipulated standards of safety and accountability standards.

In the case of 6G, the legally obligating TIA protocol should be created, which includes both horizontal and sector-specific requirements. Such a protocol must incorporate: (i) a legal risk mapping of the implications of rights (privacy, non-discrimination, autonomy), (ii) the assessment of algorithmic and systemic biases, (iii) consultation with the stakeholders and community, and (iv) a plan of remedies in contingency situations of harm. Notably, the TIA process is an ongoing, iterative process which should be updated during the lifecycle of the technology, not just in the pre-deployment phase. An example is the application of XR in healthcare or smart tourism, two areas that experts believe could be a source of 6G growth due to concerns raised regarding informed consent, accessibility, and digital divide. A TIA would test whether the XR system unreasonably gathers or processes sensitive biometric or neurological data and has such controls as opt-out, pseudonymization, or third-party audits in place. In such a way, the embedding of TIAs as an explicit legal measure would impose a procedural gate and precondition to 6G technologies commercialization and societal adoption [35].

4.6.3 Transparent Algorithmic Governance

Transparency is still among the core principles of digital rights jurisprudence and the practical legitimization of 6G-enabled systems. With so many real-time, life-and-death decisions increasingly being mediated by algorithms, the legal system needs to ensure algorithmic transparency and explainability, as well as that algorithms are publicly answerable. This convergence in the legal literature and policy debates can be seen in the rise of the concept of explainable AI (XAI) and algorithm registries, particularly when applied to high-stakes and publicly facing applications. As an example, the New York City Automated Decision Systems Law (2017) appeared that now the algorithms employed by the municipal agencies require auditing. This is the same case with the EU AI Act which seeks the documentation, traceability, and awareness of the end users on high-risk AI systems. Applying these advances to a 6G perspective, the need arises to create publicly available registries of high-impact systems revealing the root information of the functionality and purpose of real-time decision-making systems integrated into a 6G network. This step increases the level of democratic control and gives aggrieved parties a chance to go to court in order to seek redress or clarification [36].

In addition, legislative requirements defining explainability as a mandatory requirement should be formalized, especially in a situation when the outcomes of the work of AI systems involve an individual in the exercise of rights. By way of example, when a wearable system aided by 6G deprives an individual of access to a smart healthcare facility based on a black-box behavioral profiling algorithm, that natural person must be entitled to challenge such a decision to the extent that it will have a legally cognizable claim to know the reasons underlying such an operation. This complies with Article 22 of the GDPR according to which a right not to be subject to decisions based exclusively on automated processing without meaningful human intervention exists. Only the transparency of algorithms should not simply be code or the weights of models. It also demands situational legal frameworks that force the exposure of the rationale, the purpose, and the possible bias of the system, with the right to human intervention and appeal. Finally, transparent algorithmic governance, including publicly accessible registries, explanatory mandates, and user redress systems, makes up a precarious legal infrastructure of ethical application of 6G technology. It makes sure that the pace of innovation does not undermine basic rights and that the technological systems are not exempted to the democratic and legal responsibility.

4.7 TOWARD A BINDING INTERNATIONAL 6G GOVERNANCE CHARTER

4.7.1 Establishing Legal Obligations in 6G Systems

A binding international 6G Governance Charter is not purely imaginary work, since the formalization of essential legal commitments to transparency, informed user consent, and algorithmic accountability should be its first implication. These commitments play a critical role in showing that the introduction of 6G-enabling technologies, including real-time platforms in the context of XR and nanonetworks, does not interfere with individual rights and democratic values. Transparency must not only be linked to the access of technical documentation but also extended to substantial algorithm explainability. AI systems present in 6G networks need to allow users access to how decisions that impact them are made. The mechanisms to ensure consent need to be optimized to the context of a pervasive environment with wearables, implants, or ambient systems continuously capturing data—moving to an active, revocable and granular permission model beyond passive terms-and-conditions approaches [37]. Moreover, the neutrality of legal responsibility should prescribe to all high-risk AI or XR systems the mandate to perform TIAs prior to deployment, as risk-preventive governance. Collectively, these commitments can harmonize digital due process transnationally and cross-platform.

4.7.2 Defining 6G Fundamental Digital Rights

The Charter ought to define a list of the 6G Fundamental Digital Rights to deal with the new kinds of risks caused by ultra-fast communication ecosystems. These would be adaptations of the established international human rights to be suitable in an immersive, interconnected, and predictive 6G technology environment. Future rights being suggested are the algorithmic explanation right, the cognitive sovereignty right (against subliminal manipulation by XR), the right to digital anonymity in hyperconnected world, the right to bodily integrity in neurointerfaces or nanobots systems. Such rights would provide legal reference for interpreting the validity of AI-based governance, content moderation, biometric surveillance, and cross-border data flows in a 6G world.

4.7.3 Enabling Enforcement: National and Global Mechanisms

These principles can only be realized when accompanied by strong enforcement mechanisms. Removing the exemption from information on the having and the doing identification should come with giving independent AI regulators authority 6G technologies must have independent auditors with the authorization to sanction and certify the same. They need to give support to such national bodies by also establishing a supranational institution like a Global Data Oversight Council, which is established as a multilateral agreement. Such a world organization would regulate cases that concern transnational flows of data, systemic algorithmic injustices, or breaches of digital rights between different jurisdictions. It might also have a centralized registry of high-risk 6G systems and liaise between regulators to help coordinate enforcement so as to harmonize it and fill gaps in governance that exist on the basis of all national laws [38].

4.7.4 The Role of Public-Private Collaboration and Regulatory Innovation

States cannot control the situation with 6G technologies by themselves. Ethical innovation is also the responsibility of private entities that create and implement most of 6G infrastructure. The Charter ought to promote organized Public-Private Partnerships (PPPs) to bind the tech companies to standards with contractual obligations and require the companies to comply with human rights standards. Such collaborations may enable due diligence innovation in terms of joint compliance auditing, reporting obligations, and ethics-conscious designing [39]. Moreover, providing legal support for using regulatory sandboxes, the controlled background within which new high-risk technologies can be tested in the legal framework of the legislation, is another measure that has to be outlined. Such sandboxes would enable regulators to monitor in reality the effects of AI/XR systems and make sure regulatory measures are put in place before they become commonplace [40].

Finally, the legal and ethical framework can be established to control the next digital regrowth, so an International 6G Governance Charter with legal power would be a binding document. The Charter can make sure that the deployment of the 6G technologies is not only technologically superior but also ensures the maintenance of human dignity, autonomy, and justice in smart societies by: establishing universal obligations, codifying fundamental digital rights, establishing multilevel enforcement mechanisms, and promoting cooperative innovation [41].

4.8 CONCLUSION

4.8.1 Legal Reflections on 6G Governance: An Emerging Imperative

The new frontier of 6G technologies represents a radical change in the world of global communication networks because it brings ultra-high-speed connectivity, real-time AI decisions, the hologram-like presence, and ubiquitous XR experiences. Nonetheless, as has been shown in this chapter, the legal frameworks that exist now are structurally ill-equipped to address the multidimensional risks of such advancements [42]. In the global landscape, 6G regulation has become fractured and uneven-diverse approaches to policymaking around 6G range from risk-based regimes such as the AI Act and Digital Services Act in the EU, the ethical guidelines and enforcement shortcomings of the U.S. AI Bill of Rights, to India that pioneered a notional data sovereignty regime in its recently enacted Digital Personal Data Protection Act, 2023. This comparative legal analysis reveals a problematic tendency: national frameworks tend to be reactive rather than forward-looking, and may be more oriented to past considerations of data protection than to particular juxtapositions of autonomy, consent, surveillance, and algorithmic accountability that 6G presents [43].

In investigating the use of governance models in Estonia and China, the case studies demonstrated two opposing developments of smart societies, one that moves toward transparency, citizen trust and legal integration, and the hitch that looks into state surveillance and algorithmic profiling. The two cases exemplify the cautionary story to the entire world, further validating the necessity to establish legally binding frameworks for young technologies. The development of 6G would otherwise create an ever-digitized world where the violation of rights is facilitated, inequality is widened, and democratic checks and balances are diminished, unless active measures are taken to curb it [44].

4.8.2 Global Alignment as a Legal Necessity, Not an Option

The transnational and cross-border character of 6G will not allow anything less than a coordinated legal response. The pace and volume of such technologies imply that none of the countries can handle the implications in their own way. Such international legal instruments like the Budapest Convention on Cybercrime or the General Comment No. 25 (2021) of the UN Human Rights Committee on the right to privacy in the digital age provide initial points of departure, but they are not enough until updated, binding provisions are drafted, reflecting the very particular legal risks that AI-infused and real-time immersive systems bring into focus today [45]. Border questions like transnational data flow, digital jurisdiction, algorithmic decision-making, and transnational law enforcement need to be settled with harmonization of legal processes and similar regulatory norms. International bodies such as ITU, G20, and UN cannot afford to be left behind in defining such norms and here, innovation should lag behind regulation [46].

4.8.3 Regulating Speed: A Call for Precaution, Rights, and Legal Vision

In the end, control over the pace of innovation does not mean the imposition of arbitrary constraints, but the establishment of legal awareness and professional ethics in the context of technology development [47]. With market adoption of the 6G enabled by AI and driven by the proactive movement by smart society to deploy in sectors they manage such as in healthcare, transportation, tourism, and governance, the legal profession needs to keep pace. In this chapter, it is demanded that the precautionary principle and compulsory TIAs become a prerequisite to any large-scale implementation of 6G. It also suggests the development of the International 6G Governance Charter that is based on user rights, transparency, algorithmic responsibility, and legal enforceability across borders [48].

To sum up, the future of 6G cannot be dictated by the hopes of speed and effectiveness only, but by adherence to law, equality, and the dignity of a human person [49]. The legal community enjoys, thus, both a duty and an opportunity: to create a digital era in which rights are not a luxury but a premise—and in which innovation prospers within a context of justice.

REFERENCES

1. Ahokangas, P., Gisca, O., Matinmikko-Blue, M., Yrjölä, S., & Gordon, J. (2023). Toward an integrated framework for developing European 6G innovation. *Telecommunications Policy, 47*(9), 102641. https://doi.org/10.1016/j.telpol.2023.102641.
2. Kalodanis, K., Papapavlou, C., & Feretzakis, G. (2025). Enhancing security in 5G and future 6G networks: machine learning approaches for adaptive intrusion detection and prevention. *Future Internet, 17*(7), 312. https://doi.org/10.3390/fi17070312.
3. Wilczek, B., Thäsler-Kordonouri, S., & Eder, M. (2024). Government regulation or industry self-regulation of AI? Investigating the relationships between uncertainty avoidance, people's AI risk perceptions, and their regulatory preferences in Europe. *AI & Society.* https://doi.org/10.1007/s00146-024-02138-0.
4. Xu, J., Lee, T., & Goggin, G. (2024). AI governance in Asia: policies, praxis and approaches. *Communication Research and Practice, 10*(3), 275–287. https://doi.org/10.1080/22041451.2024.2391204.
5. Yang, M., Qu, Y., Ranbaduge, T., Thapa, C., Sultan, N., Ding, M., Suzuki, H., Ni, W., Abuadbba, S., Smith, D., Tyler, P., Pieprzyk, J., Rakotoarivelo, T., Guan, X., & M'rabet, S. (2024a). From 5G to 6G: a survey on security, privacy, and standardization pathways. *arXiv (Cornell University).* https://doi.org/10.48550/arxiv.2410.21986.
6. Aliyu, A., Ogbuju, E., Kolajo, T., Imoize, A. L., Abari, O. J., Masari, M. S., & Oladipo, F. (2024). Legal frameworks regulating computational models in wireless communication systems. In: *Computational Modeling and Simulation of Advanced Wireless Communication Systems*, edited by F. Oladipo, A. L. Imoize, and A. Aliyu, CRC Press, pp. 231–259. https://doi.org/10.1201/9781003457428-11.
7. Arnal, J. (2025). AI at risk in the EU: it's not regulation, it's implementation. *European Journal of Risk Regulation, 1*, 1–10. https://doi.org/10.1017/err.2025.19.
8. Sinha, S., Lambrechts, J. W., Bimana, A., & Ashipala, A. (2025). Augmented artificial intelligence in 5G, 6G, and beyond: a quantum leap. *Computer, 58*(1), 24–32. doi:10.1109/MC.2024.3445532.
9. Bareh, C. K. (2024). Reviewing the privacy implications of Indias digital personal data protection act (2023) from library contexts. *DESIDOC Journal of Library & Information Technology, 44*(1), 50–58. https://doi.org/10.14429/djlit.44.1.18410.
10. Bhattacharya, N., & Bansal, D. (2024). Enabling secure and efficient healthcare delivery in smart cities through 6G infrastructure. In: *Building Tomorrow's Smart Cities With 6G Infrastructure Technology*, edited by S. K. Singh, B. Virdee, R. B. Jadeja, S. Tanwar and A. Khanna, IGI Global, pp. 365–398. https://doi.org/10.4018/979-8-3693-8029-1.ch013.
11. Dass, P., Ujjwal, S., Novotny, J., Zolotavkin, Y., Laaroussi, Z., & Köpsell, S. (2024). Addressing privacy concerns in joint communication and sensing for 6G networks: challenges and prospects. *arXiv (Cornell University).* https://doi.org/10.48550/arxiv.2405.01742.
12. De Almeida, P. G. R., & Júnior, C. D. D. S. (2025). Artificial intelligence governance: understanding how public organizations implement it. *Government Information Quarterly, 42*(1), 102003. https://doi.org/10.1016/j.giq.2024.102003.
13. Ahokangas, P., & Aagaard, A. (2023). *The Changing World of Mobile Communications: 5G, 6G and the Future of Digital Services.* Palgrave Macmillan. https://doi.org/10.1007/978-3-031-33191-6.
14. Dotan, R. (2024). US regulation of artificial intelligence. In: *Handbook on Public Policy and Artificial Intelligence,* edited by R. Paul, E. Carmel, and J. Cobbe, Edward Elgar, pp. 153–178. https://doi.org/10.4337/9781803928241.00015.
15. Ghosh, O., & Kumar, B. (2025). The impact of 6G technology on environmental ecosystems and human life. In: *6G Impacts on Natural Habitats and Human Life*, edited by S. K. Haque and S. K. Singh, IGI Global, pp. 255–290. https://doi.org/10.4018/979-8-3373-2220-9.ch010.
16. Goyal, N. K. K. (2025). AI-powered policy calibration: a framework for dynamic regulation compliance in LLM applications. *International Journal of Scientific Research in Computer Science, Engineering and Information Technology, 11*(3), 854–860. https://doi.org/10.32628/cseit25113352.
17. Gurugopinath, R., & Venugopal, B. (2024). Policy and regulatory challenges in implementing beyond 5G and 6G networks in rural India. In: *Advances in Wireless Technologies and Telecommunication,* edited by M. A. Matin, IGI Global, pp. 59–88. https://doi.org/10.4018/979-8-3693-7788-8.ch003.
18. Cui, Q., You, X., Wei, N., Nan, G., Zhang, X., Zhang, J., Lyu, X., Ai, M., Tao, X., Feng, Z., Zhang, P., Wu, Q., Tao, M., Huang, Y., Huang, C., Liu, G., Peng, C., Pan, Z., Sun, T., Nayato, D., Chen, T., Khurram Khan, M., Jamalipour, A., Guizani, M., & Yuen, C. (2025). Overview of AI and communication for 6G network: fundamentals, challenges, and future research opportunities. *Science China Information Sciences, 68*(7), 171301. https://doi.org/10.1007/s11432-024-4337-1.

19. Chauhan, D., Mewada, H., Gondalia, V., Almalki, F. A., Patel, S., Modi, H., Kavaiya, S., Trivedi, Y., & Mujlid, H. M. (2024). Balancing technological innovation and environmental sustainability: a lifecycle analysis of 6G wireless communication technology. *Sustainability, 16*(15), 6533. https://doi.org/10.3390/su16156533.
20. Imoize, A. L., Adedeji, O., Tandiya, N., & Shetty, S. (2021). 6G enabled smart infrastructure for sustainable society: opportunities, challenges, and research roadmap. *Sensors, 21*(5), 1709. https://doi.org/10.3390/s21051709.
21. Naithani, P. (2025). Analysis of India's digital personal data protection act. *International Journal of Law and Management, 67*(5), 543–553. doi:10.1108/IJLMA-05-2024-0174.
22. Jahankhani, H., Kendzierskyj, S., & Hussien, O. (2023). Approaches and methods for regulation of security risks in 5G and 6G. In: *Advanced Sciences and Technologies for Security Applications,* edited by H. Jahankhani, S. Kendzierskyj, and O. Hussien, Springer, pp. 43–70. https://doi.org/10.1007/978-3-031-33631-7_2.
23. Jha, A. V., Appasani, B., Khan, M. S., Zeadally, S., & Katib, I. (2024). 6G for intelligent transportation systems: standards, technologies, and challenges. *Telecommunication Systems, 86*(2), 241–268. https://doi.org/10.1007/s11235-024-01126-5.
24. Kapoor, U., Shinde, S. S., Kanwer, B., Duggal, S., Kansal, L., & Dhanraj, J. A. (2024). Strategic insights. In: *Advances in Wireless Technologies and Telecommunication,* edited by X. Xu, IGI Global, pp. 99–119. https://doi.org/10.4018/979-8-3693-2931-3.ch007.
25. Khan, S. K., Shiwakoti, N., Stasinopoulos, P., & Warren, M. (2023). Cybersecurity regulatory challenges for connected and automated vehicles – State-of-the-art and future directions. *Transport Policy, 143*, 58–71. https://doi.org/10.1016/j.tranpol.2023.09.001.
26. Kumar, A., Masud, M., Alsharif, M. H., Gaur, N., & Nanthaamornphong, A. (2025). Integrating 6G technology in smart hospitals: challenges and opportunities for enhanced healthcare services. *Frontiers in Medicine, 12.* https://doi.org/10.3389/fmed.2025.1534551.
27. Letaief, K. B., Chen, W., Shi, Y., Zhang, J., & Zhang, Y. A. (2019). The roadmap to 6G: AI empowered wireless networks. *arXiv (Cornell University).* https://doi.org/10.48550/arxiv.1904.11686.
28. Maduranga, M. W. P., Tilwari, V., Rathnayake, R. M. M. R., & Sandamini, C. (2024). AI-enabled 6G Internet of Things: opportunities, key technologies, challenges, and future directions. *Telecom, 5*(3), 804–822. https://doi.org/10.3390/telecom5030041.
29. Mulders, S. (2023). The relationship between the principle of effectiveness under Art. 47 CFR and the concept of damages under Art. 82 GDPR. *International Data Privacy Law, 13*(3), 169–181. https://doi.org/10.1093/idpl/ipad012.
30. Navaie, K. (2024). Personal data protection in AI-native 6G systems. *arXiv (Cornell University).* https://doi.org/10.48550/arxiv.2411.03368.
31. Olugbade, O. (2025). In search of a global governance mechanism for Artificial Intelligence (AI): a collective action perspective. *Global Public Policy and Governance.* https://doi.org/10.1007/s43508-025-00113-z.
32. Paul, R. (2024a). The politics of regulating AI technologies: towards AI competition states. In: *Handbook on Public Policy and Artificial Intelligence*, edited by R. Paul, E. Carmel, and J. Cobbe, Edward Elgar Publishing, pp. 261–279. https://doi.org/10.4337/9781803922171.00028.
33. Reddy, K. H. K., Rao, M. G., & Roy, D. S. (2024). Health-care cyber-physical system for smart cities. In: *Cyber-Physical System 2.0: Communication and Computational Technologies*, edited by A. V. Jha and B. Appasani, CRC Press, pp. 98–122. https://doi.org/10.1201/9781003559993-5.
34. Saurabh, S. (2024). The digital personal data protection act of 2023: strengthening privacy in the digital age. *International Journal of Law in Changing World, 3*(2), 77–94. https://doi.org/10.54934/ijlcw.v3i2.84.
35. Skaria, G., Reshmi, B., Mathew, S. K., Shetty, S., & Rao, V. L. (2023). Development and testing of an electronic diabetes diary integrated with hospital information system for individuals with type 2 diabetes mellitus-study protocol. *JMIR Research Protocols, 13*, e50732. https://doi.org/10.2196/50732.
36. George, A. S., & George, A. S. H. (2024). Towards a super smart society 5.0: opportunities and challenges of integrating emerging technologies for social innovation. *Partners Universal International Research Journal, 3*(2), 1–29. https://doi.org/10.5281/zenodo.11522048.
37. Taddeo, M., McCutcheon, T., & Floridi, L. (2019). Trusting artificial intelligence in cybersecurity is a double-edged sword. *Nature Machine Intelligence, 1*(12), 557–560. https://doi.org/10.1038/s42256-019-0109-1.
38. Anh, V. T. K. (2024). The rise of AI in 6G networks: a comprehensive review of opportunities, challenges, and applications. *IEEE Conference Publications, 1*, 261–279. https://ieeexplore.ieee.org/abstract/document/10908115.
39. Moya Osorio, D. P., Ahmad, I., Vega Sánchez, J. D., Gurtov, A., Scholliers, J., Kutila, M., & Porambage, P. (2022). Towards 6G-enabled internet of vehicles: security and privacy. *IEEE Vehicular Technology Magazine, 17*(3), 82–105. https://ieeexplore.ieee.org/abstract/document/9681822.

40. Tripathi, P., & Malik, A. A. (2025). Ethical and social dimensions of 6G deployment. In: *Advances in Wireless Technologies and Telecommunication,* edited by X. Xu, IGI Global, pp. 401–434. https://doi.org/10.4018/979-8-3373-2220-9.ch014.
41. Wang, S., Qureshi, M. A., Miralles-Pechuaán, L., Huynh-The, T., Gadekallu, T. R., & Liyanage, M. (2021). Applications of explainable AI for 6G: technical aspects, use cases, and research challenges. *arXiv (Cornell University).* https://doi.org/10.48550/arxiv.2112.04698.
42. Weck, T. (2024). AI and competition policy. *AI and Ethics Review, 1*(4), 440–447. https://doi.org/10.21552/aire/2024/4/8.
43. Yrjola, S., Ahokangas, P., & Matinmikko-Blue, M. (2022). Value creation and capture from technology innovation in the 6G era. *IEEE Access, 10,* 16299–16319. https://doi.org/10.1109/access.2022.3149590.
44. Porambage, P., Gür, G., Moya Osorio, D. P., Liyanage, M., Gurtov, A., & Ylianttila, M. (2021a). The roadmap to 6G security and privacy. *IEEE Journals & Magazine, 2,* 1094–1122. https://ieeexplore.ieee.org/document/9426946.
45. Timmers, P., & Serentschy, G. (2023). Sovereignty and 6G. In: *Changing World of Mobile Communications: 5G, 6G and the Future of Digital Services,* edited by P. Ahokangas and A. Aagaard, Springer, pp. 253–282. https://doi.org/10.1007/978-3-031-33191-6_10.
46. Zuo, Y., Guo, J., Gao, N., Zhu, Y., Jin, S., & Li, X. (2023). A survey of blockchain and artificial intelligence for 6G wireless communications. *arXiv (Cornell University).* https://doi.org/10.48550/arxiv.2305.08604.
47. Ahokangas, P., Matinmikko-Blue, M., & Yrjölä, S. (2023). Envisioning a future-proof global 6G from business, regulation, and technology perspectives. *IEEE Journals & Magazine, 61,* 72–78. https://ieeexplore.ieee.org/abstract/document/9982304.
48. Wu, W., Zhou, C., Li, M., Wu, H., Zhou, H., Zhang, N., Xuemin, S., & Zhuang, W. (2021). AI-native network slicing for 6G networks. *arXiv (Cornell University).* https://doi.org/10.48550/arxiv.2105.08576.
49. Gao, R., Yu, D., Gao, B., Hua, H., Hui, Z., Gao, J., & Yin, C. (2025). Legal regulation of AI-assisted academic writing: challenges, frameworks, and pathways. *Frontiers in Artificial Intelligence, 8.* https://doi.org/10.3389/frai.2025.1546064.

Unleashing the Future with 6G Revolution *A Cybersecurity Approach*

5

Nabil Idi, Hajar Moudoud, and Zakaria Abou El Houda

5.1 INTRODUCTION

The quick development of mobile communication systems from 1G to 5G has completely changed how people talk to each other, do business, and engage with each other [1]. Each generational advancement has been characterized by distinct technological breakthroughs: 2G introduced digital voice capabilities, 3G enabled mobile data services, 4G delivered broadband internet access and streaming capabilities, while 5G promises ultra-low latency, high bandwidth, and support for massive Internet of Things (IoT) deployments [2,3].

Yet as digital ecosystems continue expanding and diversifying, the constraints of 5G technology are becoming increasingly evident. The emerging demands of digital twins, holographic telepresence, immersive extended reality (XR), tactile internet applications, and brain–computer interfaces (BCIs) require a fundamentally new connectivity paradigm. Sixth-generation (6G) networks represent this next evolutionary step, designed to provide not merely enhanced throughput and reliability, but to embed intelligence, adaptability, and trustworthiness as core architectural principles.

How societies interact, communicate, and do business has been completely transformed by the quick development of mobile communication systems from 1G to 5G [1]. Certain technological innovations have defined each generational leap: Digital voice was made possible by 2G, mobile data by 3G, broadband internet and streaming services by 4G, and ultra-low latency, high bandwidth, and massive IoT deployments by 5G [2,3]. However, the limitations of 5G are becoming apparent as digital ecosystems continue to grow and diversify. A new paradigm of connectivity is needed to meet the unprecedented demands of digital twins, holographic telepresence, immersive extended reality (XR), tactile internet, and brain–computer interfaces (BCIs). These qualities will support game-changing use cases like industrial automation, autonomous mobility, precision healthcare, and global-scale IoT ecosystems, propelling social progress and economic expansion [4–7]. But 6G's disruptive potential also brings with it new risks and complications, especially when it comes to managing trust, security, and privacy.

The development of the Radio Access Network (RAN) is one of 6G's most important features. The RAN is a crucial layer for both performance optimization and security enforcement since it serves as the initial point of contact between users and the network core [8]. Cloud-native, software-defined, and

DOI: 10.1201/9781003683599-5

AI-driven RAN architectures will provide previously unheard-of flexibility and efficiency when paired with cutting-edge technologies like cell-free massive MIMO, reconfigurable intelligent surfaces (RIS), terahertz communications, and joint communication and sensing (JCAS). By creating new weaknesses in the distributed intelligence, virtualized functions, and air interface, these innovations also increase the attack surface. As a result, protecting the RAN in 6G calls for a comprehensive and flexible strategy that goes beyond conventional perimeter-based defenses.

The Open RAN (O-RAN) framework's introduction in this context poses both a challenge and an opportunity. O-RAN encourages interoperability, innovation, and vendor diversity by breaking down the RAN into logical components and establishing standardized interfaces like A1, E2, O1, and O2. The Non-Real-Time RAN Intelligent Controller (Non-RT RIC) and the Near-Real-Time RIC (Near-RT RIC) are two examples of AI-based controllers that enable operators to integrate with multi-vendor ecosystems and dynamically optimize resource allocation, policy enforcement, and service delivery. Strong trust and access management systems are necessary because O-RAN's openness and modularity, which make it appealing, also increase its susceptibility to attacks.

This chapter concentrates on integrating Zero Trust Architecture (ZTA) into the O-RAN framework in order to address these issues [9]. ZTA goes beyond static perimeter-based trust by enforcing the "never trust, always verify" principle, which makes sure that each access request is regularly authorized, validated, and authenticated based on behavioral analytics, device posture, and context. To protect policy exchanges, enrichment data, and machine learning model management from manipulation and abuse, ZTA can be mapped onto crucial interfaces in the O-RAN environment, especially the A1 interface, which connects the Non-RT RIC and Near-RT RIC. Building robust, secure, and adaptable 6G RANs appears to be possible with this combination of ZTA principles and O-RAN's modular architecture.

The chapter also examines the role of machine learning and artificial intelligence (AI/ML) in protecting 6G networks, emphasizing intrusion detection systems (IDS) designed for extremely dynamic environments. We show how IDS models trained on benchmark datasets like CICIDS2018 can adapt across domains, offering strong defenses against changing threats, by utilizing transfer learning, adversarial domain adaptation (ADA), and CORrelation ALignment (CORAL). This experimental viewpoint emphasizes how important it is to combine cutting-edge RAN architectures with AI-driven, security-by-design frameworks to make sure that 6G is reliable and resilient in addition to being quick and intelligent.

5.2 THE 6G ARCHITECTURE LANDSCAPE

5.2.1 Core Network Transformation and Infrastructure

A fundamental change in the way telecommunications infrastructure is designed, implemented, and maintained, the transition to 6G networks signifies a paradigm shift in core network architecture [10]. Beyond small adjustments, this shift entails a total redesign of network architectures based on cloud-native concepts, AI integration, and previously unheard-of levels of automation. To build a truly intelligent and adaptable infrastructure that can support cutting-edge applications like extended reality (XR), digital twins, and brain–computer interfaces that require extremely low latency and massive computational resources, the 6G core network will make use of cutting-edge technologies like network slicing, edge-cloud convergence, and quantum-safe security protocols.

6G networks will be supported by a distributed, disaggregated infrastructure that uses network function virtualization (NFV), software-defined networking (SDN), and open radio access networks (Open RAN) to build adaptable, interoperable systems. In addition to combining cutting-edge materials like metamaterials and intelligent reflecting surfaces to maximize signal propagation, this new architectural paradigm unifies terrestrial, aerial, and satellite networks into a single three-dimensional

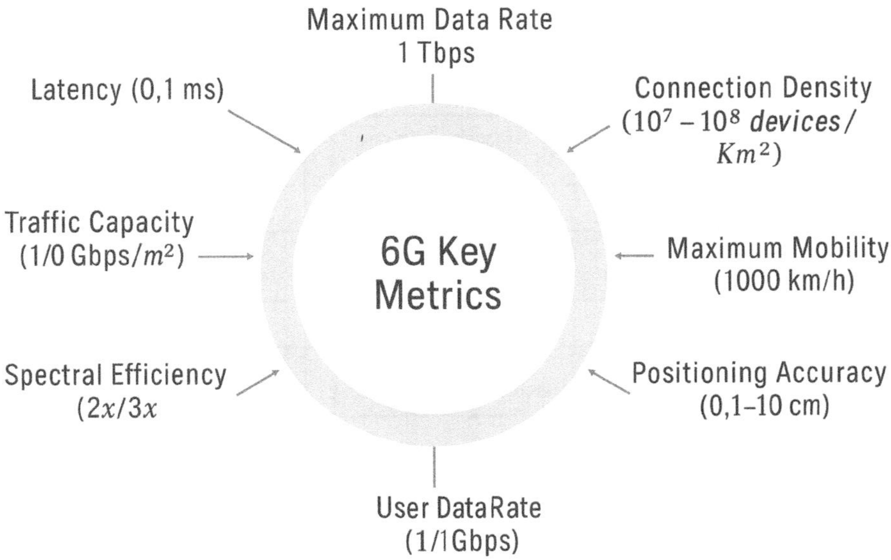

FIGURE 5.1 Overview of 6G key metric.

coverage model. Additionally, the core network transformation prioritizes energy efficiency and sustainability by utilizing renewable energy sources and green networking principles to address the environmental issues brought on by higher network densities and computational demands. The end result is an infrastructure that can dynamically allocate resources across multiple service domains, predict and prevent failures, and adapt to traffic patterns on its own. Figure 5.1 overviews some key metrics, which characterize 6G networks.

5.2.2 6G Architecture Landscape

The 6G architecture landscape is structured into five interconnected layers: the Physical Layer, which leverages advanced technologies such as THz, VLC, RIS, and Massive MIMO to deliver ultra-high data rates and reliable connectivity; the Network Layer, powered by AI-native cores, edge computing, and network slicing to provide flexible, intelligent, and customized resource allocation; the Data Layer, which introduces semantic communications, federated learning, and blockchain to ensure meaning-aware transmission, privacy-preserving intelligence, and trust; the Service Layer, enabling immersive experiences through XR, holographic communications, and the tactile internet for real-time multisensory interactions; and finally, the Application Layer, where use cases such as smart cities, autonomous systems, healthcare, and IoT materialize, showcasing how 6G integrates infrastructure, intelligence, and human-centric services into a unified ecosystem (see Figure 5.2).

In addition to its layered design, the 6G architecture emphasizes seamless integration and intelligence across all levels. Unlike previous generations that primarily focused on faster connectivity, 6G introduces a holistic ecosystem where communication networks are context-aware, adaptive, and capable of learning in real time. The synergy between physical infrastructure, AI-driven network management, secure data handling, and immersive services enables new socio-economic opportunities—from precision healthcare and autonomous mobility to sustainable smart cities and global-scale IoT systems. This vision positions 6G not only as a communication technology, but as a foundation for digital transformation in the coming decade.

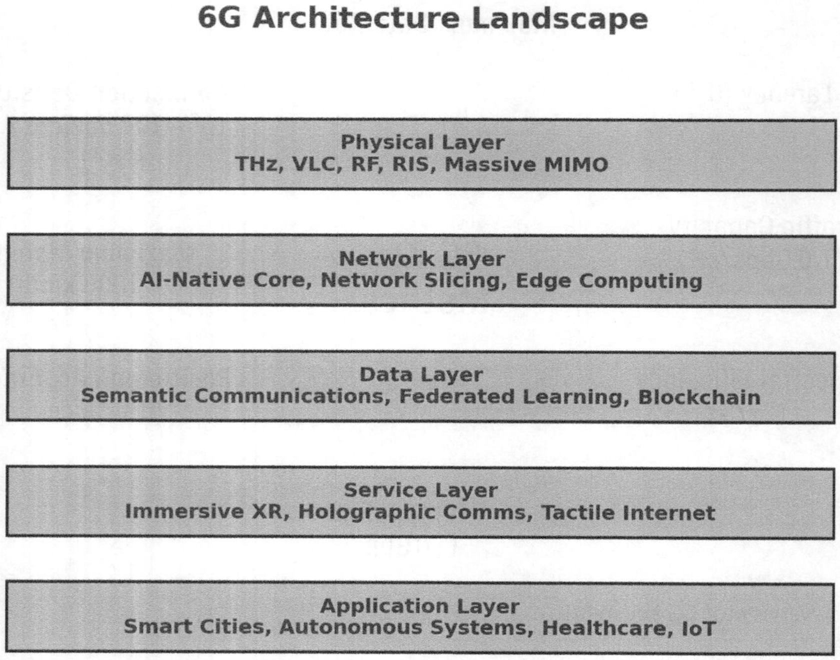

FIGURE 5.2 Multi-layered architecture of 6G networks.

5.2.3 Edge Computing Integration and Distributed Architecture

Edge computing integration represents a fundamental architectural evolution that brings computational resources, data storage, and network intelligence closer to end users and connected devices, dramatically reducing latency while improving application performance and user experience [5]. This distributed computing paradigm shifts processing capabilities from centralized cloud data centers to a mesh of edge nodes deployed at cell towers, enterprise premises, and strategic network locations, enabling real-time decision-making for latency-sensitive applications such as autonomous vehicles, industrial automation, and augmented reality systems. The integration involves sophisticated orchestration mechanisms that can dynamically distribute workloads across the edge-cloud continuum, utilizing artificial intelligence and machine learning algorithms to optimize resource allocation based on factors including network conditions, computational requirements, user proximity, and service-level agreements. The distributed architecture of edge computing creates pyramid-like layers of processing power. Small edge devices and tiny data centers are located at the bottom, close to where users are. Large cloud data centers are at the top, followed by regional facilities that serve larger areas. A connected network that can seamlessly deliver services at all levels is produced by this tiered approach.

This system works so well because it is adaptable. Modern techniques like microservices and containerization allow applications to be swiftly deployed to the locations where they are most needed. The system has the ability to automatically reallocate resources to areas with high demand. Applications can move to better locations without affecting users if network conditions change.

The networking aspect is just as advanced. Like a smart traffic control system, technologies like NFV and SDN automatically manage service connections and make sure resources go where they're needed. Edge analytics and distributed databases, on the other hand, maintain data locally when necessary, which is essential for complying with various nations' data protection regulations.

The outcome is remarkable: a strong infrastructure with lightning-fast response times that can manage millions of connected devices. Whether augmented reality applications need immediate feedback or autonomous cars need to make snap decisions, this distributed system keeps the network as a whole fast and reliable.

5.2.4 Hyperconnectivity and the Internet of Everything

True hyperconnectivity—where billions of devices, sensors, and intelligent systems are seamlessly connected in real time—is what 6G networks are intended to enable. By combining data, people, and processes into a single digital fabric, this vision transcends the conventional IoT and moves toward the Internet of Everything (IoE) [11]. Through technologies such as terahertz communications, AI-native networking, semantic communications, and edge intelligence, 6G provides the ultra-low latency, reliability, and context-aware services needed to support massive-scale interactions. Hyperconnectivity ensures that smart cities, autonomous vehicles, healthcare systems, and industrial environments can interoperate dynamically, while IoE transforms connectivity into a driver of innovation, enabling immersive experiences, sustainable infrastructures, and globally intelligent networks.

5.3 RADIO ACCESS NETWORK EVOLUTION IN 6G

5.3.1 Introduction to RAN in 6G

The Radio Access Network (RAN) represents one of the most critical components in the evolution toward 6G mobile networks, as it directly determines how users, devices, and machines connect to the core infrastructure. Unlike previous generations, 6G RAN is envisioned to integrate sub-THz spectrum bands, RIS, massive ultra-MIMO, and cell-free architectures to achieve ultra-high capacity, extreme data rates, and ubiquitous coverage. This evolution is not only a response to the exponential growth in mobile traffic but also a response to the need for ultra-reliable low-latency communication (URLLC) in applications such as autonomous vehicles, the tactile internet, digital twins, and large-scale IoT ecosystems. Additionally, 6G's RAN will depend more and more on cloud-native, virtualized, and AI-driven designs, allowing for intelligent orchestration of heterogeneous access technologies, dynamic spectrum allocation, and real-time optimization. However, because adversaries can take advantage of flaws in the air interface, RIS control channels, or virtualized network functions, this change also broadens the attack surface, making 6G RAN security a fundamental challenge. Therefore, comprehending the development of the RAN in 6G necessitates not only a technical viewpoint on performance but also an all-encompassing strategy that incorporates security-by-design, resilience, and trust into its architecture.

5.3.2 Architectural Transformations

The move to cloud-native RAN (CN-RAN) is one of the biggest changes in 6G RAN design. Cloud-native, software-defined, and fully virtualized infrastructures are replacing traditional hardware-centric deployments. Network slicing, elastic scalability, and dynamic resource allocation are made possible by this transformation in distributed computing environments. Such adaptability is essential in 6G, where a wide range of traffic profiles, from immersive XR applications to low-power IoT devices, must coexist harmoniously. Simultaneously, the traditional idea of cell boundaries is anticipated to be replaced by Cell-Free Massive MIMO. This method lowers inter-cell interference and increases spectral and energy efficiency by implementing dispersed antenna arrays that collaboratively serve users. For ultra-dense IoT deployments and safety-critical applications, coordinated beamforming across multiple access points guarantees more consistent coverage and enhanced reliability for all users. The development of integrated access and backhaul (IAB) is another fundamental component of the architecture. Fiber deployment for each small cell is not feasible or economical, as 6G networks get denser. IAB offers a solution by using multi-hop wireless backhauling, which makes use of cutting-edge relay technologies to increase network capacity and coverage. This speeds up the rollout of 6G, particularly in urban and rural areas, by enabling operators to achieve flexible densification while lowering infrastructure costs.

5.3.3 Advanced Technologies

The 0.1–10 THz spectrum is emerging as a cornerstone of 6G radio access networks. With its enormous bandwidth, terahertz (THz) technology promises data rates ranging from hundreds of gigabits to several terabits per second—enough to power demanding applications such as holographic telepresence and immersive digital twins. However, the path to real-world implementation is not straightforward. Significant challenges still exist, such as high atmospheric absorption, the challenge of maintaining exact beam alignment, and the limitations of existing hardware. In order to get past these obstacles, 6G systems are anticipated to use advanced beamforming techniques and intelligent reflecting surfaces (IRS), which can both enhance signal propagation and guarantee more dependable connectivity.

Equally central to 6G is the rise of the AI-native Radio Access Network (RAN). Here, artificial intelligence is no longer treated as an optional layer but as a core capability built into every part of the network. Machine learning will drive enhancements at the physical layer, improve traffic forecasting, and enable autonomous resource allocation as well as predictive maintenance. With networks that can continuously learn and adjust to changing conditions, the AI-native RAN is expected to simplify operations while delivering reliability and efficiency, even in highly dynamic environments.

Another transformative feature is JCAS, which integrates environmental sensing with wireless communications. This dual-purpose use of RAN infrastructure enables capabilities such as object detection, localization, and situational awareness, which are critical for autonomous vehicles, drone management, and smart city infrastructures. In 6G, the RAN will thus evolve into a platform that not only connects users but also senses the environment in which they operate.

5.4 INTEGRATING ZERO TRUST ARCHITECTURE INTO O-RAN

5.4.1 Background

The evolution of RAN architectures toward openness and virtualization has revolutionized the telecommunications industry. O-RAN promotes cloud-native deployments, modular network functions, and vendor interoperability as a key facilitator of this paradigm shift. However, because its components are dispersed and heterogeneous, the flexibility and openness it brings also increase security risks.

Static perimeter-based security measures that presume implicit trust within network boundaries are the foundation of traditional RAN deployments. The dynamic, service-based O-RAN environment is not a good fit for such models. ZTA, on the other hand, maintains that trust needs to be constantly verified, independent of location or credential history. ZTA fits in nicely with the threat landscape of contemporary mobile networks by assuming breach and implementing least privilege access.

An experimental and architectural viewpoint on integrating ZTA into the O-RAN architecture is provided in this chapter. We focus on the A1 interface, which connects the Non-RT RIC to the Near-RT RIC and is essential to intent-based RAN optimization. Our goal is to create a scalable and reliable model by mapping ZTA components onto A1 operations and protecting the control plane with trust scoring and policy enforcement.

5.4.2 Zero Trust Architecture

In cybersecurity, the adoption of ZTA marks a fundamental change in how networks are secured. Traditional perimeter-based models focused on keeping threats outside and assumed that anything within the network boundary could be trusted. That approach is no longer sufficient in today's environment, shaped by cloud computing, mobile workforces, IoT devices, and the distributed services of 6G. Modern attack vectors exploit weaknesses such as lateral movement across networks, misconfigured cloud resources, or stolen identities—making reliance on static defenses ineffective.

ZTA addresses these challenges through its guiding principle: "never trust, always verify." Trust is not granted simply because of a user's location or prior access. Instead, every request—whether it originates inside or outside the network—must be continuously authenticated, authorized, and protected through encryption.

ZTA is a security model that substitutes context-aware, adaptive access control for static, perimeter-based trust. ZTA requires verification of all access requests, regardless of their source or prior trust state, in accordance with NIST SP 800-207. The foundation of the architecture is:

- Policy enforcement points (PEPs)
- Policy engines (PEs)
- Policy administrators (PAs)
- **Supporting Systems**: CDM, IdPs, SIEM, etc.

ZTA operates via a service-based control plane where requests are enriched with contextual data such as device posture, user identity, location, and behavior analytics. Trust levels are continuously updated.

ZTA also relies heavily on a range of supporting systems that feed contextual data into the decision-making process. Continuous Diagnostics and Mitigation (CDM) systems provide insights into the security posture of endpoints, ensuring that only compliant devices are granted access. Identity Providers (IdPs) handle user and device authentication through strong, often multi-factor mechanisms, while ensuring federated identity management across distributed platforms. Security Information and Event Management (SIEM) solutions collect and analyze logs to detect anomalies, threats, or policy violations, feeding intelligence back into the policy engine for improved decision accuracy. Other complementary systems, such as Endpoint Detection and Response (EDR) and User and Entity Behavior Analytics (UEBA), further enhance visibility and adaptability in a ZTA environment.

The implementation of ZTA in 6G networks is particularly critical, given the ultra-dense device connectivity, dynamic service orchestration, and highly heterogeneous ecosystem of users and applications. For example, in a 6G-enabled smart city, billions of IoT sensors, autonomous vehicles, and edge devices continuously request access to cloud and edge resources. A perimeter-based trust model would not scale or withstand evolving attacks. ZTA ensures that each request—whether from a sensor reporting environmental data, a drone offloading video streams, or a user initiating XR communication—is contextually verified in real time based on the trustworthiness of both the entity and its communication path.

5.4.3 O-RAN and the A1 Interface

O-RAN's architecture decomposes the RAN into logical entities and interfaces. The A1 interface connects the Non-RT RIC (analytics/policy) to the Near-RT RIC (real-time control). It supports:

- **A1-P**: Policy creation and updates
- **A1-EI**: Enrichment data exchange
- **A1-ML**: ML model management

The A1 interface uses REST APIs with JSON over HTTPS/mTLS but lacks dynamic trust evaluation—opening the door for ZTA.

5.4.4 Design of a ZTA-Enabled O-RAN Architecture

The design of a ZTA-enabled O-RAN must begin with clear objectives that balance security, interoperability, and performance [12–15]. Unlike traditional perimeter-based approaches, where trust is assumed once access is granted, ZTA requires that every request, entity, and communication be continuously validated. When mapped onto the open and modular O-RAN environment, this introduces both opportunities and challenges. O-RAN's decomposition of RAN functions and reliance on standardized interfaces like A1, E2, O1, and O2 provide natural integration points for ZTA, but they also expose the control and data planes to new attack vectors. As such, the design goals must ensure that robust security mechanisms are embedded without undermining latency, scalability, or interoperability.

The first goal is to enforce continuous verification across all components and interactions. This aligns with the ZTA principle of "never trust, always verify." In practice, this means that every API call, data exchange, and policy update within O-RAN must be authenticated, authorized, and encrypted in real time. The A1 interface, which facilitates communication between the Non-Real-Time RIC and the Near-Real-Time RIC, is particularly critical, as it carries policies, enrichment data, and machine learning model updates. By embedding continuous verification at this interface, the architecture ensures that malicious or compromised entities cannot inject falsified policies or manipulate learning models, which could degrade performance or destabilize the network.

Maintaining compatibility with A1 standards is the second objective. Any ZTA integration must adhere to the open specifications that O-RAN is based on, which are intended to promote vendor interoperability. This necessitates creating standards-aligned, lightweight security overlays that may be added without compromising compatibility, such as certificate pinning, OAuth2-based identity management, and mutual TLS (mTLS). The architecture prevents vendor lock-in and guarantees that Zero Trust measures can be broadly implemented across heterogeneous installations by following A1's current requirements.

The third goal focuses on enabling distributed, low-latency enforcement. In 6G, latency-sensitive applications such as autonomous driving, remote surgery, and XR demand response times in the sub-millisecond range. Centralized security checks could introduce bottlenecks, undermining these requirements. Instead, ZTA enforcement must be distributed across edge nodes, RIC instances, and O-Cloud resources, ensuring that verification occurs as close as possible to where data and control interactions happen. Leveraging edge computing and AI-driven local decision-making, the architecture can achieve both rapid enforcement and resilience, even under high traffic volumes and dense device connectivity.

The fourth objective is to provide visibility by means of observability and telemetry. Continuous system behavior monitoring is necessary in a dynamic O-RAN context in order to identify abnormalities, evaluate risk, and modify access controls. AI/ML-driven behavioral analytics engines can use telemetry gathered from interfaces like E2 (real-time RAN telemetry) to spot suspect activity like malware xApps, unusual traffic patterns, or unauthorized configuration modifications. In order for the Non-RT RIC's policy engine to update trust ratings and make dynamic policy adjustments, observability frameworks must not only gather data but also transform it into actionable information. This establishes a closed-loop system in which ongoing learning and feedback reinforce ZTA principles.

When combined, these design objectives serve as the cornerstone of a secure and flexible ZTA-enabled O-RAN architecture. The suggested approach improves visibility, distributes enforcement mechanisms,

TABLE 5.1 Implementation strategy for ZTA in O-RAN

ZTA COMPONENT	*O-RAN INTEGRATION POINT*	*EXAMPLE STRATEGY*
Identity and access mgmt	xApp onboarding, RIC access	PKI-based identities, OAuth2 flows
Micro-segmentation	O-Cloud, RIC network zones	Kubernetes namespaces+Calico policies
Behavioral analytics	RIC telemetry (E2)	AI/ML-driven threat detection from user & slice KPMs
Least privilege	API gateways for A1, E2, O1	Role-based access, deny-all-default policy
Secure communication	All interfaces (A1, E2, O1, O2)	mTLS, IPsec tunnels, certificate pinning
Continuous verification	Policy engine in non-RT RIC	Real-time context-based access adjustment

ensures standard compliance, and integrates continuous verification to make 6G RAN installations more resilient. More significantly, it accomplishes this without sacrificing the flexibility and openness that characterize O-RAN, providing a workable blueprint for creating reliable and future-proof 6G networks.

Table 5.1 shows how the O-RAN (Open Radio Access Network) framework can incorporate the concepts of ZTA.

5.5 EVALUATION OF THE INTEGRATION OF MACHINE LEARNING TO SECURE 6G NETWORKS

A revolutionary change from static, rule-based intrusion detection to adaptive, context-aware defense systems is represented by the incorporation of machine learning (ML) into 6G security frameworks. Unlike previous generations, 6G is envisioned as an AI-native network, where intelligence is embedded at every layer of the architecture, from the physical transmission plane to the service and application layers. This makes ML not merely a support tool but a core enabler of trust, resilience, and real-time protection. However, the heterogeneity, hyperconnectivity, and massive scale of 6G environments create challenges that require models capable of generalizing across domains and adapting to evolving threats. To address this, our evaluation focuses on transfer learning techniques applied to the widely recognized CICIDS dataset, which provides a realistic representation of modern cyberattacks and benign traffic patterns.

Transfer learning is particularly relevant in the context of 6G security because of the domain shift problem: models trained in one environment (e.g., a specific slice, IoT scenario, or geographical deployment) may underperform when deployed in another. To mitigate this, we employ ADA and CORAL to align feature distributions between source and target domains. ADA ensures that the learned representations become domain-invariant by using an adversarial loss to "fool" a domain discriminator, while CORAL minimizes the difference between covariance statistics of source and target feature spaces. Together, these methods enhance the transferability of intrusion detection models, allowing them to maintain high accuracy even when deployed in new or evolving 6G scenarios.

The CIC-IDS dataset was designed to support research in the field of cybersecurity against cyber threats (see Table 5.2). It was developed by the University of New Brunswick and includes seven types of attacks: Botnet, Brute Force, DDoS, DoS, Infiltration, PortScan, and Web Attack [16]. In addition to these malicious data samples, the dataset also contains benign traffic data. However, it is important to note that this dataset is imbalanced, which presents a significant challenge for training intrusion detection algorithms aiming for high accuracy [17,18].

Figure 5.3 presents the classification report of Logistic Regression applied on the CICIDS-17 dataset, highlighting the precision, recall, and F1-scores across the two target classes (0=Normal, 1=Attack). With precision values of 0.96 for class 0 and 0.89 for class 1, the results demonstrate a reasonably balanced

TABLE 5.2 Breakdown of data by label (percentage and number) of CICIDS dataset

CATEGORY	*PERCENTAGE (%)*
Benign	58
DoS	25
DDoS	16
Brute force	1
Web attack	0
PortScan	0
Bot	0
Infiltration	0

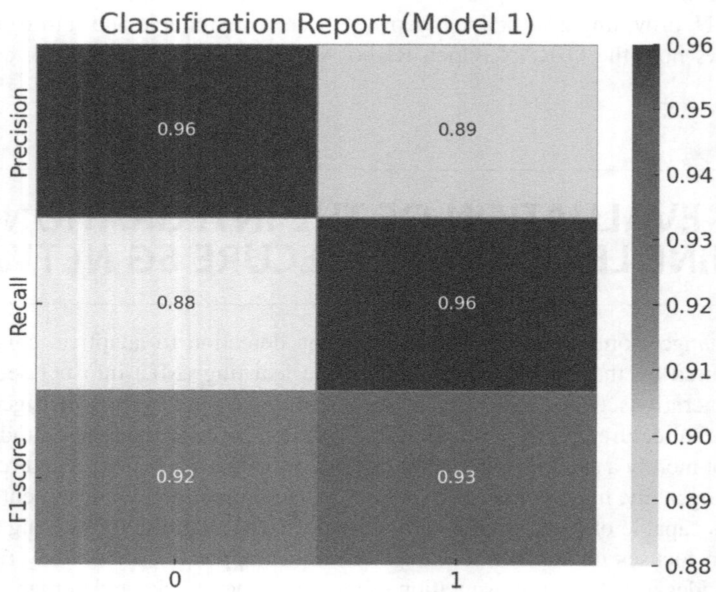

FIGURE 5.3 Evaluation of logistic regression based on CICIDS-17 dataset.

performance, suggesting that normal traffic is more consistently recognized than attack traffic. The recall values, on the other hand, show the opposite pattern, with attack detection achieving a higher recall (0.96) than typical traffic (0.88). This implies that although the model is very good at detecting attack events, it is marginally less good at reliably spotting benign flows. The overall robustness of the Logistic Regression model when trained under this scenario is confirmed by the F1-scores, which balance precision and recall, which are 0.92 and 0.93 for the normal and attack classes, respectively.

The suggested algorithm (Algorithm 5.1) improves intrusion detection performance in 6G-enabled environments by utilizing transfer learning with ADA and CORAL [19]. The model learns discriminative features from the source domain and aligns them with the target domain using adversarial feature invariance (ADA) and second-order statistical CORAL after training on the CICIDS dataset, which offers realistic traffic patterns and a variety of attack scenarios. Due to the dynamic and heterogeneous nature of 6G networks—which introduce significant domain shifts due to massive device connectivity, ultra-low latency, and distributed architectures—this dual adaptation guarantees robustness. Because preserving precise and flexible IDS capabilities is essential to safeguarding next-generation infrastructures, the algorithm is thus well-suited for real-time threat detection in 6G IoT and tactile internet applications.

Algorithm 5.1: Transfer Learning with ADA and CORAL

Input: Source dataset $\mathcal{D}_s = \left\{\left(x_i^s, y_i^s\right)\right\}_{i=1}^{n_s}$, Target dataset $\mathcal{D}_t = \left\{x_j^t\right\}_{j=1}^{n_t}$, feature extractor g_θ, classifier h_ψ, domain discriminator d_ϕ
Output: Adapted model M^*

Initialize parameters θ, ψ, ϕ. Pre-train (g_θ, h_ψ) on $\mathcal{D}_s$ using classification loss $\mathcal{L}_{\text{cls}} = \text{CE}\left(h_\psi\left(g_\theta\left(x_i^s\right)\right), y_i^s\right)$
for *each* training *epoch* **do**

1. Sample mini-batches $B_s \subset \mathcal{D}_s$, $B_t \subset \mathcal{D}_t$ and extract features $F_s = g_\theta(B_s)$, $F_t = g_\theta(B_t)$
2. Compute classification loss on source: $\mathcal{L}_{\text{cls}} = \text{CE}\left(h_\psi(F_s), B_s^y\right)$
3. Compute adversarial domain loss (with GRL): $\mathcal{L}_{\text{adv}} = \text{BCE}\left(d_\phi(F_s), 0\right) + \text{BCE}\left(d_\phi(F_t), 1\right)$
4. Compute CORAL loss: $\mathcal{L}_{\text{coral}} = \left(1/\left(4d^2\right)\right) \times \left\|C(F_s) - C(F_t)\right\|_F^2$
5. Compute total loss: $\mathcal{L}_{\text{total}} = \mathcal{L}_{\text{cls}} + \lambda_{\text{adv}}\mathcal{L}_{\text{adv}} + \lambda_{\text{coral}}\mathcal{L}_{\text{coral}}$
6. Update parameters (θ, ψ) by minimizing $\mathcal{L}_{\text{total}}$
7. Update ϕ by minimizing $\mathcal{L}_{\text{adv}}$ (without GRL)

end for
Select the best model M^* based on validation performance on $\mathcal{D}_t$
return M^*

Table 5.3 provides a more detailed comparison between ADDA (Adversarial Discriminative Domain Adaptation) and CORAL (CORrelation ALignment) when integrated with Logistic Regression on the CICIDS2018 dataset. Interestingly, both adaptation methods achieve perfect precision for the attack class (1.00), which means no benign traffic was misclassified as malicious. However, the recall values for the attack class (0.86 for ADDA and 0.84 for CORAL) indicate that some attacks were missed. The normal traffic class was not detected at all (precision and recall=0), reflecting a dataset imbalance where the model becomes biased toward the majority attack class. Consequently, the macro-averaged F1-scores remain low (0.46), showing the difficulty of handling imbalanced datasets even with transfer learning strategies. Despite this, the weighted averages are relatively high (0.92 for ADDA and 0.91 for CORAL), which reflects the dominance of the attack samples in the dataset and the strong model performance on this class.

Table 5.4 reports the performance of the Random Forest classifier on the CICIDS2018 dataset, evaluated across multiple attack categories as well as benign traffic. The results highlight both the strengths and weaknesses of Random Forest when applied to highly imbalanced datasets. For the benign class, the model achieves moderate performance, with a precision of 0.61, recall of 0.74, and an F1-score of 0.67, indicating that normal traffic is relatively well detected but still prone to false positives and false negatives. In contrast, the model performs extremely poorly across most attack categories, with precision, recall, and F1-scores close to zero for Bot, Brute Force, DDoS, Infiltration, PortScan, and Web Attacks. Only the DoS class shows minimal recognition, with precision of 0.13, recall of 0.16, and an F1-score of 0.14, which remains far from acceptable for practical use.

TABLE 5.3 Detailed classification report: ADDA vs CORAL (Logistic Regression on CIC-IDS2018)

CLASS	*ADDA+LR (PRECISION/ RECALL/F1–SCORE)*	*CORAL+LR (PRECISION/ RECALL/F1–SCORE)*
Normal	0.00/0.00/0.00	0.00/0.00/0.00
Attack	1.00/0.86/0.92	1.00/0.84/0.91
Macro avg	0.50/0.43/0.46	0.50/0.42/0.46
Weighted avg	1.00/0.86/0.92	1.00/0.84/0.91

TABLE 5.4 Classification report with Random Forest on CICIDS2018

CLASS	*PRECISION*	*RECALL*	*F1-SCORE*	*SUPPORT*
Benign	0.61	0.74	0.67	91,767
Bot	0.00	0.00	0.00	287
Brute_force	0.00	0.00	0.00	1,830
DDoS	0.00	0.00	0.00	25,603
DoS	0.13	0.16	0.14	38,751
Infiltration	0.00	0.00	0.00	7
PortScan	0.00	0.00	0.00	391
Web_attack	0.00	0.00	0.00	429
Accuracy			0.47	159,065
Macro avg	0.09	0.11	0.10	159,065
Weighted avg	0.39	0.47	0.42	159,065

The overall accuracy of 0.47 and weighted F1-score of 0.42 reflect this imbalance in detection capability. The macro-averaged results (precision=0.09, recall=0.11, F1=0.10) are particularly concerning, as they expose the model's inability to generalize across minority classes. This strongly suggests that Random Forest, in its standard form, is heavily biased toward the dominant class (benign traffic) and fails to effectively capture patterns associated with rarer attack types.

These results emphasize the challenges of using traditional ML models for intrusion detection in realistic, imbalanced traffic environments such as those represented by CICIDS2018. In the context of 6G networks, where traffic heterogeneity and attack sophistication will be even greater, such limitations become critical. The inability to detect low-frequency yet high-impact attacks (e.g., infiltration or zero-day exploits) can have devastating consequences. To overcome class imbalance and improve robustness, this emphasizes the need to implement cutting-edge strategies like transfer learning, adversarial domain adaptation, ensemble methods, and cost-sensitive learning.

5.6 CONCLUSION

In addition to offering previously unheard-of possibilities for connectivity, intelligence, and service diversity, the transition to 6G networks also makes it easier for sophisticated cybercriminals to target users. This chapter has examined the integration of cutting-edge security paradigms like machine learning-driven defense mechanisms, O-RAN-based architectural flexibility, and ZTA in order to address these issues. Through the integration of real-world benchmark datasets like CICIDS2017/2018 with adaptive transfer learning techniques like ADDA and CORAL, we have illustrated the potential of intelligent intrusion detection frameworks that can dynamically adjust to diverse traffic patterns and hostile environments.

The evaluation's findings show the current methods' potential as well as their shortcomings. The need for hybrid strategies and deeper model generalization is highlighted by the fact that, although adversarial and correlation-based alignment methods enhance attack detection in transfer learning contexts, their performance still varies across attack classes. Additionally, continuous verification, fine-grained control, and secure orchestration of 6G services are guaranteed by the integration of distributed intelligence and AI-native RAN with ZTA principles.

In the end, a route to creating robust and secure 6G infrastructures is provided by the combination of cloud-native RAN, AI-driven automation, and sophisticated trust models. But realizing this goal necessitates tackling important unresolved problems like minimizing model bias, guaranteeing energy-efficient AI training, and creating international security interoperability standards. In order to provide proactive, flexible, and long-lasting security frameworks, future research should concentrate on integrating quantum-resilient cryptography, federated learning, and transfer learning in the context of 6G.

REFERENCES

1. S. Dang, O. Amin, B. Shihada, et al. "What should 6G be?" *Nature Electronics* 3(1), 20–29, 2020.
2. L. Jiao, Y. Shao, L. Sun, et al. "Advanced deep learning models for 6G: Overview, opportunities, and challenges." *IEEE Access* 12, 133245–133314, 2024.
3. Z. Xiao and Y. Zeng. "An overview on integrated localization and communication towards 6G." *Science China Information Sciences* 65(3), 131301, 2022.
4. C. Yeh, G. D. Jo, Y.-J. Ko, and H. K. Chung "Perspectives on 6G wireless communications." *ICT Express* 9(1), 82–91, 2023.
5. H. Moudoud, L. Khoukhi and S. Cherkaoui. "Prediction and Detection of FDIA and DDoS Attacks in 5G Enabled IoT." *IEEE Network* 35(2), 194–201, 2021. doi:10.1109/MNET.011.2000449.
6. H. Moudoud and S. Cherkaoui. "Empowering security and trust in 5G and beyond: A deep reinforcement learning approach." *IEEE Open Journal of the Communications Society* 4, 2410–2420, 2023. doi:10.1109/OJCOMS.2023.3313352.
7. P. Porambage, G. Gur, D. P. M. Osorio, et al. "The roadmap to 6G security and privacy." *IEEE Open Journal of the Communications Society* 2, 1094–1122, 2021.
8. V.-L. Nguyen, P. C. Lin, B. C. Cheng, et al. "Security and privacy for 6G: A survey on prospective technologies and challenges." *IEEE Communications Surveys & Tutorials* 23(4), 2384–2428, 2021.
9. H. Moudoud, Z. A. E. Houda and B. Brik. "Zero trust security architecture for 6G open radio access networks (ORAN)." *IEEE Networking Letters* 6(4), 272–275, 2024. doi:10.1109/LNET.2024.3514357.
10. K. Trichias, A. Kaloxylos and W. Colin. "6G global landscape: A comparative analysis of 6G targets and technological trends." In: *2024 Joint European Conference on Networks and Communications & 6G Summit (EuCNC/6G* Summit*)*, Gothenburg, Sweden. IEEE, 2024.
11. N. Janbi, I. Katib, A. Albeshri, and R. Mehmood. "Distributed artificial intelligence-as-a-service (DAIaaS) for smarter IoE and 6G environments." *Sensors* 20(20), 5796, 2020.
12. H. Moudoud and S. Cherkaoui. "Enhancing open RAN security with zero trust and machine learning." In: *GLOBECOM 2023-2023 IEEE Global Communications Conference*, Kuala Lumpur, Malaysia, pp. 2772–2777, 2023. doi:10.1109/GLOBECOM54140.2023.10437043.
13. Z. A. El Houda, H. Moudoud and L. Khoukhi. "Blockchain meets O-RAN: A decentralized zero-trust framework for secure and resilient O-RAN in 6G and beyond." In: IEEE *INFOCOM 2024-IEEE Conference on Computer Communications Workshops (INFOCOM WKSHPS)*, Vancouver, BC, Canada, 2024, pp. 1–6. doi:10.1109/INFOCOMWKSHPS61880.2024.10620803.
14. Y. He, D. Huang, L. Chen, Y. Ni, and X. Ma. "A survey on zero trust architecture: Challenges and future trends." *Wireless Communications and Mobile Computing* 2022(1), 6476274, 2022.
15. V. Stafford, "Zero trust architecture." *NIST Special Publication* 800(207), 800–207, 2020.
16. A. Kumar and D. Pandey. "Enhancing intrusion detection with ML and deep learning: A survey of CICIDS 2017 and CSE-CIC-IDS2018 datasets." *AIP Conference Proceedings* 3168, 1, 2024.
17. M. Ouaissa and A. Rhattoy. "A secure model for machine to machine device domain based group in a smart city architecture." *International Journal of Intelligent Engineering & Systems* 12(1), 151–164, 2019.
18. M. Ouaissa and A. Rhattoy. "A new scheme of group-based AKA for machine type communication over LTE networks." *International Journal of Electrical & Computer Engineering* 8(2), 2088–8708, 2018.
19. Z. Zhao, L. Alzubaidi, J. Zhang, Y. Duan, and Y. Gu. "A comparison review of transfer learning and self-supervised learning: Definitions, applications, advantages and limitations." *Expert Systems with Applications* 242, 122807, 2024.

6G-Enabled Blockchain for Authentication and Access Control

6

Raghad Jamal Abunasser and Qasem Abu Al-Haija

6.1 INTRODUCTION

The growing number of Internet of Things devices and drone usage makes secure authentication and access control necessary for modern surveillance systems. These technologies transform our ability to monitor and control physical areas, but simultaneously create major security challenges. The modern digital environment is causing traditional security frameworks to fall short in its rapid transformation. Most security systems suffer from outdated architecture that depends on centralized management, which is inflexible and incapable of efficient access rule modifications in dynamic situations [1].

Traditional authentication methods, including password protection and biometric verification, remain common, yet present important restrictions. The security of passwords deteriorates because users forget them, while hackers and unauthorized access make them vulnerable to attack. The risk associated with biometric data is that any compromised iris scan or fingerprint remains unalterable after the breach. Surveillance systems that use only these authentication methods face high risks because they have weak points that attackers can exploit [2].

The decentralized operation with tamper-resistant characteristics of blockchain technology makes it an attractive solution. Blockchain is distributed throughout its entire network, thus making attacks on a single point ineffective because they cannot cause widespread damage. The system provides more advanced Identity and Access Management (IAM) capabilities, which align better with modern surveillance environment requirements [3].

Blockchain records information so that alterations require agreement from all network participants to occur. Such tracking capabilities become essential for documenting which users access specific resources at specific times for specified reasons. The system permanently records all login successes, blocked attempts, and policy modifications. It detects suspicious behavior and complies with data privacy regulations [4].

Blockchain provides security monitoring capabilities that traditional systems cannot provide. Authorized users can view all access changes and policy updates, leading to increased trust and accountability.

DOI: 10.1201/9781003683599-6

TABLE 6.1 Traditional authentication and access control vs. blockchain-based solutions

ASPECT	*TRADITIONAL METHODS*	*BLOCKCHAIN SOLUTIONS*
Centralization	The central server is a single point of failure	Decentralized architecture lowers systemic risk
Scalability	Limited by outdated infrastructure	Easy scales to accommodate large IoT ecosystems
Policy enforcement	Static and slow to update	Dynamic and automated via smart contracts
Auditability	Incomplete or easily manipulated logs	Immutable, transparent records stored on distributed ledgers
User control	Data owned by central entities	Users retain ownership and control over personal data

Smart contracts push this even further. These automated systems use conditions to activate rule changes. Blockchain technology enables lockdown restrictions to be automatically enforced to allow only first responders during emergencies, whereas conventional systems would experience delays in response [5].

Several weaknesses in traditional systems prove that changes are needed to maintain secure operations. First is the danger of centralization. Legacy systems commonly depend on a single server to perform all login operations and to handle permission control functions. The entire system becomes vulnerable to risk whenever the central server falls victim to an attack. Security breaches targeting centralized servers have resulted in the disclosure of millions of user data records throughout history [6]. Then there is the issue of scale. Modern systems were not designed to handle the substantial number of newly connected devices now populating the market. System performance deteriorates when combined with system crashes, and increased security risks emerge because the system cannot handle excessive authentication requests [7]. The access control models from earlier times maintained rigid rules that do not change. Security breaches or operational emergencies cause traditional access control systems to fail in their ability to adapt swiftly. The delay in updating access rules enables unauthorized users to enter while blocking authorized personnel from gaining access [8,9]. The essential differences between traditional methods and blockchain-based approaches are illustrated through the summary in Table 6.1.

The comparison proves that blockchain technology establishes entirely new security principles instead of providing minor enhancements to existing security systems. Blockchains distribute trust across multiple network nodes, which reduces the opportunities for attackers to identify one vulnerable point. The approach disrupts traditional system designs because it distributes control, making them easy targets for attackers. The comparison proves that blockchain technology establishes entirely new security principles instead of providing minor enhancements to existing security systems. Blockchains distribute trust across multiple network nodes, which reduces the opportunities for attackers to identify one vulnerable point. The approach disrupts traditional system designs because it distributes control, making them easy targets for attackers.

Blockchain technology enables organizations to prove their current privacy measures effectively. The system keeps a traceable log of user interactions that enables verification of unauthorized access attempts. The traceability system builds trust and helps system administrators and regulators who must validate correct data handling procedures [10]. The adoption of blockchain for surveillance systems aligns with the global trend toward security models that place users first and use decentralized frameworks. Various governmental and private organizations aim to find alternatives to control systems operating centrally to deliver better transparency, resilience, and enhanced accountability. Organizations must address the current threat landscape, including complex cyberattacks and IoT botnet campaigns, since these threats keep evolving. The digital world's progression through AI, 5G, and edge computing technology requires defensive infrastructure to protect against increasing risks [11].

Blockchain technology serves an essential purpose beyond current problem-solving because it develops enduring security frameworks. Blockchain's future success depends on its ability to predict future threats and develop security systems at the same pace as new threats emerge. This chapter analyzes traditional authentication and access control system vulnerabilities through blockchain implementation for core surveillance system component transformation. The discussion begins with blockchain basics and explains its relationship to identity and access management principles. The following section presents concrete blockchain implementations in operational surveillance systems. The analysis will present current system weaknesses before revealing continuing obstacles and critical aspects for development in this developing field.

The following chapter provides an exhaustive analysis of blockchain technology applications for constructing secure surveillance infrastructure. The analysis proves blockchain technology satisfies current security demands by developing systems that accommodate present-day digital network complexities.

The following background section investigates how surveillance systems have evolved access control and authentication methods to reveal their limitations. The historical context demonstrates the need to immediately adopt blockchain technology to replace outdated systems alongside modern methods.

6.2 BACKGROUND: AUTHENTICATION AND ACCESS CONTROL IN SURVEILLANCE

The combination of technological progress with increasing IoT device adoption has caused significant transformations in surveillance system development and operation during the last ten years. Today's surveillance infrastructure surpasses its historical forms because it incorporates advanced technology and interconnected systems serving private and commercial domains, urban areas, and critical infrastructure. Traditional surveillance methods experience transformative change through smart cameras, drones, and networked sensors, enabling real-time data acquisition and environmental response capabilities [12]. Drones have brought revolutionary surveillance capabilities through their ability to provide aerial perspectives that boost situational understanding. Technology has become vital for disaster response operations, environmental tracking, and public security applications. IoT devices have established connectivity links by enabling immediate device-to-device data exchange and real-time information transfer. Enhanced system connections boost operational efficiency while enabling rapid, informed decisions [13]. The expanding ecosystem demands advanced authentication and access control systems that adapt to its growing complexity and scale. Integrating numerous devices into one system creates a critical security requirement, ensuring authorized personnel maintain access to sensitive information and system control capabilities. The combination of extensive network distribution and continuous live data transfer requires security solutions that offer greater flexibility and intelligence than previous solutions [14].

The core authentication methods of surveillance systems from the past suffer from various limitations in their operation. The verification tool of passwords stands as one of the leading methods for identity verification even today. The basic nature of passwords makes them vulnerable to their most significant disadvantage. The ease of password creation leads users to select or reuse weak choices. These weaknesses make passwords highly susceptible to threats, including phishing attacks, brute-force assaults, and social engineering techniques [15]. Using biometric authentication through fingerprint, facial recognition, and iris scanning systems provides better security than passwords because each person possesses distinct biological characteristics. The implementation of biometric solutions comes with multiple challenges to consider. Biometric authentication data faces serious privacy risks because of its sensitive nature and because it can be stolen or spoofed when it is stored [16]. A combination of three verification components in Multi-Factor Authentication (MFA) provides enhanced security: something you know (password), something you have (device), and something you are (biometric). Implementing MFA enhances security yet causes obstacles in situations where fast access is critical and delays would create disruptions [17].

Implementing traditional security methods in large-scale surveillance systems often proves insufficient because of their inherent limitations. The systems' scalability reveals their operational constraints while requiring more adaptable solutions. Modern security architecture needs blockchain technology

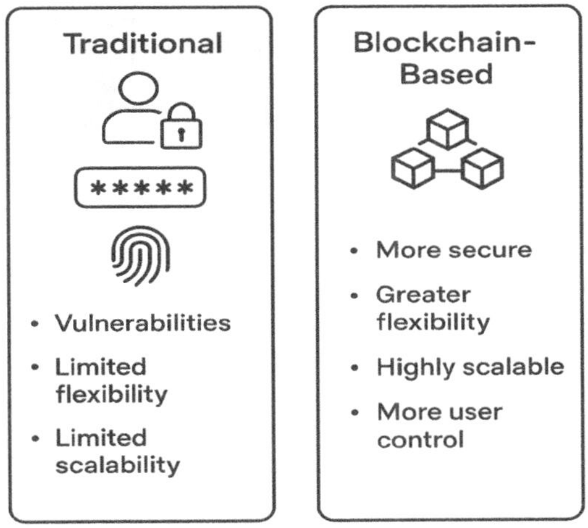

FIGURE 6.1 Traditional vs. blockchain-based authentication and access control models.

solutions because it addresses evolving security requirements better than previous methods. The illustration in Figure 6.1 compares blockchain security requirements, including passwords, biometrics, and MFA.

Surveillance systems depend on access control systems to establish which personnel can access assets with what permission level. System administrators have established multiple access control models that adapt to specific system requirements. Role-Based Access Control (RBAC) is a widely implemented model that assigns access permissions according to established organizational roles. The security personnel maintain greater surveillance access compared to reception staff members at the front desk. The security approach helps organizations better control access in structured systems and protect their overall systems. The strict nature of this model hinders operational flexibility in dynamic work environments where job responsibilities change often [18].

Attribute-Based Access Control (ABAC) presents an alternative access control method with higher flexibility. ABAC differs from traditional role-based systems because it establishes access decisions through multiple user attributes, including identity, resource types, and contextual elements. The system grants access permission only when users authenticate through authorized locations during authorized periods. The security features of this approach allow for detailed control management and better defense measures. System maintenance and administrative supervision became more complex because of the intricate details in policy management [19]. The security approach of Mandatory Access Control (MAC) serves environments with extremely sensitive information, including military operations and government institutions. This system operates in the direction of central authority and uses fixed rules that cannot be altered. The strict nature of MAC ensures high security, but its rigid structure prevents its usage in dynamic surveillance systems that need flexible, rapid decision-making [20].

Surveillance systems benefit from different access control methods, but the evolving operational requirements and security threats require models that respond automatically to system changes. Real-time adaptation becomes essential for distributed surveillance architectures that handle diverse devices and geographic regions with minimal latency. Evolution introduces multiple essential priorities for consideration. First is confidentiality, assuring that sensitive data is only available to those authorized to see it. Organizations must implement powerful encryption systems, secure communication channels, and strict authorization standards to achieve this goal [21]. Integrity is another essential concern. The surveillance data must stay untouched from capture time until analysis completion. Combining cryptographic hashing and digital signatures helps verify that information remains unchanged during transmission [21].

Equally important is availability. Surveillance systems must keep operating continuously while facing hardware breakdowns, natural disasters, and cyberattacks. High availability provides authorized users with uninterrupted access to important data during emergencies [21]. The need for auditability stands as a fundamental element in the current surveillance infrastructure. The combination of detailed logs that track data access along with time stamps, user identities, and purposes ensures both accountability and supports legal and regulatory compliance and facilitates forensic investigations [21]. Real-time data processing represents an essential requirement for modern surveillance systems. Surveillance systems must swiftly handle enormous amounts of input to respond promptly to emerging threats [21]. The growing connection between these systems makes them more susceptible to cyber threats. Unauthorized access remains a significant security issue because outdated authentication systems lack sufficient protection against credential stuffing, phishing, and brute-force attacks [21]. Biometric data faces the most critical security threat from data breaches because compromised biometric information cannot be replaced with new credentials. Since biometric data cannot be changed after an attack, protecting this information must be absolute [21].

The denial of service (DoS) attack presents a common system threat. It enables attackers to overwhelm networks with traffic until systems fail to respond to genuine threats when needed most. In distributed surveillance systems, data transmissions between devices face the risk of man-in-the-middle (MitM) attacks, breaking confidentiality and integrity by allowing attackers to modify or extract information. At the same time, it is transmitted [21]. The risk associated with malware remains significant. Malicious programs can enter surveillance networks through software vulnerabilities or hardware misconfigurations to cause disruptions, steal data, and gain complete system control [21].

Security should no longer be considered an afterthought in system design because these risks exist. The development of surveillance platforms requires designers to implement robust security mechanisms throughout the system-building process. The system requires three main security features: strong authentication protocols, access control systems, continuous software and firmware updates, real-time system activity monitoring, and quick incident response protocols. The next part explores the essential components of blockchain technology after analyzing the problems with conventional authentication and access control frameworks. The following section will introduce blockchain technology as an adaptable solution to enhance surveillance system security. Understanding blockchain core principles allows us to assess its potential solution for the mentioned limitations and future surveillance system requirements.

6.3 FUNDAMENTALS OF BLOCKCHAIN TECHNOLOGY

Blockchains introduce an innovative security paradigm because trust and data integrity remain essential. Technology provides an entire change in basic assumptions by introducing innovative techniques for transparency, decentralized governance, and tampering protection. The system relies on distributed ledgers, cryptographic hashing, and consensus systems as its fundamental principles. This section explains blockchain fundamentals, smart contracts, the classification between public and private networks, and their integration into IoT-based surveillance systems.

The core blockchain system operates under the Distributed Ledger Technology (DLT) designation, contrasting with the traditional centralized databases of older systems. The blockchain network distributes its data across multiple peer-to-peer connections instead of in a single centralized location. Each node holds a copy of the entire ledger, and all participants maintain identical ledger states. The system design eliminates single vulnerable entry points, making unauthorized modifications impossible. The entire network can observe transactions, so tampering becomes much simpler to detect. The ledger prevents alterations because network members need to control most of the nodes to make changes, and this scenario is impossible to occur [22].

Blockchain implements consensus mechanisms that guarantee network-wide agreement before ledger entries become permanent. The systems work to establish network-wide agreements about valid content. Each blockchain network operates with its distinct consensus methods. Different blockchain systems prioritize security features, energy efficiency, or operational speed [23]. Proof of Work (PoW) serves as Bitcoin's validation system. The security features are high, but nodes must resolve complex mathematical problems, resulting in high energy consumption. Proof of Stake (PoS) decreases energy consumption by picking validators according to their network investment amount. The approach allows both speed and sustainability of operations [23].

There are other approaches, too. The permissioned blockchain system employs Practical Byzantine Fault Tolerance (PBFT) for fast node agreement when most participants support the process. The DPoS system enhances scalability because users can select representatives to make decisions through voting. System requirements determine the optimal choice between speed, efficiency, and security [23].

The main advantage of blockchain technology stems from its immutable properties. Once information enters the system, it becomes permanent and unchangeable. Implementing Merkle trees and hash chains enables cryptographic tools to achieve this. The connection between consecutive blocks ensures that any attempt to modify the blockchain would require the network to rehash everything, which occurs only during exceptional circumstances. The surveillance industry benefits from this feature because it ensures accuracy and trust in its operations. Audits and investigations become more efficient by creating logs that cannot be altered secretly. The system becomes more accountable because users cannot deny their actions through cryptographic guarantees [24].

The primary advantage stems from the implementation of smart contracts. Code pieces stored on blockchain networks execute automatically after predefined conditions are satisfied—surveillance benefits from real-time access management through automatic systems that function without human supervision. The system rules would analyze the origin location and current time during decision-making processes [25]. A smart contract will only grant drone access to restricted airspace after validating drone ID information and mission specifications matching pre-approved criteria. The system needs multiple personnel to validate before accessing sensitive documents. If a threat is detected? Self-executing contracts enable the system to shut down access or send alerts immediately because of their automatic execution capabilities [25].

DApps function as an additional layer that operates with the other features. Users can interact with blockchain systems through these apps, while traditional apps require central servers. These applications become resistant to censorship and more resilient against hacking attempts and system crashes. DApps enable surveillance systems to manage identity verification for drones and IoT sensors through decentralized authentication protocols. Through their live access logs, admins can monitor real-time requests made by devices. Because these apps maintain decentralized data storage, they become resistant to major breaches. The decentralized management of surveillance software through DApps provides better privacy and reliability [26].

Figure 6.2 shows the main blockchain components reviewed, including its distributed ledger framework, consensus mechanisms, and cryptographic methods, which sustain it.

The blockchain networks exist in two main types that serve distinct operational requirements. The blockchain systems of Bitcoin and Ethereum operate as open networks where anyone can access them. Users who join the network can verify transactions and monitor all network activities. The open nature of

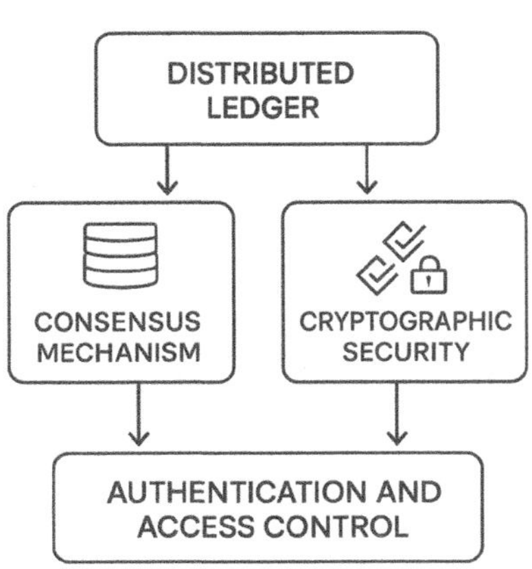

FIGURE 6.2 Blockchain fundamentals and components.

these systems makes them suitable for applications requiring extensive public oversight, such as smart city surveillance. The broad network openness creates multiple negative consequences. The decentralized nature of public blockchains leads to speed and scaling challenges, which limit their use in real-time surveillance applications [27].

Private blockchains operate through distinct operational mechanisms. These systems function with permission-based access, which requires authorized participants to enter the network. The enterprise-focused platforms Hyperledger Fabric and R3 Corda serve organizations requiring strict management and data protection features. Private blockchains operate faster because they do not need consensus from numerous unknown participants, thus enabling Practical Byzantine Fault Tolerance (PBFT) and Raft protocols. Private blockchains provide fast transaction confirmation and improved system performance through this method. Through selective visibility, organizations achieve better security benefits from blockchain since they can conceal sensitive data. Private blockchain technology proves essential for government and military surveillance applications because these organizations require absolute privacy and control [27].

A combination of different approaches often produces the optimal solution. The hybrid blockchain design unites public and private model features to achieve transparency alongside data confidentiality. A system uses private chains to protect sensitive surveillance data while publishing public access logs on a different chain for auditing purposes. The hybrid blockchain model attracts healthcare and law enforcement interest because it fulfills privacy standards while meeting transparency requirements [28].

The implementation of blockchain technology in IoT surveillance systems faces various challenges. The continuous data stream from drones and smart cameras demands systems that quickly and securely process high volumes of information. Traditional blockchains face difficulties because their consensus approaches tend to operate at slow speeds while consuming substantial system resources. Some systems address this challenge by implementing Proof of Authority (PoA) consensus algorithms, decreasing computational requirements while maintaining network trustworthiness. Off-chain processing is a smart approach that records essential events like access permission modifications on the blockchain while handling ordinary data independently to maintain system efficiency [29].

Device identity management has become stronger because of the implementation of blockchain technology. Creating distinct blockchain identities for every IoT device makes it impossible for attackers to create fake devices or perform unauthorized access. A private blockchain network enables smart city drones to authenticate themselves while smart contracts determine which drones can access restricted zones. The established system prevents unauthorized drones from passing through, thus substantially increasing security measures [29].

Implementing blockchain technology in surveillance systems delivers multiple advantages besides access control functions. The system enables immediate policy modifications while generating forensic-quality audit trails and reduces dependence on central authorities, decreasing insider threat exposure. The immutable logs provide an absolute tamper-proof history of all actions, which helps organizations meet legal requirements and facilitates incident investigations [29].

The following section will examine the practical implementation of blockchain technology to enhance authentication after introducing fundamental blockchain concepts. The section demonstrates how blockchain technology creates secure, decentralized frameworks that enhance identity management and establish trust within surveillance systems.

6.4 BLOCKCHAIN FOR AUTHENTICATION: PRINCIPLES AND MODELS

Blockchain authentication represents a major change in basic assumptions because it moves away from centralized management of identities. The Self-Sovereign Identity (SSI) concept allows users to maintain complete control of their credentials because blockchain eliminates the need for trusted intermediaries such as Identity Providers (IdPs) and Certificate Authorities (CAs). This section examines the fundamental concepts of blockchain authentication, including federated identity models and enhanced MFA and Decentralized Identifiers (DIDs). It compares them to traditional methods [30].

Traditional identity management systems experience multiple weaknesses because they depend on single points of failure, create opportunities for data breaches, and restrict user control capabilities. Blockchain uses consensus protocols together with cryptographic proofs to distribute trust throughout networks. A blockchain grants each participant distinctive digital identities, including drones, IoT devices, and people. The cryptographic protection and independent verification of these identities exist without a central authority. The Sovrin and Hyperledger Indy platforms were designed to manage decentralized identities and implement Verifiable Credentials (VCs), which protect privacy and resist tampering [31].

The main advantage of blockchain identity management lies in its ability to defend against Sybil attacks and identity theft. Attackers find creating fake or duplicate identities difficult because network members must verify each request. The blockchain ledger ensures complete transparency when tracing identity origins because it remains unalterable. This feature is valuable for surveillance systems needing strong device authentication against impersonation attempts. Blockchain functions to authenticate drones before airspace entry by verifying their identities to ensure only approved units gain access [32].

Decentralized Identifiers (DIDs) are essential for authentication systems built on blockchain technology. These represent a fresh type of identifier that operates without centralized registration systems. DIDs operate independently from distributed ledgers, and their owners maintain full control of these identifiers, which differ from conventional identifiers such as usernames and email addresses. The World Wide Web Consortium (W3C) has created standards for DIDs, enabling different blockchains to work together. A DID is a pointer to documents containing public keys, authentication procedures, and service locations that verify identities [30].

The SSI concept relies on DIDs to give users complete authority and mobility when handling their identity information. Users and devices can disclose identity information through selective methods without third-party involvement. IoT devices and drones can maintain privacy protection through trusted issuers such as manufacturers or regulators when they prove their credentials. A smart city camera demonstrates blockchain-backed credential verification to control centers to establish legitimacy without disclosing additional information [30].

Users can independently manage their consent and data deletion through SSI without central administrative oversight. Organizations benefit from SSI because it supports GDPR privacy compliance, which remains crucial in surveillance environments, emphasizing data reduction and user privacy standards. According to the Decentralized Identity Foundation (DIF) research, SSI reduces identity fraud by 80% compared to traditional federated systems [33].

The traditional version of MFA remains vulnerable to phishing attacks because it depends on centralized servers. Blockchain technology improves MFA operations through distributed authentication processes and decentralized removal of potential failure points. A blockchain MFA system can achieve security by combining hardware tokens under smart contracts and geospatial oracles for location verification and biometric data stored as zero-knowledge proofs within the blockchain. Multiple security layers protect the system from complete breaches because attackers must overcome each protection step individually [34].

Enterprises use federated authentication extensively, but blockchain introduces genuine advantages through its decentralized trust system. Blockchain enables decentralized federations through which different organizations can establish trust relationships instead of depending on a single identity provider like Google or Microsoft. Smart contracts maintain smooth authentication rules between diverse groups. This approach benefits cross-border surveillance because law enforcement agencies use permissioned blockchains to validate each other's credentials without exchanging personal information [35].

OAuth, Active Directory, and LDAP represent centralized systems without protection against major system failures. The SolarWinds hack revealed how attacks on central identity providers can impact millions of users. The decentralized nature of Blockchain authentication distributes trust among numerous nodes, preventing any single entity from dominating the authentication process, thus providing enhanced security [36].

TABLE 6.2 Centralized authentication vs. blockchain-based authentication

ASPECT	*CENTRALIZED AUTHENTICATION*	*BLOCKCHAIN-BASED AUTHENTICATION*
Trust model	Resolutions on a central authority (CA, IdP)	Decentralized, consensus-based verification
Data ownership	Controlled by service providers	User-controlled (SSI principles)
Resilience to attacks	Vulnerable to single-point breaches	Tamper-proof due to cryptographic immutability
Auditability	Logs can be altered or deleted	Immutable audit trails on the blockchain
Interoperability	Often sold within organizations	Cross-domain compatibility via DIDs

The following Table 6.2 compares centralized and blockchain-based authentication systems based on important aspects.

The National Institute of Standards and Technology (NIST) and other research organizations have discovered that decentralized authentication systems reduce identity-related breaches by more than 60% compared to traditional centralized systems. Blockchain maintains an unalterable and visible record of all login attempts. The permanent ledger system enhances security incident investigations by providing easy access to historical records [37].

Blockchain authentication systems face multiple difficulties during implementation. The public Ethereum blockchain network experiences performance degradation and increased costs when user traffic increases, creating scalability problems. Private blockchain networks operate at higher speeds but sometimes compromise their decentralized structure. The key management process becomes complex because users need a secure yet easy-to-use method to recover their private keys after losing them. The implementation of Hardware Security Modules (HSMs) for secure key storage and Shamir's Secret Sharing (SSS) for key part distribution helps resolve this issue [38].

Regulatory compliance presents a major challenge for organizations. The distributed nature of blockchain technology creates conflicts with surveillance systems that must comply with strict data sovereignty regulations. The solution to this issue involves testing hybrid blockchain systems. The system uses off-chain storage for sensitive data alongside on-chain verification to satisfy legal requirements without compromising security [39]. The discussion about blockchain authentication changes leads to examining access control systems. The following section examines blockchain-based flexible decentralized access control solutions that transform traditional surveillance network access control systems.

6.5 BLOCKCHAIN-BASED ACCESS CONTROL MECHANISMS

Introducing blockchain technology to large surveillance networks transforms access control systems by adopting decentralized, flexible access instead of rigid centralized systems. Real-time updates to permissions exist throughout extensive IoT and drone networks since they operate independently from centralized systems. The built-in capabilities of blockchain, including transparency, unalterable records, and powerful encryption, enable this process [29].

Implementing smart contracts within blockchain technology demonstrates how it implements modern security standards to traditional Role-Based Access Control (RBAC) systems. Blockchain maintains permission data through cryptographic coding rather than storing it in a centralized database. The blockchain permanently logs all role modifications in a way that functions as an impenetrable historical record. A traffic officer in a smart city possesses authorized access to public cameras but lacks permission to access residential cameras. The automated system executes these rules to provide precise and uniform access control according to [29]. The simple management of RBAC remains accessible, but a central failure point becomes impossible in this configuration. The smart contracts' organizational structure enables organizations to modify roles and permissions rapidly during their development. The system maintains importance when personnel roles shift frequently, or security requirements produce new positions [29].

Blockchain systems enhance the capabilities of ABAC authorization systems. Smart contracts enable the execution of complex conditional rules to determine authorization decisions. Drone authorization for specific locations requires the system to verify its identification, position, mission specifications, environmental conditions, and time of day. The blockchain system confirms that all nodes share the same security checks, thus blocking unauthorized access and permission discrepancies [29]. Organizations gain better control through ABAC by implementing specific access rules. Drone operations are limited to daylight hours and specific weather conditions in designated locations. Such access management enhances security standards while preventing unauthorized access in areas with changing environmental conditions [29].

Blockchain technology provides advantages to Discretionary Access Control (DAC), particularly in surveillance operations that involve multiple groups working together. The blockchain platform gives users complete visibility when assigning and revoking access permissions. Security companies can grant police temporary emergency camera access, which automatically expires when the emergency ends. The permanent blockchain record is an everlasting proof against disagreements and establishes trust between parties [29]. Blockchain-based DAC systems incorporate automatic expiration functionality for these temporary rights, which benefits drone mission operations that require immediate access termination after completion [29].

The implementation of blockchain technology delivers more accurate and dependable policy management systems. The execution of smart contracts follows their pre-programmed instructions, which reduces human mistakes. Blockchain technology instantly distributes updated policies throughout all network nodes, enabling immediate enforcement of current rules. The system requires this functionality because security threats evolve rapidly, and policies must adapt correspondingly [29]. The financial district monitoring system implements camera restrictions during peak periods and additional authentication procedures when cyber threats become more probable. The automated rules update function through smart contracts that depend on real-time data [29]. The decision-making process for access control becomes faster through blockchain implementation. Smart contracts operate at a millisecond speed, enabling instant permission adjustments in response to emerging risks. The combination of smart contracts with AI-based analytics produces automated security policy updates. The system grants immediate access to authorized personnel when cameras detect suspicious activities while maintaining complete records of every access attempt for subsequent review [29].

The process of revoking access has become more efficient. Traditional authorization systems take time to terminate credentials, thus creating temporary vulnerabilities for attackers to exploit. Blockchain technology lets the network update smart contracts, immediately blocking compromised credentials. Several advanced systems employ temporary keys that automatically expire at specific points. Security-intensive environments need multiple authorization steps for access revocation to prevent unauthorized activities [29]. Blockchain access control establishes smart cryptographic methods to protect user privacy. Users demonstrate their rights through zero-knowledge proofs, which protect their additional information from disclosure. The application of rules happens directly on encrypted data through homomorphic encryption techniques. Different methods conceal resource access information from users or maintain system accountability features. The system requires the right balance between privacy and transparency for public surveillance operations and sensitive personal data handling [29].

The audit trail function of blockchain technology provides complete visibility through its features. The system permanently documents all access events through cryptographic proof, including granted, denied, and revoked access. The trustworthy log enables organizations to conduct audits, perform system improvements, and conduct forensic investigations. System analytics help detect unusual behavior patterns, policy violations, and key compromise incidents. Security teams obtain complete records, sometimes revealing the precise contract logic behind decision-making processes, thus enabling faster and more

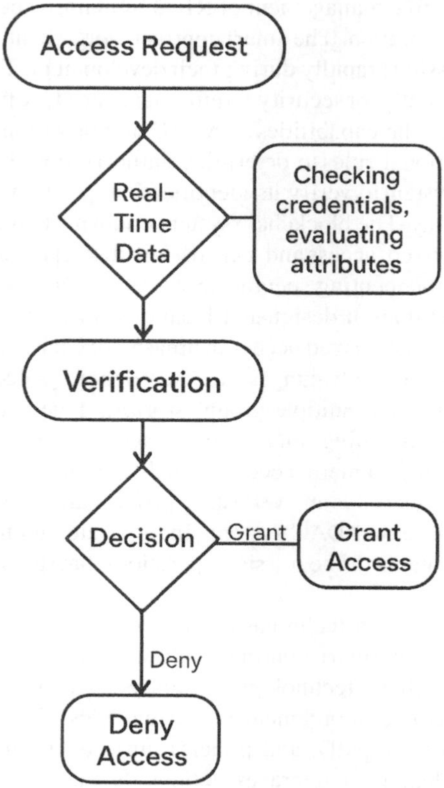

FIGURE 6.3 Smart contract access control workflow.

precise incident response [29]. The diagram in Figure 6.3 demonstrates the process of smart contracts used for access control management.

The complete understanding of blockchain advanced access control requires knowledge of its fundamental system architectures and protocols. The following section describes the essential architectural components required to implement blockchain technology effectively within surveillance systems.

6.6 SYSTEM ARCHITECTURES AND PROTOCOLS

The construction of blockchain-based surveillance networks demands a thorough evaluation of the advantages and disadvantages of blockchain. Traditional speed-based and low-delay systems operate together with blockchain technology during points of maximum value delivery in various resilient setups. The blockchain network foundation uses consensus protocols that fulfill the precise surveillance requirements of its use case. Private or consortium blockchain networks with Practical Byzantine Fault Tolerance (PBFT) and Raft consensus algorithms deliver the optimal balance between system decentralization and operational speed [40].

These systems use different operational layers to create a unified system. The fundamental component of the system is the blockchain network, which maintains records of all access control events. System requirements regarding privacy, scalability, and transparency determine which blockchain type to select between public, private, and consortium options. Private blockchain systems function best when data needs protection from disclosure, but public blockchain networks work best for systems requiring open auditing processes. The selected blockchain model strongly affects the operational security and the system's effectiveness [40].

Smart contracts sit directly above the base layer, which hosts the access control rules. The access control rules and authorization processes within smart contracts determine user access levels and their corresponding conditions. The modular structure allows organizations to update flexible permissions, preventing system disruption. The version control system enables effortless updates while enabling developers to create contracts that adapt to specific roles and context requirements. The access control system uses these contracts to enforce policies and verify identities while managing all activities according to predefined rules. This approach delivers enhanced security through rapid condition changes while simplifying system management [40].

The integration layer serves as the following component, uniting physical surveillance equipment with blockchain components. The system converts device protocol data into blockchain-compatible transactions while determining which access decisions should be processed locally to enhance speed and which need blockchain verification. Techniques like caching help maintain system responsiveness during high-traffic periods, and secure APIs with encryption protect data as it transfers between systems [40].

At the device level, components link the cameras, sensors, or drones to the blockchain network. The blockchain network contains lightweight client devices and powerful gateway devices that function as full blockchain nodes. All devices need to execute access rules and maintain secure operations for cryptographic authentication protocols. Specialized firmware must be implemented because different devices have unique capabilities to handle blockchain interactions properly and enable smooth authentication and communication [40].

Security functions as a vital priority that extends throughout the entire system design. Protecting cryptographic keys and operations relies on Hardware Security Modules (HSMs) and Trusted Platform Modules (TPMs) that function as hardware-based security measures. Network segmentation provides security benefits by splitting blockchain nodes from public-facing surveillance equipment, which enables safe connections and minimizes potential attack entry points [3]. The network encounters scalability issues when it expands because transaction quantities rise. The state channels of Layer-2 solutions enable numerous access control operations to occur outside the blockchain while maintaining blockchain-level security. The main chain receives occasional updates from sidechains specializing in video analytics functions. The system operates efficiently and remains responsive when device numbers and data volumes increase because of these methods [40].

Interoperability also matters for smooth integration with existing systems. The application logic becomes independent from blockchain implementations through abstraction layers, which enable different platforms to function together. The W3C Verifiable Credentials standard enables better access control implementation on blockchain networks while providing users with portable identity features. Transitioning to blockchain transactions can be done gradually through adapters that convert RADIUS and LDAP protocols [41].

The architecture must include adaptability features that enable the future integration of new technologies, including machine learning (ML) and artificial intelligence (AI). The system becomes more intelligent and responsive through automated decision-making capabilities, which detect anomalies and provide predictive insights for surveillance applications. AI systems can dynamically modify access controls using real-time information to maintain strong security measures against developing threats. The architecture of a blockchain-based surveillance system is presented through the visual representation shown in Figure 6.4.

The architectural foundations are now well understood, so we can move on to real-world case studies and applications of blockchain in surveillance. The following examples demonstrate how blockchain technology is used in different sectors to improve security.

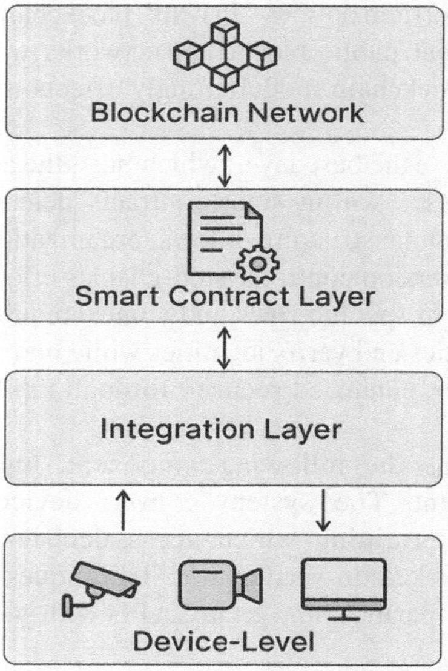

FIGURE 6.4 Blockchain architecture for surveillance systems.

6.7 CASE STUDIES AND IMPLEMENTATIONS

Blockchain technology introduces decentralized trust systems to critical infrastructure security alongside smart cities and drone operations, thus transforming their operational security dynamics. A smart city functions through blockchain authentication of sensors, traffic lights, and cameras, significantly minimizing cyber threats. BCAuthEN systems employ immutable ledgers to protect edge devices while ongoing monitoring with biometric checks maintains privacy protection [42]. The infrastructure sector has received substantial improvements. Combining blockchain's tamper-proof records with automated smart contracts for access management has reduced security incidents in power grids, water supplies, and public transit systems. Every system action gets recorded and verified, proving against any security breach [42].

Drones benefit, too. Blockchain is an authentication method that protects drone identities across IoT networks while maintaining data reliability and preventing identity theft. Blockchain-based authentication reduces unauthorized drone threats, making deliveries more dependable and enhancing disaster response coordination [43]. Technology exists as an actual system. The integration process of blockchain with existing systems demands custom implementation. However, this technology has already begun accelerating energy microgrid transactions, securing healthcare patient information, granting patients control, and fighting supply chain fraud through product authentication verification [44].

Of course, challenges remain. Some blockchain systems require significant computational resources while regulatory frameworks continue to develop. Blockchain provides a new paradigm for securing essential systems through distributed trust mechanisms that eliminate centralized system vulnerabilities. The expansion of blockchain technology creates networks that unite infrastructure systems with improved security measures. The implementation of blockchain for IoT security testing requires real-world environments. The development of pilot programs allows these concepts to advance their performance levels before they become widespread applications. MedAccessX is a blockchain system that combines hybrid access control to protect medical IoT networks. A private Ethereum network test demonstrated that the system performs better with lower expenses than existing alternatives [45]. A zero-trust access system serves as an illustration for 6G IoT environments. The system employs blockchain and advanced encryption to establish detailed decentralized security protocols. Research studies indicated that early implementations produced faster operations alongside improved encryption performance [46]. The pilots create conditions for industrial acceptance of modern technologies. Hyperledger Fabric and Kafka managed high-stress testing of thousands of requests when decentralized storage (IPFS) was integrated to address blockchain scalability issues [47].

The MedAccessX system developed improved smart contracts for access control automation and cost reduction through flexible attribute-based and simple role-based control tests with up to twenty nodes. The open-source code and detailed performance data allow other developers to build upon their work. The zero-trust system fulfills 6G requirements for swift and protected network connections as MedAccessX enhances healthcare information sharing capabilities [45]. Before broad implementation, these pilots demonstrate how to establish regulations and standards by conducting attack resistance tests against identity fraud and data leaks. Financial operations undergo continuous changes due to rapid technological advancements. Blockchain and traditional systems each bring strengths. Blockchain technology's decentralized and transparent nature works well for applications that require high trust, such as smart contracts and international payment transactions. Traditional financial systems provide fast transaction processing and regulatory compliance that adhere to current banking standards [48].

The traditional banking platforms successfully manage massive transaction volumes because they operate on established platforms, yet blockchain technology reduces fraud through its cryptographic mechanisms. The selection between these systems depends on which aspects are more important: fast scalability alongside familiar systems versus secure, tamper-proof operations with transparency. The decentralized blockchain framework eliminates vulnerabilities because it does not depend on any node for operation, enhancing cybersecurity resilience. Smart contracts on this system enable automated process execution, thus eliminating intermediaries and delivering faster operations at reduced expenses. Despite its advantages, Bitcoin demonstrates limited scalability because it handles fewer transactions per second than Visa or SWIFT. The use of many intermediaries in traditional systems leads to delayed operations and additional costs despite their fast transaction speeds [49].

Decentralized Finance (DeFi) represents another blockchain application that enables unbanked individuals to use financial services through blockchain technology without needing conventional banking institutions. Although established systems benefit from well-defined legal frameworks, ambiguous regulatory frameworks restrict the industry's expansion. The conflict emerges from the push for modern technologies to maintain stability [50]. Blockchain technology achieves excellent success in uniting disparate networks through its functionality. The cross-chain transaction capabilities of platforms like Polkadot and Cosmos create better opportunities for data sharing and liquidity. The integration of legacy systems proves difficult because it results in operational difficulties [51]. Blockchain systems face two major difficulties, which stem from complicated regulations and Proof-of-Work systems that consume excessive energy. Traditional platforms maintain foreseeable energy consumption together with established regulatory frameworks. NexGen Networks' Blockchain Interconnect Exchange presents hybrid solutions that unify blockchain benefits, such as scalability and security, to establish an infrastructure for financial operations [52].

The most effective strategy for financial ecosystem development involves merging traditional system speed and compliance with blockchain transparency and automation capabilities. The case studies demonstrate blockchain's potential, yet successful implementation requires overcoming its existing obstacles to achieve its complete potential for surveillance and additional applications.

6.8 CHALLENGES AND LIMITATIONS

6.8.1 Scalability, Latency, and Resource Limits

The process of expanding blockchain systems for access control across extensive IoT and drone networks proves to be challenging. Proof of Work (PoW) blockchains experience performance degradation when handling numerous transactions because of their inability to process them efficiently. Surveillance systems require thousands of simultaneous authentications, which exceed the transaction capacity of Bitcoin and Ethereum networks. Private blockchains that use PBFT consensus methods encounter performance constraints when their nodes are distributed across various locations. Many IoT devices' limited power and memory capabilities prevent them from operating as full blockchain nodes or performing complex cryptographic operations. Implementing sidechains, state channels, and off-chain computing through Layer-2 solutions helps address these issues, but typically reduces security and decentralization. The speed benefits of off-chain work remain limited because it needs periodic on-chain updates, which can slow down access control operations. Storage is another problem. The generation of surveillance data produces massive logs and access records that need permanent storage. Blockchains become impractical for smart city and factory applications when they grow too large because they lack efficient data trimming and sharding methods.

6.8.2 Vulnerabilities: 51% Attacks and Contract Bugs

Blockchains are not bulletproof. The control of most mining or staking power by someone enables attackers to rewrite historical data and perform token double-spending. The risk of a 51% attack remains low for large public chains, yet smaller private surveillance blockchains become more vulnerable to attacks when a few nodes control the network. Smart contracts add risk, too. Implementing access rules through smart contracts creates security vulnerabilities, allowing hackers to gain unauthorized access while blocking authorized users. The 2016 DAO hack demonstrates that contract flaws can result in major security breaches. Formal audits help, but are not foolproof and need specialized skills. The immutability of blockchain data requires disruptive hard forks to correct mistakes, damaging trust in the system. The system faces insider threats because bad actors who have administrative access can control nodes and modify rules. The system requires strong governance, constant security checks, and emergency safeguards to function properly.

6.8.3 Legal and Compliance Challenges

The permanent records of blockchain systems conflict with privacy regulations such as GDPR's "right to be forgotten" provision. Deleting personal data on blockchain systems requires a modification that would compromise the system's integrity. Implementing data storage off-chain and using zero-knowledge proof methods helps, but does not guarantee complete compliance. The implementation of cross-border data rules creates additional challenges. Implementing surveillance systems across multiple countries with different data storage and sharing regulations makes blockchain usage challenging. Identity management also faces issues. The SSI model gives users control over their data, but this approach may violate regulations that demand centralized verification through KYC. Implementing hybrid systems, which store sensitive data off-chain while maintaining access logs on-chain, becomes necessary. Blockchain adoption in sensitive fields will remain slow until regulatory frameworks evolve to match technological advancements.

6.8.4 Usability and Adoption Hurdles

Blockchain systems present challenges to both users and administrators. The process of managing cryptographic keys proves more complicated than handling passwords. Users who lose their keys will experience a permanent lockout because no central recovery mechanism exists. Non-tech users avoid using multi-signature wallets and key sharing solutions because they increase system complexity. Organizations must handle different blockchain platforms because they implement distinct standards, which require managing multiple systems. The administration requires specialized knowledge to construct protected contracts, operate nodes, and resolve consensus problems. Users encounter complex payment procedures through "gas" fees and unpredictable waiting periods, creating difficulties during urgent situations such as emergency drone authentication. The adoption of blockchain technology will remain limited until user interfaces become more straightforward.

6.8.5 Emerging Threats

Blockchain security will encounter fresh obstacles in the future. The current cryptographic systems can be compromised by quantum computers, revealing encryption keys and enabling signature forgery. The development of post-quantum cryptography continues, but its complete implementation requires additional time. Analyzing blockchain data through AI systems presents a risk of weaponization, enabling predictions of system weaknesses and contract manipulation. Exploiting minor consensus flaws by AI systems could result in delayed access transactions. Open systems remain vulnerable to Sybil attacks because attackers can create numerous fake identities to flood the network. Using external data sources creates oracle attacks, allowing false information to deceive contracts into granting unauthorized access. The connection of blockchains will create potential bugs and exploits that span across different chains. The protection of blockchain surveillance requires three essential defensive measures: decentralized oracles, AI-powered audits, and rotating quantum-resistant keys. Blockchain surveillance security requires continuous innovation because attackers rapidly develop their methods.

6.9 FUTURE TRENDS AND RESEARCH OPPORTUNITIES

Emerging technologies advance quickly, and changing security requirements push the creation of blockchain-based access control and authentication systems. The following section presents research directions for developing decentralized security systems with surveillance operations, Internet of Things (IoT), and drone networks. The widespread adoption of blockchain in surveillance operations faces major barriers because of limited scalability, high energy consumption, and latency issues. Researchers must focus on developing consensus protocols requiring fewer computational resources while providing strong security features for upcoming research. The Proof of Stake (PoS) and Delegated PoS and Proof of Authority protocols demonstrate promising capabilities as substitutes for the energy-consuming Proof of Work (PoW) consensus mechanism. Developers should implement off-chain transaction processing with data partitioning techniques such as sharding to enhance transaction throughput.

Public and private blockchain networks should use a hybrid architecture to improve the overall blockchain architecture. The combined approach protects sensitive information through private ledgers yet uses public chains for authentication proof verification to achieve data privacy and system transparency. Developing unified access control protocols for surveillance infrastructure networks requires interoperability between different blockchain systems, and Polkadot and Cosmos serve as successful examples. Combining Artificial intelligence (AI) and machine learning (ML) technologies will enhance the functionality of blockchain-based access control systems. AI analytical systems examine blockchain transaction data to identify security threats when unauthorized IoT devices or drone access occur. Smart contracts can execute adaptive policies automatically through AI-driven permission modifications based on environmental factors, user activities, and threat assessments, thus creating an exciting development. Artificial Intelligence systems with predictive security functions identify potential attack vectors, including Sybil attacks and vulnerabilities in smart contracts, to enable defensive measures in advance. Integrating blockchain technology with AI systems generates problems concerning processing requirements and protecting personal information. The security challenges of combining blockchain with AI can be addressed through federated learning because this method trains models using decentralized data without revealing sensitive information.

Future studies must examine decentralized identity federation mechanisms because modern surveillance systems consist of government agencies, security providers, and hardware manufacturers as entities. Secure trustless identity sharing between organizations becomes possible through Verifiable Credentials and Decentralized Identifiers while avoiding centralized control. Implementing smart contracts for multi-party agreement enforcement should include automatic legal compliance features that permit drone operations across international territories. Combining blockchain with zero-trust security models eliminates implicit trust by performing continuous identity verification and always enforcing least-privilege access. Quantum computing threatens cryptographic algorithms that secure blockchain systems, including RSA and ECDSA. Implementing post-quantum cryptographic solutions, including hash-based signature schemes and lattice cryptography, must become a research priority for blockchain authentication protocols. The development of quantum-resistant decentralized methods to create and distribute keys is essential for defending against quantum-enabled attacks. Developing hybrid cryptographic models will be necessary to support a transition to quantum-resistant blockchains, which should work with current infrastructures.

Adopting blockchain technology across surveillance operations requires complete standardization and regulatory synchronization. Implementing frameworks is necessary to validate blockchain access control systems against data privacy regulations, including GDPR, HIPAA, and NIST. Standards bodies like IEEE and W3C should develop interoperability protocols that enable seamless connection between decentralized identifiers and smart contract platforms. Implementing certification and benchmarking procedures enables organizations to evaluate and improve the security posture of blockchain authentication systems against new cyber threats. This chapter investigates the blockchain revolution in surveillance security and presents directions for building adaptive, resilient security systems.

6.10 CONCLUSION

Blockchain technology transforms surveillance security operations by abandoning centralized systems, which remain at risk of cyberattacks. The system operates through decentralized ledgers using smart contracts and cryptography to address single points of failure, limited scalability, and strict access rules. Users and devices maintain full control over their credentials through SSI, thus decreasing instances of fraud. Modern smart contracts enforce immediate access rules and provide quick adaptation capabilities that older systems lack. Blockchain provides permanent, transparent records that help organizations meet audit requirements and maintain compliance standards. Despite these benefits, there are still certain challenges. The public blockchain model faces scalability issues while being incompatible with privacy regulations in certain cases. Hybrid blockchains and zero-knowledge proofs promise to solve these problems, although key management remains challenging for numerous users.

The upcoming years will demand attention to quantum-resistant cryptography, cross-chain interoperability, and AI solutions. The combination of blockchain activity monitoring through AI and smart contracts enables automatic permission adjustments when system conditions evolve. The development of quantum-resistant encryption algorithms becomes essential because current encryption methods will be vulnerable to quantum computing threats. The interoperability capabilities of Polkadot and Cosmos protocols enable blockchains to communicate, which facilitates drone and IoT device integration as well as connection to existing systems. Decentralized Autonomous Organizations (DAOs) enable users to maintain direct governance control through automated rule-making systems.

The adoption of blockchain faces challenges from regulatory uncertainty and public chain delay issues that impact real-time surveillance operations. To achieve widespread blockchain adoption, standardized frameworks and user-friendly interfaces are needed. The application of blockchain technology delivers a substantial transformation in how trust operations and security management work in surveillance systems. The system eliminates central vulnerabilities by implementing immediate rule enforcement and maintaining permanent audit trails. Blockchain technology presents an excellent solution for securing sensitive infrastructure within smart cities and critical facilities. The complete realization of this concept demands that developers create simpler interfaces and comply with changing regulations and faster consensus methods. The increasing drone and IoT surveillance market requires blockchain technology to provide a scalable, transparent security solution that replaces outdated, vulnerable systems.

This book's fifth chapter evaluates blockchain technology's impact on access control and authentication systems for contemporary surveillance networks that utilize drones and IoT devices. According to the author, decentralized blockchain systems demonstrate superior strength to traditional centralized systems because they avoid scalability and security issues.

The chapter connects theoretical concepts with concrete examples through discussions about decentralized identities alongside smart contract-based access control automation. Real-world examples in case studies establish practical connections between theoretical concepts so readers can develop secure surveillance systems that scale effectively. The chapter's concluding section demonstrates blockchain's potential to create a better security infrastructure.

REFERENCES

1. Y. Jiao, K. Sato, and B. Wu, "A Walk-through Type Authentication System using Real-Time Gaze & Motion Detection," In: *2024 IEEE International Conferences on Internet of Things (iThings) and IEEE Green Computing & Communications (GreenCom) and IEEE Cyber, Physical & Social Computing (CPSCom) and IEEE Smart Data (SmartData) and IEEE Congress on Cybermatics,* Copenhagen, Denmark, 2024, pp. 38–43, doi:10.1109/iThings-GreenCom-CPSCom-SmartData-Cybermatics62450.2024.00030.
2. A. Abu Zaid, M. Aljaidi, Q. Al-Naamneh, and B. Qadoumi, "Advancements and Challenges on the Internet of Drones Security Issues: A Comprehensive Review," In: A. Abu Zaid, M. Aljaidi, Q. Al-Naamneh, and B. Qadoumi (Eds.), *Machine Intelligence Applications in Cyber-Risk Management,* IGI Global, 2024, doi:10.4018/979-8-3693-7540-2.ch001.
3. D. Commey, B. Mai, S. G. Hounsinou, and G. V. Crosby, "Securing Blockchain-Based IoT Systems: A Review," *IEEE Access*, vol. 12, pp. 1–15, 2024, doi:10.1109/ACCESS.2024.3428490.
4. H. Bari and N. Patel, "Generalized Immutable Ledger (GILED) using Blockchain Technology," In: *2023 IEEE International Students' Conference on Electrical, Electronics and Computer Science (SCEECS),* Bhopal, India, 2023, pp. 1–9, doi:10.1109/SCEECS57921.2023.10062983.
5. A. Nigmatov, A. Pradeep, and N. Musulmonova, "Blockchain Technology in Improving Transparency and Efficiency in Government Operations," In: *2023 15th International Conference on Electronics, Computers and Artificial Intelligence (ECAI),* Bucharest, Romania, 2023, pp. 01–06, doi:10.1109/ECAI58194.2023.10194154.
6. F. A. M. Do Nascimento and F. Hessel, "Decentralized Federated Learning for Intrusion Detection in IoT-based Systems: A Review," In: *2022 IEEE 8th World Forum on Internet of Things (WF-IoT),* Yokohama, Japan, 2022, pp. 1–6, doi:10.1109/WF-IoT54382.2022.10152174.
7. V. Hu, D. Ferraiolo, D. R. Kuhn, and K. Scarfone, *Guide to Attribute-Based Access Control (ABAC) Definition and Considerations*, National Institute of Standards and Technology (NIST), 2014.

8. H. Jiang, M. Liu, S. Zhang, R. Zheng, and X. Lan, "DNLight: Learning Efficient Evaluation for Traffic Signal Control," In: *2024 43rd Chinese Control Conference (CCC),* Kunming, China, 2024, pp. 6526–6531, doi:10.23919/CCC63176.2024.10662517.
9. G. Shi, M. Qi, Q. Zhong, N. Li, W. Gao, L. Zhang, and L. Gao, "MedAccessX: A Blockchain-Enabled Dynamic Access Control Framework for IoMT Networks," *Sensors*, vol. 25, no. 6, Art. no. 1857, 2025, doi:10.3390/s25061857.
10. M. Shuaib, S. Alam, M. S. Alam, and M. S. Nasir, "Compliance with HIPAA and GDPR in Blockchain-Based Electronic Health Record," *Materials Today: Proceedings,* 2021, doi:10.1016/j.matpr.2021.03.059.
11. M. J. C. S. Reis, "AI-Driven Anomaly Detection for Securing IoT Devices in 5G-Enabled Smart Cities," *Electronics*, vol. 14, no. 12, Art. no. 2492, 2025, doi:10.3390/electronics14122492.
12. N. S. Labib, M. R. Brust, G. Danoy, and P. Bouvry, "The Rise of Drones in the Internet of Things: A Survey on the Evolution, Prospects, and Challenges of Unmanned Aerial Vehicles," *IEEE Access*, vol. 9, pp. 118750–118776, 2021, doi:10.1109/ACCESS.2021.3104963.
13. A. Barbosa, H. Mubarak, F. Mohammadi, M. J. Sanjari, and M. Saif, "A Real-Time IoT-Based Data Acquisition and Monitoring System for Photovoltaic Applications," In: *2024 3rd International Conference on Power Systems and Electrical Technology (PSET),* Tokyo, Japan, 2024, pp. 667–670, doi:10.1109/PSET62496.2024.10808965.
14. A. Alshaeri and M. Younis, "Efficient Distributed Authentication for Intelligent Transportation Systems Using Mobile Devices," *IEEE Transactions on Intelligent Transportation Systems,* vol. 25, no. 8, pp. 8865–8880, 2024, doi:10.1109/TITS.2024.3376517.
15. S. Sarkar and M. Nandan, "Password Strength Analysis and its Classification by Applying Machine Learning Based Techniques," In: *2022 Second International Conference on Computer Science, Engineering and Applications (ICCSEA),* Gunupur, India, 2022, pp. 1–5, doi:10.1109/ICCSEA54677.2022.9936117.
16. K. Yliana, A. Arina, and P. Anastasia, "Vulnerability of Biometric Protection," In: *2023 International Russian Smart Industry Conference (SmartIndustryCon),* Sochi, Russian Federation, 2023, pp. 275–280, doi:10.1109/SmartIndustryCon57312.2023.10110772.
17. R. A. Grimes, *Hacking Multifactor Authentication,* Wiley, 2021, pp. 122–140, doi:10.1002/9781119672357.ch5.
18. Z. M. Iqal and A. Selamat, "A Comprehensive Analysis of Risk-Based Access Control Models for IoT: Balancing Security, Adaptability, and Resource Efficiency," In: *2024 IEEE International Conference on Computing (ICOCO), Kuala Lumpur,* Malaysia, 2024, pp. 344–349, doi:10.1109/ICOCO62848.2024.10928193.
19. S. Jadhav and N. Pise, "Secure and Transparent Blockchain Donations: An Attribute-Based Access Control (ABAC) Framework for Enhanced Donor Control," In: *2024 IEEE International Conference on Blockchain and Distributed Systems Security (ICBDS),* Pune, India, 2024, pp. 1–6, doi:10.1109/ICBDS61829.2024.10837149.
20. T. Ferrer, S. Céspedes, and A. Becerra, "Review and Evaluation of MAC Protocols for Satellite IoT Systems Using Nanosatellites," *Sensors*, vol. 19, no. 8, Art. no. 1947, 2019, doi:10.3390/s19081947.
21. J. M. Kizza, *Guide to Computer Network Security*, 5th ed., Springer, 2017, doi:10.1007/978-3-319-55606-2.
22. A. K. Tyagi, S. Tiwari, and K. Naithani, "Distributed Systems and Distributed Ledger Technology: An Introduction," In: M. K. Shukla, S. Tiwari and S. Ahmad (Eds.), *Artificial Intelligence-Enabled Digital Twin for Smart Manufacturing,* Wiley, 2024, pp. 103–123, doi:10.1002/9781394303601.ch6.
23. B. S. Anupama and N. R. Sunitha, "Analysis of the Consensus Protocols used in Blockchain Networks: An overview," In: *2022 IEEE International Conference on Data Science and Information Systems (ICDSIS),* Hassan, India, 2022, pp. 1–6, doi:10.1109/ICDSIS55133.2022.9915929.
24. V. Ali, A. A. Norman, and S. R. B. Azzuhri, "Characteristics of Blockchain and Its Relationship With Trust," *IEEE Access*, vol. 11, pp. 19750–19765, 2023, doi:10.1109/ACCESS.2023.3243700.
25. P. Ravisankar, "Smart Contracts: Understanding the Technology Behind Self-Executing Blockchain Agreements," *International Journal of Computer Engineering and Technology*, vol. 16, no. 1, pp. 3288–3300, 2025, doi:10.34218/IJCET_16_01_229.
26. S. Verma, S. Dash, A. Joshi, and Kavita, "A Detailed Study of Blockchain and dapps," In: *2022 International Conference on Cyber Resilience (ICCR),* Dubai, United Arab Emirates, 2022, pp. 1–5, doi:10.1109/ICCR56254.2022.9996003.
27. N. Horvat, D. B. Gajić, P. Trifunović, V. B. Petrović, D. Dragan, and V. A. Katić, "Performance Evaluation of a Distributed Ledger-Based Platform for Renewable Energy Trading," *IEEE Access*, vol. 12, pp. 65234–65247, 2024, doi:10.1109/ACCESS.2024.3416612.
28. O. Can, T. Dag, and M. Kantarcioglu, "A Blockchain-Based Hybrid Architecture for Auditable Consent Management," *IEEE Access*, vol. 12, pp. 70102–70117, 2024, doi:10.1109/ACCESS.2024.3431292.
29. U. Majeed, L. U. Khan, I. Yaqoob, S. M. A. Kazmi, K. Salah, and C. S. Hong, "Blockchain for IoT-Based Smart Cities: Recent Advances, Requirements, and Future Challenges," *Journal of Network and Computer Applications*, vol. 181, Art. no. 103007, 2021, doi:10.1016/j.jnca.2021.103007.

30. A. Goel and Y. Rahulamathavan, "A Comparative Survey of Centralised and Decentralised Identity Management Systems: Analysing Scalability, Security, and Feasibility," *Future Internet*, vol. 17, no. 1, Art. no. 1, 2024, doi:10.3390/fi17010001.
31. P. Nidyanantham, P. M. Rex Aantonny, and P. M. Ponnavaikko, "Blockchain-Based Validating Documents by Self-Sovereign Identity Using Hyperledger Indy," *Gedrag & Organisatie*, vol. 33, no. 3, 2020, doi:10.37896/GOR33.03/493.
32. K. Moriyama and A. Otsuka, "Sybil-Resistant Self-Sovereign Identity Utilizing Attested Execution Secure Processors and Zero-Knowledge Membership Proofs," *IEEE Access*, vol. 13, pp. 8812–8825, 2025, doi:10.1109/ACCESS.2025.3533877.
33. C. Bartolini, G. Lenzini, and L. Robaldo, "The DAta Protection REgulation COmpliance Model," *IEEE Security & Privacy*, vol. 99, 2019, doi:10.1109/MSEC.2019.2937756.
34. J. A. B. Moya, J. Ayoade, and M. A. Uddin, "A Zero-Knowledge Proof-Enabled Blockchain-Based Academic Record Verification System," *Sensors*, vol. 25, no. 11, Art. no. 3450, 2025, doi:10.3390/s25113450.
35. S. S. M. Rahman, N. Hossain, M. Hossain, and M. H. I. Sohag, "OAuth 2.0: A Framework to Secure the OAuth-Based Service for Packaged Web Application," In: S. Paiva (Ed.), *Innovative Perspectives on Interactive Communication Systems and Technologies,* IGI Global, 2020, doi:10.4018/978-1-7998-3355-0.ch005.
36. M. A. Thakur and R. Gaikwad, "User Identity and Lifecycle Management Using LDAP Directory Server on Distributed Network," In: *2015 International Conference on Pervasive Computing (ICPC),* Pune, India, 2015, pp. 1–3, doi:10.1109/PERVASIVE.2015.7086970.
37. P. A. Grassi, J. L. Fenton, E. M. Newton, R. A. Perlner, A. R. Regenscheid, W. E. Burr, J. P. Richer, N. B. Lefkovitz, J. M. Danker, Y.-Y. Choong, K. K., Greene, and M. F. Theofanos, "Digital Identity Guidelines: Authentication and Lifecycle Management," NIST Special Publication 800-63B, 2017. https://doi.org/10.6028/NIST.SP.800-63b
38. T. A. Alghamdi, R. Khalid, and N. Javaid, "A Survey of Blockchain-Based Systems: Scalability Issues and Solutions, Applications and Future Challenges," *IEEE Access*, vol. 12, pp. 57890–57912, 2024, doi:10.1109/ACCESS.2024.3408868.
39. S. Han and S. Park, "A Gap between Blockchain and General Data Protection Regulation: A Systematic Review," *IEEE Access*, vol. 10, pp. 90550–90566, 2022, doi:10.1109/ACCESS.2022.3210110.
40. G. Yang, K. Lee, K. Lee, Y. Yoo, H. Lee, and C. Yoo, "Resource Analysis of Blockchain Consensus Algorithms in Hyperledger Fabric," *IEEE Access*, vol. 10, pp. 69160–69174, 2022, doi:10.1109/ACCESS.2022.3190979.
41. R. Belchior, L. Riley, T. Hardjono, A. Vasconcelos, and M. Correia, "Do You Need a Distributed Ledger Technology Interoperability Solution?" *Distributed Ledger Technologies: Research and Practice*, vol. 2, no. 1, Art. no. 1, pp. 1–37, 2024, doi:10.1145/35645.
42. A. Iftikhar, K. N. Qureshi, F. B. Hussain, M. Shiraz, and M. Sookhak, "A Blockchain-Based Secure Authentication Technique for Ensuring User Privacy in Edge-Based Smart City Networks," *Journal of Network and Computer Applications*, vol. 233, Art. no. 104052, 2025.
43. A. Koulianos and A. Litke, "Blockchain Technology for Secure Communication and Formation Control in Smart Drone Swarms," *Future Internet*, vol. 15, no. 10, Art. no. 344, 2023, doi:10.3390/fi15100344.
44. M. Choobineh, A. Arabnya, B. Sohrabi, A. Khodaei, and A. Paaso, "Blockchain Technology in Energy Systems: A State-of-the-Art Review," *IET Blockchain*, vol. 2, no. 1, pp. 1–17, 2022, doi:10.1049/blc2.12020.
45. G. Shi, M. Qi, Q. Zhong, N. Li, W. Gao, L. Zhang, and L. Gao, "MedAccessX: A Blockchain-Enabled Dynamic Access Control Framework for IoMT Networks," *Sensors*, vol. 25, no. 6, Art. no. 1857, 2025, doi:10.3390/s25061857.
46. S. Nie, J. Ren, R. Wu, P. Han, Z. Han, and W. Wan, "Zero-Trust Access Control Mechanism Based on Blockchain and Inner-Product Encryption in the Internet of Things in a 6G Environment," *Sensors*, vol. 25, no. 2, Art. no. 550, 2025, doi:10.3390/s25020550.
47. F. Wen, Z. Wang, L. Qu, X. Hu, and Y. Liu, "Enhancing Secure Multi-Group Data Sharing Through Integration of IPFS and Hyperledger Fabric," *PeerJ Computer Science*, vol. 10, Art. no. e1962, 2024, doi:10.7717/peerj-cs.1962.
48. K. Sathish, J. G. Thatipudi, P. Manikandan, N. Kanthimathi, T. S. Rao, and P. Alexander, "Blockchain-based Enhancement of Digital Revolution in Financial Sector," In: *2023 International Conference on Sustainable Computing and Data Communication Systems (ICSCDS),* Erode, India, 2023, pp. 1283–1286, doi:10.1109/ICSCDS56580.2023.10104724.
49. M. M. Islam and H. P. In, "An Auditable, Privacy-Preserving, Transparent Unspent Transaction Output Model for Blockchain-Based Central Bank Digital Currency," *IEEE Open Journal of the Computer Society*, vol. 5, pp. 1483–1497, 2024, doi:10.1109/OJCS.2024.3486193.

50. S. Dos Santos, J. Singh, R. K. Thulasiram, S. Kamali, L. Sirico, and L. Loud, "A New Era of Blockchain-Powered Decentralized Finance (DeFi): A Review," In: *2022 IEEE 46th Annual Computers, Software, and Applications Conference (COMPSAC)*, Los Alamitos, CA, USA, 2022, pp. 1286–1292, doi:10.1109/COMPSAC54236.2022.00203.
51. K. Košťál, D. Morháč, and J. Mečír, "Building Interoperability: A Decentralized Bridge Connecting Polkadot and Cosmos Ecosystems," In: *2025 37th Conference of Open Innovations Association (FRUCT)*, Narvik, Norway, 2025, pp. 136–144, doi:10.23919/FRUCT65909.2025.11008190.
52. W. Y. Leong, Y. Z. Leong, and W. S. Leong, "Blockchain Technology in Next Generation Energy Management System," In: *2024 7th International Conference on Green Technology and Sustainable Development (GTSD)*, Ho Chi Minh City, Vietnam, 2024, pp. 15–19, doi:10.1109/GTSD62346.2024.10675034.

6G-Enabled Edge Intelligence and Blockchain Integration for Secure Real-Time Surveillance

7

Nada Mohammad Altwaiq and Qasem Abu Al-Haija

7.1 INTRODUCTION

In the last few years, real-time surveillance systems have seen increased demand due to rising security threats and the need for public safety. Traditional surveillance systems introduce several challenges that hinder them from responding immediately to critical threats, besides the need for manual control. These limitations may lead to missing opportunities for preventing crimes, responding to emergencies, or managing crowded public events effectively. The growth of urban cities and Internet of Things (IoT) technologies raised the expectations for surveillance systems to operate as intelligent, responsive infrastructure, and passive observers. Real-time surveillance features provide immediate control over incidents, allowing quicker emergency response and automated decision-making. These capabilities are essential where risks are common, especially in disaster-prone areas where seconds make a critical difference [1,2].

IoT devices generate enormous amounts of data, such as data collected from smart cameras, drones, and environmental sensors. Traditional surveillance systems would transmit these data to centralized cloud servers for analysis and storage. This process increases delayed connection, consumes bandwidth, and may also raise privacy concerns. Edge computing technology provides a solution to decrease latency by reducing the distance between the data source and the processing center. It allows real-time analytics, making it a suitable solution for critical situations and emergencies. This technology was applied to surveillance applications to benefit from the low-latency feature, which has systems that support real-time decision-making. At the same time, blockchain technology provides a secure, decentralized way to store and verify data because of its cryptographic nature and resistance to tampering. By this, blockchain ensures security, integrity, and transparency.

DOI: 10.1201/9781003683599-7

The intersection between edge computing and blockchain technology provides a complementary role. While edge enhances fast responses, blockchain ensures data security and transparency. Their integration into the IoT surveillance frameworks moves toward decentralized, secure, and intelligent monitoring infrastructures that can respond to modern challenges such as cyber threats, regulatory compliance, and operational scalability.

In this chapter, we focus on multiple objectives: to explore the integration of edge computing and blockchain technologies, especially in IoT surveillance systems, and to explore the limitations of traditional systems in real-time surveillance. This chapter aims to illustrate how combining the low-latency feature of edge computing with the secure and immutable nature of blockchain would lead to smart, responsive, and trustworthy surveillance infrastructures. The key contributions of this chapter include:

- This chapter provides an overview of edge computing and blockchain fundamentals in surveillance terms.
- An insight into several architectures of various system architectures (edge, cloud, and hybrid edge-cloud), advantages, and limitations.
- Presentation of middleware and frameworks for an effective combination of edge-blockchain.
- Examine the role of AI in supporting edge computing, exploring edge AI technology while maintaining resources, and presentation of AI applications at the edge.
- Analysis of blockchain-based authentication, access control, and privacy-preserving techniques. We will also discuss the role of smart contracts in automated policy enforcement.
- This chapter provides insight into practical implementation and pilot projects, and illustrates the real-world application of edge-blockchain systems.
- Determine key challenges of edge-blockchain systems, including latency, scalability, security vulnerabilities, usability, and regulatory concerns.
- Explore future sites for research areas of the next-generation development of edge-blockchain systems.

This chapter is structured into several sections to provide a logical and comprehensive presentation:

1. **Introduction**: present the motivation for real-time surveillance and highlight the features of edge computing and blockchain in IoT.
2. **Background**: **Surveillance, Edge Computing, and Blockchain**: Overview of traditional and IoT-based surveillance, exploration of the edge architecture and blockchain technology, and their role in edge-blockchain surveillance.
3. **System Architectures for Edge-Blockchain Surveillance**: Discusses typical architecture, the role of IoT and drones, network topology, and on-chain/off-chain approaches in blockchain.
4. **Real-Time Edge Processing at the Edge**: Focuses on edge AI, its contribution, and its application in video analytics and anomaly detection.
5. **Blockchain for Security and Integrity**: Representation of authentication, access control, distributed consensus, smart contracts, and privacy-preserving techniques in blockchain.
6. **Integration Models and Frameworks**: Covers frameworks of edge blockchain, middleware solutions, APIs, security at integration points, and performance strategies.
7. **Case Studies and Applications**: Real-world implementations, comparisons with conventional systems, and best practices.
8. **Challenges and Open Issues**: Addresses technical, legal, ethical, and usability barriers in deploying edge-blockchain systems.
9. **Future Trends and Research Directions**: Explores advanced developments such as federated learning, quantum-resistant cryptography, and next-generation blockchain protocols.
10. **Conclusion**: Summarization of the key findings and confirmation of the potential of edge-blockchain integration for surveillance

7.2 BACKGROUND: SURVEILLANCE, EDGE COMPUTING, AND BLOCKCHAIN

The rapid advancement of technological innovations has led to a comprehensive growth of surveillance systems, transforming them from simple, traditional surveillance methods to extremely advanced and complex smart systems. A comprehensive invention of Closed-Circuit Television (CCTV) was in the 20th century, expanding aggressively in various sectors such as banking, retail, and government facilities. However, a notable shortcoming has occurred, including human monitoring to detect anomalies and poor image quality. Besides CCTV, analog cameras were an option in some cases, with the limitation of a steady scene. In the 21st century, surveillance systems integrate with advanced technologies such as artificial intelligence (AI), IoT, and cloud computing. Smart surveillance systems transform manual monitoring techniques into more efficient, faster, and intelligent automated observation techniques. By taking advantage of AI, IoT, and cloud computing, these methods remarkably enhance the capabilities of surveillance systems. This automation enables continuous data collection, real-time analysis, and intelligent decision-making, which decreases human error and improves response times. As a result, smart surveillance systems increase operational efficiency and ensure proactive threat detection, improved situational awareness, and adaptive security management in dynamic environments. However, the wide use of smart surveillance systems, data security, and privacy issues will also need to be addressed [3,4].

7.2.1 Fundamentals of Edge Computing

The adoption of IoT and cloud computing has transformed our lives into a more modern, urban society. This transformation is a significant step toward the deployment of smart cities. The development of more effective safety and security strategies, the creation of strong connections between various systems and devices to create a more comfortable and sustainable urban environment, and the provision of a foundation for efficient resource, infrastructure, and public service management are just a few of the benefits that these technologies offer. Nevertheless, latency, network congestion, and privacy concerns remain challenges for smart cities and other applications; therefore, edge computing was a promising approach to enhancing data security in smart cities by reducing latency and setting data processing close to the data source [5].

7.2.1.1 An overview and concepts of edge computing

Edge computing is a modern paradigm that brings data **processing and analysis closer to data generation sources** such as sensors, devices, or local nodes. Unlike traditional cloud computing that relies on centralized data centers, edge computing reduces latency and network traffic by processing data locally. It enhances distributed resources like micro data centers, cloudlets, or edge nodes near users or devices. Definitions may vary, but they commonly affirm three core features: **proximity to data sources, real-time processing, and computing and network bandwidth integration**. These features reduce latency, improve security concerns, and enable real-time analytics. IoT devices produce a massive volume of data. At the same time, a study showed a production of about 180 petabytes of data per day in a city of 1 million in 2019 and a projection of over 29 billion IoT-connected devices by 2030 [6]. This data volume overwhelms centralized cloud systems in terms of processing, storage, and bandwidth. The shift from centralized cloud computing toward decentralized edge computing addresses these needs by enabling real-time processing and efficient bandwidth usage.

7.2.1.2 Architecture of edge computing

The edge computing architecture is a federated network topology that brings cloud services to the network's edge. The structure of cloud-edge computing contains an edge device layer, an edge layer, and a cloud computing layer. An outline of each layer's components and roles in the edge computing architecture is provided below.

1. **Edge Devices Layer**: This layer comprises all devices connected to the edge (e.g., cameras, sensors, smartphones, drones, etc.). In this layer, devices consume and provide data, which they collect and capture from the physical environment through their sensors. Although the upper layers possess better computational power, the devices in this layer can perform trivial computing.
2. **Edge Layer**: This layer is the core of the three-tier topology. Consists of edge nodes distributed between the upper layer, the cloud layer, the lower layer, or the device layer. Thus, through this distribution of servers, the edge layer addresses latency concerns in cloud computing by processing data closer to the data source. Data is sent to the cloud for further processing after being processed and analyzed in the edge layer. Furthermore, the closeness between the source and edge layer ensures data security.
3. **Cloud Layer**: Although edge computing is a powerful technology, cloud computing still has its own data processing power. This layer consists of high-performance servers and storage devices with powerful computing and storage capabilities, making it suitable for large data analysis tasks like maintenance and business decision support. Additionally, this layer stores data reported from the edge layer and performs analysis tasks that the edge computing layer cannot handle effectively.

7.2.2 Blockchain Technology

Overview and Principles of Blockchain Technology: Blockchain is a distributed, decentralized network that connects participants in a secure, transparent, immutable way, where data is stored in blocks, which are distributed ledgers with no failure points. The main goal of blockchain is to offer an **immutable** transaction across a network without the need for centralized authorities [7]. At its core, blockchain is a distributed database composed of sequentially linked blocks, each block containing a list of transactions. Accordingly, the distribution of data on multiple blocks decentralizes authority. Blockchain ensures that once data is recorded, it cannot be edited or deleted without consensus from the network. This immutable behavior enhances the transparency of this technology. Besides immutability, participants can view historical transaction logs in the public blockchain, which is beneficial for **traceability** in surveillance systems.

Consensus Algorithm: At its core, blockchain depends on duplicating data in blocks. However, consensus algorithms ensure that every duplicate is synchronized and always matches. Consensus algorithms rely on the algorithm design's core principles: fault tolerance, resilience, and privacy preservation [8].

Smart Contracts: A computer-based transaction protocol to lower fraud loss and harmful costs. When stockholders agree on a contract that has been added to the blockchain, transactions can be conducted if certain requirements are fulfilled [8]. Integrating edge computing and blockchain technology introduces a transformative approach to smart surveillance systems. The low proximity of data processing offered by edge computing enables real-time analytics, reduces latency, limits data tampering, lowers network traffic, and enhances data integrity, transparency, and traceability through its decentralized and distributed nature. These capabilities create a cooperative framework where surveillance data can be processed locally and securely. This combination of technologies addresses key challenges in cloud-based surveillance, such as centralized points of failure and vulnerability to data tampering. Surveillance systems benefit from the decreasing latency property to make rapid decisions in time-critical events and improve privacy control, making them more efficient and resilient in critical environments like smart cities, healthcare, and disaster-prone areas.

7.3 SYSTEM ARCHITECTURES FOR EDGE-BLOCKCHAIN SURVEILLANCE

For the Edge-Blockchain surveillance system, it is a significant choice when it comes to architecture. Latency, data privacy, and computational efficiency are critical in this choice. Cloud-based systems rely on centralized servers for data analysis, processing, and storage. However, these capabilities have a strong impact on enabling powerful analytics; they suffer from high latency, bandwidth consumption, and privacy risks, especially in real-time applications. In contrast, local data processing is applied in an edge-based architecture. Data is often processed on the edge layer. The low-latency edge computing capabilities are crucial for time-critical surveillance tasks that require immediate response. Combining edge computing and cloud technologies in one hybrid architecture achieves both best. The type of tasks determines which technology will enable the best solution. Basic analytics tasks like motion detection can be implemented at the edge, while advanced processing like behavior prediction and backup storage can be loaded on the cloud. Further integration with blockchain will enhance the features of secure access and data storage in the edge-cloud hybrid system. See Table 7.1 [9,10] for a comparison between these three architectures.

7.3.1 Role of Drones and IoT Sensors in Edge-Blockchain Surveillance

IoT plays a compatible role in surveillance systems in real-time data analysis, enabling real-time tracking and data collection. A data source serves as the eyes and ears in surveillance systems, gathering information for event analysis. Uncrewed Aerial Vehicles (UAVs), or drones, are highly effective for data collection and transmission, using video cameras to capture visuals and IoT sensors to detect motion, audio, thermal signals, and more. IoT sensors are smart sensors capable of transmitting collected data through a network and are a fundamental part of smart systems such as surveillance networks.

An edge-blockchain infrastructure enables real-time data analysis on the device or edge layer of the edge computing architecture. At the same time, the processed data can be stored securely on the cloud using blockchain in the form of blocks. For instance, if a car accident happens on a street, a drone carrying

TABLE 7.1 Comparison between cloud, edge, and hybrid architectures

CRITERIA	*CLOUD-BASED SYSTEMS*	*EDGE-BASED SYSTEMS*	*HYBRID SYSTEMS*
Latency	High (due to network repetitive transmissions)	Extremely low (local processing)	Optimized (real-time at edge, intensive tasks offloaded to cloud)
Scalability	High (flexible cloud resources)	Moderate (limited by local hardware)	High, with balanced load across layers
Privacy	Low (data transmitted to centralized servers)	High (data processed locally)	Extremely high, especially with blockchain ensuring secure access and audit
Deployment cost	High (requires cloud infrastructure + bandwidth)	Low to moderate (device-level setup)	Moderate (although the setup is complex, it is cost-effective for workloads)

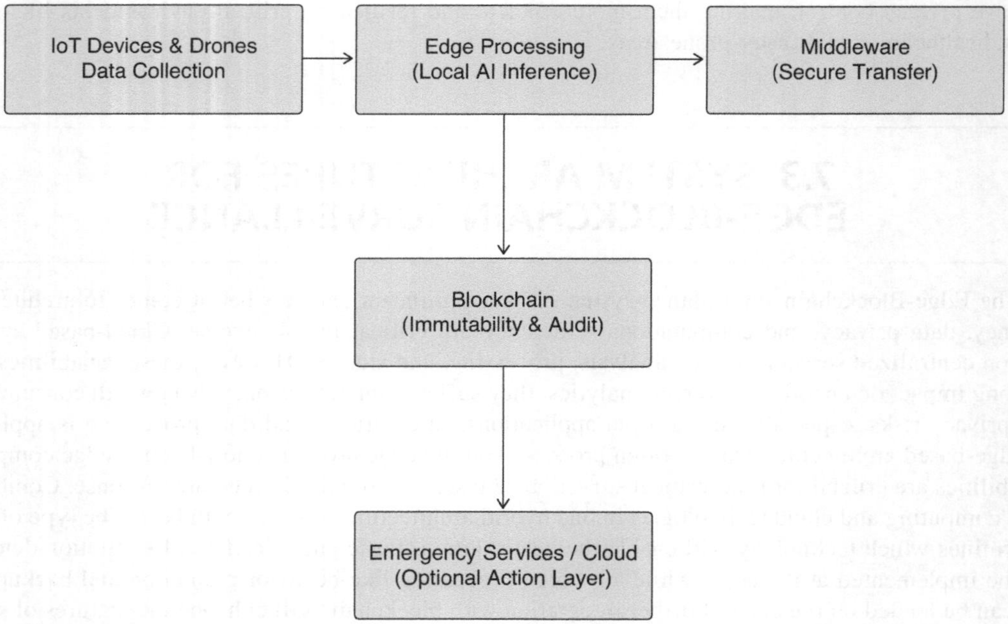

FIGURE 7.1 Data flow in edge-blockchain surveillance system.

a video camera can capture the footage, while audio sensors detect the surrounding sound. The drone and IoT sensors initiate edge-based data analysis, and AI models could be included to maintain tasks; also, the processed information can be sent to the cloud for deeper analytics. Finally, this data can be stored immutably using blockchain technology [11]. These capabilities of edge-blockchain surveillance systems offer a secure, fast, and reliable monitoring environment, especially useful in large public events and disaster-prone areas. Data flow in edge-blockchain surveillance system is illustrated in Figure 7.1.

7.3.2 Network Topology of Edge-Blockchain System

IoT-based systems that adopt edge-blockchain architecture can consist of 3–6 layers, depending on the complexity and required functionality [12]. The basic three-layer architecture is the most fundamental and widely used. It typically includes an edge computing layer, blockchain, and middleware between layers [13]. Each layer uniquely manages data flow, processing, and storage in surveillance environments. A description of the layers is provided as follows:

1. **Edge Computing System**: This layer comprises a stack of connected edge system components. IoT devices are connected to the internet and can collect and transmit data, using network protocols to ensure communication between IoT devices. A user interface (UI) enhances the visual presentation.
2. **Middleware Layer**: This layer bridges the edge computing and blockchain layers in the middle of the network architecture. Automating the connection between layers improves system communication. This layer manages data transmission using connection technologies.
3. **Blockchain Layer**: This layer is the decentralized storage environment. It offers security, privacy, immutability, and data auditability. So once a surveillance event is on the blockchain, it cannot be altered or deleted, which is essential for accountability and trust. Figure 7.2 illustrates this network topology [13].

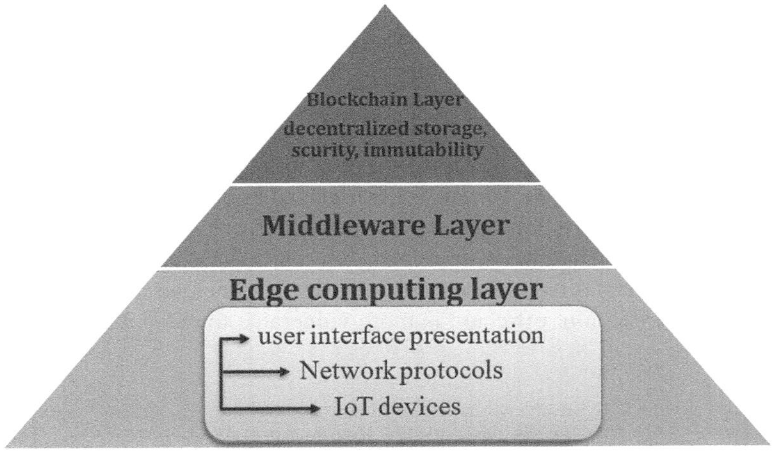

FIGURE 7.2 Three-layer network topology.

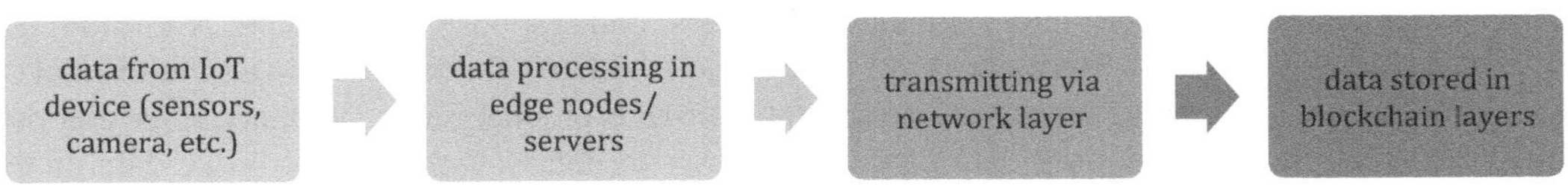

FIGURE 7.3 Data flow in edge-blockchain network.

7.3.3 Data Flow through the Network

Data flows seamlessly through the network topology. The flow begins with IoT sensors and drones, which collect data from the environment. This data is transmitted from IoT devices via network connections to edge servers and processed in real-time. After processing, the significant data is transferred through the networking layer to the blockchain layer, which is stored in a highly secure and immutable manner. Figure 7.3 shows how this data flow performs.

7.3.4 Blockchain-based Methods: On-Chain, Off-Chain

Blockchain technology has addressed several concerns in digital systems, such as security vulnerabilities and the risks associated with centralized architectures. However, blockchain itself suffers from high transaction fees and scalability concerns. IoT-based surveillance systems produce a massive amount of data (e.g., images, videos, audio streams, thermal data, etc.) in addition to real-time analysis performed at the edge. Thus, data size must be considered at the processing and storage levels.

On-Chain Approach: Each block in the blockchain stores network transactions along with metadata. Although on-chain transactions provide security and immutability by storing data directly on the blockchain, this approach may be time-consuming because the consensus mechanism requires validation and is resource-intensive. Furthermore, scalability is limited, as each transaction increases the blockchain's size, reaching hundreds of gigabytes or more. These factors can highly affect the real-time surveillance systems, causing delays and higher operational fees.

Off-Chain Approach: Storing data outside the blockchain infrastructure and maintaining data references or cryptographic hashes of transactions stored on-chain. Therefore, systems that offer larger storage capacity, lower fees, and faster processing capture actual data. The increased processing speed enables off-chain transactions to be processed immediately or within seconds, avoiding the latency of on-chain transactions, which can take minutes or even hours to confirm. Although the off-chain address constrains the on-chain approach, it introduces security and data continuity risks. The sensitive nature of the collected data and the distributed architecture of the edge-blockchain surveillance systems made security and privacy an essential feature to consider. The main principles of blockchain are immutability, decentralization, and cryptographic integrity. On-chain data processing benefits from these principles to ensure strong security. When critical information, such as the time and location of the event or analysis results, is stored on the blockchain, it becomes tamper-proof and permanently auditable. These advantages enhance trust, transparency, and resistance to data manipulation. On the other hand, off-chain offers faster computation and flexible data handling, often occurring on edge devices or cloud infrastructure. Due to the storage location outside the blockchain, off-chain is more vulnerable to cyberattacks, unauthorized access, and data loss.

7.4 REAL-TIME DATA PROCESSING AT THE EDGE

Low-latency processing manages data acquisition, analysis, and response in real time. A minimum network delay achieves a fast or approximate real-time decision, which is fundamental in a time-critical environment, for instance, detecting suspicious movement in crowds or emergency incidents. Edge computing architecture is based on positioning data processors closer to data sources to offer edge data analysis. Therefore, edge analysis and processing present a solution to reduce system latency by analyzing data at the edge nodes. This creates faster decision-making instead of transmitting it to a distant cloud server.

7.4.1 AI at the Edge

Edge AI and Its Operation: Combining edge computing technology and AI to deploy machine learning models on local edge devices, providing real-time feedback. This combination benefits from AI's automated nature and edge analysis's computational capabilities. The training process of edge AI starts with using neural networks and deep learning to train models to perform specific tasks. A centralized data center or cloud is used to process the massive amount of training data for this initial training process. AI models improve gradually, which means that if the model faces an issue, it can improve and handle it. Edge AI models have the same feature: transferring the data that caused the issue to the cloud to train the initial AI model further. This cycle of development through continuous feedback enhances model performance.

Edge AI makes automation in surveillance systems easier by allowing quick decisions on the spot and cutting down on the need to constantly send data to the cloud, unless it is for updating the model. As a result, edge AI is a better option in situations where time-critical predictions and low-latency data processing are required. On the other hand, cloud AI deploys models on cloud servers. Thus, cloud AI applications are fully developed and implemented on the cloud, benefiting from high computational power and scalability. In surveillance systems, where danger sources have high potential and an urgent need for rapid danger detection and real-time feedback, the edge AI capability of local processing reduces network connection risks. However, while cloud computing provides substantial storage capacity and high processing power, it suffers from high latency, which makes it less ideal for real-time applications like surveillance systems.

Resource Management of Edge AI: The wide usage of edge data processing technology has raised several challenges, particularly latency, bandwidth consumption, and storage concerns. These challenges urged the need for efficient resource management strategies that optimize and increase the efficiency of edge processing systems. In addition, resource management solutions require flexibility. Quick adaptation is also essential for real-time response, processing, and environmental factors. These features support fundamental requirements such as low latency and energy efficiency, which are crucial in real-time surveillance systems. Resource management plays a critical role in edge AI due to the substantial demand on system resources. Edge devices have limited computational power and processing capacity, while AI algorithms require high computational power to perform effectively and reduce delay. Tasks of AI, such as object recognition or anomaly detection, would be limited in the edge processing; for this reason, the need for deploying a lightweight model on the edge, such as MobileNetV2, has arisen to avoid bottlenecks and delays. These models can significantly reduce the process delay and perform in less than 50 ms [14]. Technology processes data locally on devices at the edge; edge AI also applies this functionality. By handling data at the source, edge computing eliminates the need to send large volumes of data to the cloud servers, directly affecting the network bandwidth. Bandwidth refers to the volume of data transmitted over a network; consequently, deploying AI at the edge requires lower consumption. Additionally, addressing storage limitations is one of the key resource constraints. Edge AI consumes memory space, and the generated results, such as logs, scores, or alerts, accumulate over time. Without smart strategies to address storage concerns, devices risk running out of available space, which can interrupt processing or cause data loss.

7.4.1.1 Applications of edge AI

1. **Video Analytics at the Edge**: The visual data analysis from video extracts meaningful insights. Many techniques involved in video analytics, such as motion detection, face recognition, and object tracking, are being processed at the edge, allowing systems to process and respond in real time. Thus, real-time decision-making and privacy can be enhanced by avoiding transmitting data to servers through the network and only processing it at the edge. Dynamic environments, such as emergencies and crowded areas, increase the need for an immediate response. For instance, in Wang et al.'s [9] study of surveillance videos prompts extraction using a deep learning CNN; their approach achieved a 5.4 times faster response and reduced bandwidth consumption by seven times less than a public cloud application.
2. **Anomaly Detection in Edge Surveillance**: To explore patterns, behaviors, or events that do not follow the expected behavior. Detecting unusual motion, unexpected crowding, unauthorized access, or abnormal audio signals. By implementing anomaly detection at the edge, surveillance systems ensure a real-time response to a threat, emergency, or situation where time is critical—utilizing machine learning or statistical models on edge devices. Models can train from normal behavior patterns and immediately mark the non-normal behavior. This minimizes the time between detection and action and reduces the need for continuous data transmission—a recent study by Song et al. [10] presented the effectiveness of edge-based anomaly detection in a high-density CCTV environment, using a hybrid approach of edge and GPU. The first phase processes data on the edge; if a critical event is detected, it transfers to the GPU. This approach minimized overall system load and enhanced response time. Table 7.2 includes a summary of edge applications.

TABLE 7.2 Edge AI applications in surveillance

USE CASE	*MODEL TYPE*	*RESPONSE TIME*	*RESOURCE CONSTRAINTS*
Video analytics at the edge	CNN-based models (e.g., motion detection, face recognition)	Up to 5.4× faster than cloud [9]	Limited compute capacity Must operate under bandwidth limits
Anomaly detection in surveillance	Hybrid edge+GPU (machine learning/ statistical models)	Near real-time (critical event triggers GPU processing) [16]	Edge device processing limited Offload only critical data

7.5 BLOCKCHAIN FOR DATA SECURITY AND INTEGRITY

Blockchain-based Strategies for Authentication, Access Control, and Auditability: Blockchain exploits its tamper-resistant ledger and distributes consensus mechanisms to perform secure authentication, access control, and data integrity within sensitive systems. The essential role of data authentication and access control is to enable secure strategies in data management, especially in blockchain-based systems. The data authentication aims to ensure integrity, security, and tamper-proofness during transmission or storage by validating the source integrity. This goal is achieved by blockchain cryptographic techniques such as digital signatures and hashing, which assure the stakeholders that data originates from a valid source and prove who accessed data or took any other action.

In contrast, access control is a technique that manages who is authorized to access data. Anyone wishing to access data can access, modify, or share data within the system. Identity verification and permission management are implemented in blockchain systems to improve access control. Using smart contracts that depend on predefined policies to automatically control access rights and ensure every party is authorized and verified. In addition to keeping data private and managing who can access it, other key features that work together are the permanent records and audit trails. This combination provides a comprehensive approach to data protection. The intersection of this combination and blockchain infrastructure ensures that identities are verified, permissions are strictly enforced, and every activity is recorded immutably and fully auditable over time [15,16].

In blockchain systems, each block is linked to the previous block using a cryptographic hash. This cryptographic architecture provides immutable logging and audit trails. Once the transaction is stored on a block, it is added to the validated blockchain. Whenever any transaction or data is stored, it cannot be altered without changing all subsequent blocks, which makes this operation computationally infeasible in distributed networks. Also, blockchain connects every transaction with a unique identifier and a signature, providing end-to-end traceability. Therefore, blockchain stores data irreversibly and provides an activity log enabling the audit trail feature, offering complete transparency by showing who accessed or modified what data, when, and under what conditions.

7.5.1 Distributed Consensus for Real-Time Systems

Distributed consensus is a mechanism that enables all devices in blockchain networks to agree on the same data and data order in which events occur via timestamps. In timestamps, therefore, real-time systems remain reliable. Consensus protocols are this mechanism's main characteristic, ensuring all participating nodes agree on the system's current state. This is critical, especially in the decentralized system with no central authority and in real-time systems, such as surveillance systems, where processing, sharing, and data validation must be done instantly to achieve integrity. For example, a situation where surveillance devices are used (e.g., drones) to capture a video; consensus protocols ensure the validation and recording of this video's metadata through the network in approximate real-time, ensuring trust, traceability, and resilience. Although traditional consensus algorithms such as Proof of Work (PoW) and Proof of Stake (PoS) ensure strong security, they may cause latency in performance, which is incompatible with time-critical applications. Practical Byzantine Fault Tolerance (PBFT), Delegated Proof of Stake (DPoS), and Proof of Authority (PoA) are lightweight consensus mechanisms that address latency concerns, providing faster validation and preserving reliability, which makes them preferred in IoT research [17].

7.5.2 Smart Contracts for Automated Policy Enforcement in Blockchain-based Systems

Smart contracts are pieces of code that are automatically executed when specific conditions of contracts are met. Contracts are written in programming languages and interact with the blockchain's decentralized ledger. Contracts become part of the blockchain network when deployed and are situated at a specific address, and create a transaction. Transactions are validated by the network through a consensus mechanism to ensure the immutability and transparency of the contracts. Activation of the contracts happens when transactions trigger a predetermined condition. This ensures trust and reliability across all nodes in the blockchain network. The execution of the contracts is recorded on the blockchain, providing an auditable and tamper-proof history of all contract actions. Smart contracts are vital in implementing automated policy enforcement by defining precise access controls and ensuring they are deployed. For instance, they can specify the conditions of permitted access of certain users and devices, which data they can access, and for how long. In edge systems, the data collected from edge devices can only be available to authorized personnel, minimizing unauthorized access or misuse.

Moreover, smart contracts support accountability and compliance by including preservation policies such as automatic data deletion after a period following the completion of a specific condition. In addition, since every transaction driven by a smart contract is recorded immutably on the blockchain, auditing becomes transparent and traceable; thus, any unauthorized change to enforcement policies is immediately detected [18]. Smart contracts have the potential to support adaptability across jurisdictions. Encoding legal frameworks such as the General Data Protection Regulation (GDPR) or the Health Insurance Portability and Accountability Act (HIPAA) into smart contracts may automate specific compliance, such as data retention limits or access control. However, legal interpretation and jurisdictional conflicts still require oversight beyond code.

7.5.3 Privacy-Preserving Techniques

Although blockchain offers strong capabilities such as data integrity, transparency, and immutability, it can be enhanced by integrating privacy-preserving techniques, ensuring that blockchain-based systems remain verifiable and auditable, and the content is kept confidential. A highly effective technique in privacy-preserving is **Zero-Knowledge Proofs (ZKPs).** ZKPs are defined as "cryptographic techniques that enable one to prove possession of specific information without revealing the information itself," based on Watanabe [19]. In the context of surveillance applications where data is sensitive to share, it can be confirmed that the user is verified to access specific data without revealing who this person is or the content of the data. This technique facilitates the validation of sensitive attributes (e.g., identity or permissions) while preserving privacy.

Homomorphic encryption is a widely used privacy-preserving technique that requires no encryption process. In this approach, data is encrypted, and computations are applied to the encrypted data without the urge to decrypt it. This ensures data privacy during processing, even if it is being processed at the edge, where local processing is performed due to latency. On the other hand, **differential privacy** is a widely used mathematical technique. The main goal of this technique is to provide statistical information about a data set with a guarantee of an individual's privacy. This approach ensures privacy by adding controlled noise to the data, which makes it very hard to detect to which individual this information refers. **Secure Multi-Party Computation (SMPC)** allows cooperative work between different entities to perform computations over their inputs while maintaining the privacy of those inputs. In Table 7.3, we provided an overview of privacy techniques.

TABLE 7.3 Summary of privacy-preserving techniques in blockchain-based systems

TECHNIQUE	*CONCEPT*	*APPLICATION IN EDGE SURVEILLANCE SYSTEMS*	*PRIVACY CONTRIBUTION*
Zero-knowledge proofs (ZKPs)	Prove knowledge of a fact without revealing the fact itself	Verifying user access without revealing identity or data content	Enables verification without data exposure
Homomorphic encryption	Perform computations directly on encrypted data	Enables processing of encrypted video/data at the edge without decryption	Enables verification without data exposure
Differential privacy	Add statistical noise to datasets to hide individual entries	Overall behavior analysis without revealing specific user data	Prevents the identification of an individual in shared datasets
Secure multi-party computation (SMPC)	Multiple parties compute functions mutually without sharing raw data	Collaborative analysis between nodes (e.g., smart cameras, edge servers) while keeping inputs private	Maintains input confidentiality in distributed environments

7.6 INTEGRATION MODELS AND FRAMEWORK

7.6.1 Frameworks for Integrating Edge Computing with Blockchain

A well-designed architecture is required to achieve seamless communication and data security in an edge-blockchain combination. These frameworks serve as blueprints that clarify how edge nodes, processing nodes, middleware, and blockchain layers interact, making a resilient and scalable surveillance system. One of the most referenced frameworks is the **Blockchain-based Edge Computing Architecture (BECA) [13].** BECA is a three-layer model; the first layer is an edge computing layer where IoT devices, such as smart cameras and drones, are included to collect data and do initial processing. The second layer is the middleware, which is the main communication line between layers, and the third layer is the blockchain layer, where secure processing is provided and immutable audit trails are kept. This structure provides low latency and security across systems.

Another example is the **EdgeMeediChain** framework proposed by Akkaoui et al. [12]. This framework was primarily developed to secure data exchange between medical records, but it can also be implemented in a surveillance system. Edge devices are used for local processing, and a blockchain secure architecture stores metadata, sensitive data, and access records. This framework is a four-layer system; the first layer is the end-user layer to manage the data-exchanging parties. The second layer is the Local Edge-Mining Layer, where data is processed off-chain after receiving it from the data collector. If this data is critical, it will be sent to the third layer, the Global Blockchain Layer, where off-chain storage provides integrity by storing the blocks' hashes while maintaining authority using a consensus mechanism. The off-chain distributed storage layer is the fourth layer that stores on-chain and off-chain data to ensure privacy and authenticity.

Additionally, **the BlockEdge** framework introduces a scalable and authorized architecture that takes care of the service edge resource in a lightweight blockchain infrastructure. This framework transfers data from the server to the user through the network by verifying mechanisms, improving integrity, and transparency [20]. These frameworks usually depend on specific principles where every network layer has its own responsibility, and the edge helps with the real-time analysis. At the same time, blockchain contributes to security and access control.

7.6.2 Middleware Solutions and APIs

Edge and blockchain networks provide solutions for many concerns; however, the connectivity needs to be addressed further, requiring a cohesive communication layer that ensures interoperability, security, and data consistency. That is where middleware comes to the surface. We can describe middleware as a bridge between the edge-blockchain network's layers, handling the smoothness of the data flow and protocol translation. Middleware manages data flows in edge-blockchain architectures, besides triggering smart contracts to serve and verify access before transferring the data to the blockchain. It also handles the data generated from edge devices by ensuring its compatibility with the verification standards required by the blockchain layer. Service protocols, such as APIs, RESTful APIs, and gRPC protocols, expose the application interface between edge services and blockchain so that they can be proposed for both. An application can be developed using APIs to serve smart contract functions, access controls, or request data records. Lightweight middleware is always preferred to ensure performance quality, especially in resource-limited environments. Furthermore, MQTT or Apache Kafka could be used to support real-time communications and reduce latency in feedback.

7.6.3 Security Mechanisms at Integration Points

Many concerns have been raised since the edge and blockchain integration; one of the most critical concerns is the security at integration points. Integration points connect the data collection, local processing, and immutable storage, making it vulnerable to several attacks, such as man-in-the-middle attacks, data tampering, identity spoofing, and unauthorized access. These threats can be addressed using many mechanisms through interface layers, such as mutual authentication protocols to validate each of the edge devices and blockchain identities before connection; however, in limited edge resource environments, lightweight cryptographic protocols are preferred, such as Elliptic Curve Cryptography (ECC) [21]. The pipelines must be encrypted through data transfer to secure data transmitted between the edge and the blockchain, such as using TLS/SSL. This prevents any possible attack through data exchange, especially if the network is connected wirelessly or in public networks. In blockchain, smart contracts often trigger access controls in automated and immutable ways to ensure only authorized users access, edit, and share data. Thus, it reduces the risk of human error and unauthorized access to sensitive data.

7.6.4 Scalability and Interoperability

In this fast-paced world where the edge-blockchain combination is widely used, there is an increasing data volume, transactions, and number of devices. This raises concerns about the ability of the systems to grow due to their critical effect on the scalability and interoperability of the systems. Blockchain architecture has a core effect on the scalability challenge, specifically when transaction throughput is limited and consensus delays occur, such as **in traditional algorithms like PoW and PoS,** which require energy and high latency time, making them unsuitable for time-critical environments or real-time decision-making. On the other hand, a new adaptation of recent algorithms, such as PBFT, DPoS, and PoA, makes them suitable for many tasks in blockchain-based systems. Furthermore, the **off-chain approach** provided storage to offload large amounts of data while maintaining integrity. Interoperability is the ability of the system to handle several edge devices, protocols, and blockchain platforms. **Middleware** has been able to pay attention to scenarios where networks have a variety of devices and edge nodes vary in hardware and communication standards; the key role of middleware is to connect these differences. Also, **API gateways** and protocol adapters are solutions to address interoperability concerns that ensure data smoothness during exchange between incompatible systems.

7.6.5 Performance Optimization Strategies

Many performance strategies have been developed to address issues of latency, bandwidth, and computation power consumption

- **Dynamic Task Offloading**: Heavy computational tasks are offloaded from edge devices to edge servers of cloud nodes. The offloading guide is based on system load or network bandwidth.
- **Local Data Filtering and Prioritization**: Transfer only relevant and critical events to the blockchain, thus reducing the storage space from the unnecessary data. For instance, cameras can detect an object locally while transferring incidents to the blockchain.
- **Transaction Batching and Asynchronous Processing**: Collect multiple events to be processed into one transaction asynchronously to reduce delay.
- **Layer-2 Solution**: Apply an off-chain technique to reduce the load on the main blockchain layer, such as state channels or rollups.
- **Edge Caching**: Temporarily store the frequently accessed data at the edge to prevent frequent access to the blockchain and enhance query speed.
- Lightweight Consensus Mechanisms.

7.7 CASE STUDIES AND APPLICATIONS

7.7.1 Real-World Implementation

Edge-blockchain integration has moved beyond conceptual designs to practical implementations in the real world across several domains. Here, we propose three state-of-the-art applications combining edge and blockchain integration technology with surveillance systems in drone surveillance, smart city traffic management, and critical infrastructure protection.

1. **Drone-based Surveillance**: Drones have been used in many surveillance scenarios. Gumaei et al.'s study proposed a framework to recognize the drone states, such as powered off, connected, active flight, hovering, and video capture, where data is collected from 5G drones, models are trained and applied to edge devices, and blockchain ensures trust and security. Flight data are being collected and sent to clouds to train DRNN models; after that, these models are dispersed to edge devices to identify drones and their states. Blockchain plays a core role at this level in immutably storing data, ensuring verifiability and trustworthiness. **This approach achieved faster detection times** and protected sensitive flight records from manipulation [22].
2. **Smart Cities and Traffic Management**: Transit in smart cities has been a critical issue; many solutions have been provided. Recently, a study by Chen et al. proposed a **Smart Traffic Management System (STMS)** framework that has the advantage of the key features of edge computing, blockchain, IoT, and reinforcement learning to monitor traffic. Edge devices are used to analyze the road traffic locally, and blockchain and reinforcement learning algorithms ensure the integrity of data transportation through the network. Also, smart contracts are applied to blockchain to enable automatic implementation of traffic policies, maintaining regulatory compliance and traceability. IoT sensors are used on streets and transport units as a data source to collect information [23].
3. **Critical Infrastructure Protection**: "Securing Critical Infrastructure with Blockchain Technology" introduces a blockchain's ability to resist threats on critical infrastructures vulnerable to cyberattacks. Using blockchain's tamper-resistance, distributed architecture, and

auditability enhances the robustness of these critical infrastructures. This study approached solutions by **iteratively refining the system** using user feedback and performance analysis, thus enhancing security resistance to threats [24].

7.7.2 Pilot Projects and Experimental Results

Experimental studies have been proposed and tested due to their vital role in evaluating existing strategies. These experimental studies are still in chapter but propose a future insight into a probable solution. A study by Awais et al. in 2024 has provided a methodology to help emergency vehicles, as these types of vehicles are usually in situations where a street with no traffic or any other obstacles can be reached as fast as possible. For this reason, this study has appeared in the contributed of detecting whether the event is an accident or non-accident. This methodology proposed a technique that depends on visual sensors and cameras that collect videos to recognize traffic accidents using a trained neural network. Once these frames are validated, blockchain provides secure storage for the critical event to ensure integrity and traceability. The system achieved high accuracy in this classification study, achieving 97% for detecting emergencies [25]. Another study presented by Quaisem and Nisa in 2025, to solve the problem of blockchain's high latency in analyzing videos, which affects the performance of real-time surveillance, such as self-driving cars, besides the low computational power of blockchain. To address this problem, this chapter proposed video surveillance techniques to achieve low-latency data transfer in a multiple access edge computing (MEC) system, combined with blockchain for video surveillance in autonomous vehicles, and applied optimization techniques. This simulation approved their method and achieved a vital performance compared to other methods, especially in self-driving cars [26]. Table 7.4 compares traditional and edge-blockchain systems regarding latency, bandwidth usage, scalability, privacy, and auditability. Table 7.4 provides a Comparative Analysis with Conventional Systems.

7.7.3 Lessons Learned and Best Practices

The implementation and experiments of edge-blockchain technology highlighted many lessons and practical recommendations for future implementations.

1. **Balancing between On-Chain and Off-Chain**: Storing all the data from the data source on-chain emerged scalability concerns and costs of transactions, while best practices suggest offloading data to edge or cloud storage and transferring only critical data on-chain for validity.
2. **AI Models at Edge**: Using lightweight AI models at the edge reduces latency and decreases the computation power to support immediate decision-making in real-time surveillance, such as motion detection, without overloading hardware.

TABLE 7.4 Comparative analysis between traditional and edge-blockchain

FEATURE	*TRADITIONAL SYSTEMS*	*EDGE-BLOCKCHAIN SYSTEMS*
Latency	High latency due to the dependency on cloud operations	Low due to local edge processing
Bandwidth usage	High network traffic due to continuous cloud uploads	Reduced bandwidth through local filtering
Security	Exposed to manipulation and unauthorized access in public clouds	Supported by blockchain immutability and smart contracts
Scalability	Limited due to the centralized architecture	Scalable due to the distributed edge nodes
Privacy	High risk due to centralized handling	Better privacy protection using access controls
Auditability	Vulnerable to manipulation	Full audit trails through blockchain

3. **Smart Contract to Deploy Access Control**: By automating smart contracts to be deployed on the access rules based on authority, data type, etc. Thus, enhancing legal and ethical compliance, especially in sensitive environments.
4. **Apply Privacy-Preserving Techniques**: Privacy-preserving techniques are critical in protecting personally identifiable information while maintaining data auditability, such as ZKPs, homomorphic encryption, and differential privacy.
5. **Achieving Interoperability through Middleware**: Middleware is a bridge between distributed edge devices and blockchain protocols. Thus, ensuring a valid connection and triggering smart contracts effectively.

7.8 CHALLENGES AND OPEN ISSUES

Edge-blockchain integration introduces a promising technological advancement; however, it concerns technical challenges in performance and scalability, which may raise questions about its suitability for certain applications. Blockchain networks are known for their limited transaction throughput and high overhead, especially the public blockchain, hindering scalability in real-time systems. This raises concerns when it comes to handling vast amounts of data generated from edge devices such as IoT devices and drones. Although edge computing is developed to minimize the distance between the processing center and the data source to reduce processing delays, blockchain, which involves time-consuming consensus protocols, leads to delays in verifying transactions. Bandwidth constraints also occurred in edge-blockchain systems. The regulatory movement on networks to transmit data between distributed edge nodes and the blockchain consumes substantial network resources, particularly in environments with limited connectivity.

7.8.1 Security Threats of Edge and Blockchain Technologies

Attacks on Edge computing: Due to edge networks' distributed and decentralized nature, edge nodes are more vulnerable to security threats. Edge devices such as IoT devices and drones are often deployed in uncontrolled environments with limited physical protection, exposing them to cyberattacks. Many attacks could be exposed to these devices, such as DDoS, tampering, unauthorized access, and malware injection. Unpatched firmware, open ports, or weak authentication mechanisms could be used to compromise an edge node, gaining access to sensitive data or manipulating device behavior [27–29]. Furthermore, the edge devices' public communication and lack of a security network expose them to man-in-the-middle attacks, eavesdropping, and spoofing. Real-time edge systems also reduce the time for traditional defense mechanisms like patching and manual oversight. In conclusion, attacks highlighted the need for decentralized security measures such as device-level encryption, secure boot processes, and AI-driven anomaly detection to protect edge environments from exploitation.

Blockchain Vulnerabilities: Although blockchain improves security when applied to a system, this does not mean it is immune to vulnerabilities. A 51% attack is a Consensus Protocol attack where an entity or group of entities takes control of more than 50% of the network's computing power (hash rate), which allows them to manipulate transaction records. This threat often occurs in small or private edge networks with limited participants. Smart contract bugs are another vulnerability of blockchain. Attackers can exploit the immutability of contracts by inserting a bug in contract code to manipulate contract behavior, which will be a permanent flaw. In addition, blockchain provides transparency, which can cause privacy issues by exposing sensitive transaction data [30]. Table 7.5 summarizes threats to edge computing and blockchain technology.

TABLE 7.5 Summary of security threats of edge and blockchain technologies

TECHNOLOGY	*THREAT TYPE*	*EXAMPLE*
Edge computing	Physical vulnerability	Device tampering
	Network attacks	Man-in-the-middle attacks, eavesdropping, spoofing
	System exploits	Malware injection, unauthorized access, and DDoS attacks
	Real-time constraints	Delayed updates
Blockchain	51% attack	rewriting transaction history
	Smart contract bugs	Logic errors in contracts, exploitation of unchecked functions
	Transparency exposes transaction data that may be sensitive	Tracking user identity through transaction analysis

7.8.2 Legal, Regulatory, and Ethical Concerns

Edge-blockchain systems can raise significant legal, regulatory, and ethical challenges. These systems collect, process, and store sensitive data, especially in edge-blockchain surveillance systems such as facial images, behavioral patterns, or location information, which are regulated by data protection laws such as GDPR in the EU and the California Consumer Privacy Act (CCPA). Meeting these regulatory requirements is complicated, as data in decentralized architectures could be stored or processed across multiple jurisdictions with varying legal standards. From an ethical insight, edge-blockchain-based systems must balance public protection objectives and individual privacy rights. Safeguards must be set to ensure that they do not support mass surveillance, profiling, or discriminatory practices. In addition, although the blockchain immutability feature ensures auditability, it may conflict with legal rights such as GDPR's law of "the right to be forgotten" [31]. These concerns require careful system design, legal review, and the insertion of privacy-preserving mechanisms.

7.8.3 Usability and Deployment Barriers

Despite edge-blockchain systems having the technical potential, from a usability perspective, they require specialized knowledge in edge computing and blockchain technologies, which could limit the usage of these systems in non-expert organizations. Managing decentralized nodes, configuring, and smart contracts are challenges limiting usability in edge-blockchain systems development. Deployment of edge-blockchain systems requires blockchain integration with various edge devices, from low-power devices to high-performance edge servers, to ensure compatibility and resource optimization. Adoption of a lightweight client interface, flexible middleware architecture, automated deployment mechanisms, and strong developer support are required to enable scalable integration of edge-blockchain systems in real-world applications.

7.8.4 Cost and Resource Limitations

Investments in infrastructure and operational resources are substantial in systems that depend on edge-blockchain architecture, which can create major challenges for adoption, especially in environments with limited resources. Edge technology, where edge devices (e.g., sensors and cameras) are implemented and managed on distributed networks, leads to hardware-related costs, energy consumption, network bandwidth, and on-site maintenance. Also, devices may require frequent updates, secure provisioning, and environmental protection, raising the full cost of deployment and operation. On the blockchain side, resource requirements can be equally high as edge technology. When maintaining full ledgers, operations are usually limited to high storage and computation overhead. Transaction fees are required in public blockchains, while private blockchains require constant management and scaling efforts. Organizations must balance cost effectiveness with system performance and consider trade-offs between decentralization, security, latency, and resource availability.

7.9 FUTURE TRENDS AND RESEARCH DIRECTIONS

7.9.1 Edge AI and Federated Learning

IoT devices are in increasing demand for growth in edge-based systems. This shifted attention toward integrating AI into these systems by using Edge AI. This allows for real-time analytics to be performed locally on devices such as smart cameras, drones, and sensors instead of offloading data to centralized servers for processing. This provides high privacy preservation by reducing data transmission, besides the features provided by edge computing, such as low latency and faster decision-making. Federated Learning has risen in this field, giving promising development. The contribution of Federated Learning in Edge AI is represented by enabling AI models to be trained across multiple edge devices without transferring sensitive data to a central server. Each one of the network's devices trains a local copy of the AI model using its own generated data, and only the updated model parameters are shared and aggregated centrally. This technique serves systems by enhancing privacy and data security, which is significant in surveillance contexts where personally identifiable (PI) information is frequently collected. In addition, this combination allows systems to be intelligent and privacy-conscious. Distributed processing reduces network load and makes the overall architecture more resilient. These advances provide decentralization and real-time responsive monitoring systems.

7.9.2 Next Generation of Blockchain Protocols for IoT

Although blockchain can address many limitations due to its features such as tamper-resistance, traceability, cryptographic protection, and fault tolerance, traditional blockchain protocols are often not ideal for integration with IoT devices, as a result of multiple limitations, including high energy consumption, slow transaction speeds, and large storage requirements. These raise concerns in resource-constrained environments where IoT devices are lightweight, low-power, bandwidth-limited, and deployed on a large scale. This motivates the development of the next-generation blockchain protocols to handle the integration with IoT [19,32]. A lightweight blockchain protocol has been developed to operate efficiently in decentralized networks with minimal computational power. Protocols such **as IoTeX** are designed for IoT environments using lightweight architecture and a DPoS consensus mechanism, enabling secure device identity, real-time data transfer, and privacy-preserving communication between IoT devices [33,34]. Innovations such as **sharding, off-chain scaling, and layer-2 solutions** enhance data throughput while maintaining security. These advancements enable low-latency validation of transactions, authentication, and secure data storage within IoT systems [35,36].

7.9.3 Quantum-Resistant Security

Many assessments today are applied to the fundamental security principles of existing cryptographic systems, including the cryptographic algorithms used in blockchain and IoT-based edge systems. **ECC,**

digital signatures, and other classical encryption algorithms are widely used algorithms in blockchain that guarantee fast, secure communication and transaction verification in most blockchain platforms. However, these algorithms are vulnerable to quantum attacks, especially from **Shor's algorithm**, which can efficiently factor large numbers or compute discrete logarithms. To address this, quantum-resistant cryptographic techniques have been developed and designed to ensure security even under quantum threats. **Post-Quantum Cryptography (PQC)** is a new set of cryptographic algorithms designed to be implemented on classical computers today, but secure against quantum attacks in the future, such as lattice-based, hash-based, and multivariate polynomial-based cryptographic schemes; however, they are not widely adopted in blockchain systems due to several constraints [37,38]. Surveillance systems, especially those based on blockchain, involve long-term data storage and audit trails, which increase the need for data to be resistant against quantum attacks and remain secure for long-term, years or even decades.

7.9.4 Integration of Edge Computing and 5G/6G Wireless Technologies

Edge computing integration with next-generation wireless technologies such as 5G and 6G transforms the IoT ecosystem. While edge computing architecture is meant to reduce latency, deployment of 5G or beyond 5G (B5G) enhances edge computing performance by producing ultra-low latency (in milliseconds), high-speed connectivity, and network slicing capabilities. This combination supports real-time decision-making and enables efficient data exchange between edge nodes and IoT devices. Network slicing in 5G and beyond 5G (B5G) creates multiple virtual networks (slices) and **multi-access (or mobile) edge computing (MEC)** capabilities, which process at the edge of the network, supporting an edge computing strategy. Thus, ensuring reliable and efficient communication between drones, sensors, and smart cameras. Researchers projected 6G to offer greater integration between AI, edge computing, and terahertz communication to achieve ultra-high data rates (up to terabits per second) and sub-millisecond or even microsecond latency. Thus, it enhances the ability to deploy ultra-intelligent and context-aware surveillance systems. We summarize future trends in Table 7.6, their benefits, and examples.

7.9.5 Open Research Questions

Remarkable progress has been made in integrating edge computing and blockchain technology in surveillance. However, several aspects of research have arisen through this work in computing and legal frameworks.

TABLE 7.6 Summary of future trends in edge-blockchain surveillance systems

FUTURE TREND	*BENEFITS FOR SYSTEMS*	*EXAMPLE TECHNOLOGIES*
Edge AI & federated learning	• Reduced latency and network load • Decentralized learning	AI models
Next-gen blockchain protocols for IoT	• Real-time secure data exchange • Lower power and bandwidth usage • Scalability in IoT	IoTeX, DPoS, Sharding, Layer-2 solutions
Quantum-resistant security	• Long-term data protection • Post-quantum secure blockchain for surveillance audit trails	PQC algorithms: lattice-based, hash-based, multivariate polynomial-based cryptographic schemes
5G/6G and edge computing integration	• Milliseconds latency • High bandwidth • Support for real-time surveillance in dynamic environments	Multi-access edge computing (MEC), Terahertz communications

1. **What best practice strategies balance latency, privacy, and computation in distributing AI processing tasks across edge and blockchain layers in dynamic environments (e.g., crowded events, emergencies)?** Although recent studies have proposed foundational work, such as heuristic-based task offloading frameworks for edge, fog, and cloud systems. Real-time adaptation to dynamic environments lacks practical validation.
2. **Explore efficient implementation of ZKPs on lightweight edge devices in surveillance systems while maintaining latency and energy efficiency.** Some studies have tested ZKPs for resource-limited environments. Most implementations are applied to financial blockchains, not real-time surveillance systems.
3. **How can smart contracts integrate real-time legal compliance checks for multi-jurisdictional environments in decentralized systems**? Available compliance frameworks are static and jurisdiction-specific.

7.10 CONCLUSION

This chapter explores the integration of edge computing and blockchain technologies in IoT surveillance systems. Addressing limitations of traditional cloud-based systems regarding latency, integrity, and trust. Edge computing increases the speed of response by maintaining low latency and enabling data processing near its source. Blockchain produces decentralized data storage, immutable audit trails, and secure transaction verification. The combination of these technologies provides real-time data processing while ensuring security and resistance to tampering. Multiple architectures (cloud, edge, and hybrid edge-cloud technology) have been reviewed, exploring the network topology of edge-blockchain systems and their data flow, and the role of IoT devices in edge-blockchain surveillance. Investigate on-chain and off-chain approaches, edge AI technology, and its application (video analytics and anomaly detection). Consensus Algorithms and Smart Contracts were reviewed, and their role in automating access control, policy enforcement, and consensus real-time decision-making was identified. Also, privacy-preserving techniques such as ZKPs were discussed to provide additional protection for sensitive data in blockchain. Despite the efficient performance of edge-blockchain systems, they face technical, ethical, and legal limitations. These limitations, which opened areas for research, are represented in scalability, response latency, and limited edge device resources. In addition, issues related to data ownership, transparency, and automated surveillance must be carefully addressed, especially in multi-jurisdictional environments. Future contributions in optimizing edge-blockchain technology can be summarized in these points:

- Enhancing edge-AI distributed strategies by designing an adaptive framework for distributing AI tasks between edge nodes and cloud servers.
- Explore the implementation of ZKPs on limited-resource IoT devices to ensure privacy without affecting performance.
- Creating adaptive smart contracts capable of responding in real-time in legal frameworks.
- Propose resource-aware strategies and lightweight edge-blockchain frameworks designed for resource-limited environments.
- Exploit ultra-low-latency and high-bandwidth features of 5G/6G to enhance real-time processing and communication between edge and blockchain layers.

This chapter enhances the book's vision of surveillance regarding the intersection of edge computing and blockchain and how it enables secure, intelligent, and real-time monitoring infrastructures, providing a significant base for surveillance. Although challenges of traditional systems still exist, this chapter addressed these challenges and offered architectural, technical, and practical insights. In conclusion, we can say that this chapter shows a clear way toward real-time surveillance.

REFERENCES

1. M. F. Nurnoby and T. Helmy, "A Real-Time Deep Learning-based Smart Surveillance Using Fog Computing: A Complete Architecture," *Procedia Computer Science*, vol. 218, pp. 1102–1111, 2023, doi:10.1016/j.procs.2023.01.089.
2. P. C. P. M. Sajjanar, "Real-Time, Low-Latency Surveillance Using Entropy-based Adaptive Buffering and MobileNetV2 on Edge Devices," *arXiv* preprint arXiv:2506.14833, 2025. doi:10.48550/arXiv.2506.14833.
3. G. Moepi, "Smart Surveillance Systems: Trends, Challenges and Future Directions," *Indonesian Journal of Computer Science*, vol. 14, no. 2, 2025, doi:10.33022/ijcs.v14i2.4802.
4. J. I. Khan, J. Khan, F. Ali, F. Ullah, J. Bacha, and S. Lee, "Artificial Intelligence and Internet of Things (AI-IoT) Technologies in Response to COVID-19 Pandemic: A Systematic Review," *IEEE Access*, vol. 10, pp. 62613–62660, 2022, doi:10.1109/ACCESS.2022.3181605.
5. Y. Zhang and J. Feng, "Towards a Smart and Sustainable Future with Edge Computing-Powered Internet of Things: Fundamentals, Applications, Challenges, and Future Research Directions," *Journal of the Institution of Engineers (India): Series B,* vol. 106, no. 2, pp. 785–804, 2025, doi:10.1007/s40031-024-01186-w.
6. D. Kolevski and K. Michael, "Edge Computing and IoT Data Breaches: Security, Privacy, Trust, and Regulation," *IEEE Technology and Society Magazine*, vol. 43, no. 1, pp. 22–32, 2024, doi:10.1109/MTS.2024.3372605.
7. M. K. I. Rahmani, "Blockchain Technology," In: P. K. Paul, P. S. Aithal, and K. K. Shukla (Eds.), *Blockchain Technology and Computational Excellence for Society 5.0*, IGI Global, 2022, pp. 16–27, doi:10.4018/978-1-7998-8382-1.ch002.
8. S. Islam, M. J. Islam, M. Hossain, S. Noor, K.-S. Kwak, and S. M. R. Islam, "A Survey on Consensus Algorithms in Blockchain-based Applications: Architecture, Taxonomy, and Operational Issues," *IEEE Access*, vol. 11, pp. 39066–39082, 2023, doi:10.1109/ACCESS.2023.3267047.
9. S. Wang, S. Yang, and C. Zhao, "SurveilEdge: Real-time Video Query based on Collaborative Cloud-Edge Deep Learning," In: *Proceedings of the IEEE Conference on Computer Communications (INFOCOM 2020)*, Toronto, ON, Canada, April 27–30, 2020.
10. J. Song and J. Nang, "Pedestrian Abnormal Behavior Detection System Using Edge–Server Architecture for Large–Scale CCTV Environments," *Applied Sciences*, vol. 14, no. 11, p. 4615, May 2024, doi:10.3390/app14114615.
11. S. K. Jagatheesaperumal, M. Rahouti, K. Xiong, A. Chehri, N. Ghani, and J. Bieniek, "Blockchain-based Security Architecture for Unmanned Aerial Vehicles in B5G/6G Services and Beyond: A Comprehensive Approach," *IEEE Open Journal of the Communications Society,* vol. 6, pp. 1042–1069, 2025.
12. R. Akkaoui, X. Hei, and W. Cheng, "EdgeMediChain: A Hybrid Edge Blockchain-based Framework for Health Data Exchange," *IEEE Access*, vol. 8, pp. 113467–113486, 2020, doi:10.1109/ACCESS.2020.3003575.
13. O. J. Ajayi, J. Rafferty, J. Santos, M. Garcia-Constantino, and Z. Cui, "BECA: A Blockchain-based Edge Computing Architecture for Internet of Things Systems," *IoT*, vol. 2, no. 4, pp. 610–632, 2021, doi:10.3390/iot2040031.
14. L. Banjanović-Mehmedović and A. Husaković, "Edge AI: Reshaping the Future of Edge Computing with Artificial Intelligence," In: I. Karabegović (Ed.), *Basic Technologies and Models for Implementation of Industry 4.0.* Academy of Sciences and Arts of Bosnia and Herzegovina, 2023, pp. 133–160, doi:10.5644/PI2023.209.07.
15. M. B. Thazhath, J. Michalak, and T. Hoang, "Harpocrates: Privacy-Preserving and Immutable Audit Log for Sensitive Data Operations," In: *Proceedings of the IEEE International Conference on Trust, Privacy and Security in Intelligent Systems and Applications (TPS-ISA)*, Atlanta, GA, USA, November 9–12, 2022.
16. A. Aßmuth, R. Duncan, S. Liebl, and M. Söllner, "A Secure and Privacy-Friendly Logging Scheme," *Proceedings of the Twelfth International Conference on Cloud Computing, GRIDs, and Virtualization (Cloud Computing 2021)*, Porto, Portugal, April 18–22, 2021, pp. 8–12.
17. N. Moosavi and H. Taherdoost, "Blockchain Technology Application in Security: A Systematic Review," *Blockchains*, vol. 1, no. 2, pp. 58–72, 2023, doi:10.3390/blockchains1020005.
18. L. Alevizos, "Automated cybersecurity compliance and threat response using AI, blockchain and smart contracts," *International Journal of Information Technology*, vol. 17, no. 2, pp. 767–781, 2025, doi:10.1007/s41870-024-02324-9.
19. D. Das, "Toward Next Generation of Blockchain Using Improvized Bitcoin-NG," *IEEE Transactions on Computational Social Systems*, vol. 8, no. 2, pp. 512–521, 2021, doi:10.1109/TCSS.2021.3049477.

20. D. Guha Roy, "BlockEdge: A Privacy-Aware Secured Edge Computing Framework Using Blockchain for Industry 4.0," *Sensors*, vol. 23, no. 5, p. 2502, 2023, doi:10.3390/s23052502.
21. K. Kaur, S. Garg, G. Kaddoum, M. Guizani, and D. N. K. Jayakody, "A Lightweight and Privacy-Preserving Authentication Protocol for Mobile Edge Computing," In: *Proceedings of the IEEE Global Communications Conference (GLOBECOM 2019)*, Waikoloa, HI, USA, 9–13 December 2019, pp. 1–6.
22. A. Gumaei, M. S. Al-Rakhami, M. M. Hassan, M. Alrashoud, and G. Fortin, "Deep Learning and Blockchain with Edge Computing for 5G-Enabled Drone Identification and Flight Mode Detection," *IEEE Network*, vol. 35, no. 1, pp. 94–100, 2021, doi:10.1109/MNET.011.2000204.
23. Y. Chen, Y. Qiu, Z. Tang, S. Long, L. Zhao, and Z. Tang, "Exploring the Synergy of Blockchain, IoT, and Edge Computing in Smart Traffic Management across Urban Landscapes," *Journal of Grid Computing*, vol. 22, no. 2, p. 45, 2024, doi:10.1007/s10723-024-09762-6.
24. J. Govea, W. Gaibor-Naranjo, and W. Villegas-Ch, "Securing Critical Infrastructure with Blockchain Technology: An Approach to Cyber-Resilience," *Computers*, vol. 13, no. 5, p. 122, 2024, doi:10.3390/computers13050122.
25. M. Awais, M. A. Khan, A. Almogren, and A. R. Alharbi, "Deep learning based enhanced secure emergency video streaming approach by leveraging blockchain technology for Vehicular AdHoc 5G Networks," *Journal of Cloud Computing*, vol. 13, no. 1, p. 130, 2024, doi:10.1186/s13677-024-00665-1.
26. M. T. Quasim and K. U. Nisa, "Blockchain-Enabled Edge Computing Techniques for Advanced Video Surveillance in Autonomous Vehicles," *Computers, Materials & Continua*, vol. 83, no. 1, pp. 1239–1255, 2025, doi:10.32604/cmc.2025.061541.
27. A. M. Sheikh, M. R. Islam, M. H. Habaebi, A. Kabbani, S. A. Zabidi, and A. R. bin Najeeb, "Securing the IoT Edge Devices Using Advanced Digital Technologies," *Asian Journal of Electrical and Electronic Engineering*, vol. 4, no. 2, pp. 52–60, 2024, doi:10.69955/ajoeee2021.v1i2.69.
28. M. Aamir and M. A. Zaidi, "DDoS Attack and Defense: Review of Some Traditional and Current Techniques," *arXiv* preprint arXiv:1401.6317, 2014.
29. G. Alandjani, "Securing Edge Devices: Malware Classification with Dual-Attention Deep Network," *Applied Sciences*, vol. 14, no. 11, p. 4645, 2024, doi:10.3390/app14114645.
30. S. E. Ali, N. Tariq, F. A. Khan, M. Ashraf, W. Abdul, and K. Saleem, "BFT-IoMT: A Blockchain-based Trust Mechanism to Mitigate Sybil Attack Using Fuzzy Logic in the Internet of Medical Things," *Sensors*, vol. 23, no. 9, p. 4265, 2023, doi:10.3390/s23094265.
31. GDPR.EU., "Everything you need to know about the 'Right to be forgotten,'" General Data Protection Regulation (GDPR). https://gdpr.eu/right-to-be-forgotten/.
32. S. Tanwar, N. Gupta, C. Iwendi, K. Kumar, and M. Alenezi, "Next Generation IoT and Blockchain Integration," *Journal of Sensors*, vol. 2022, pp. 1–14, 2022, doi:10.1155/2022/9077348.
33. Phemex Academy, *What is IoTeX: A Privacy-Centric Blockchain for the IoT*, Phemex Academy.
34. E. U. Haque, A. Shah, J. Iqbal, S. S. Ullah, R. Alroobaea, and S. Hussain, "A scalable blockchain-based framework for efficient IoT data management using lightweight consensus," *Scientific Reports*, vol. 14, no. 1, p. 7841, 2024, doi:10.1038/s41598-024-58578-7.
35. A. Gangwal, H. R. Gangavalli, and A. Thirupathi, "A Survey of Layer-Two Blockchain Protocols," *Journal of Network and Computer Applications*, vol. 209, p. 103539, 2023, doi:10.1016/j.jnca.2022.103539.
36. M. M. Al Barat, S. Li, C. Du, Y. T. Hou, and W. Lou, "SoK: Public Blockchain Sharding," In: *Proceedings of the IEEE International Conference on Blockchain and Cryptocurrency (ICBC 2024)*, Dublin, Ireland, 27–31 May 2024, pp. 766–783, doi:10.1109/ICBC59979.2024.10634422.
37. S. Almuhammadi and S. Alghamdi, "A Novel Transition Protocol to Post-Quantum Cryptocurrency Blockchains," *Frontiers of Computer Science*, vol. 7, 2025, doi:10.3389/fcomp.2025.1457000.
38. P. Nwaga and S. Idima, "Post-Quantum Cryptographic Algorithms for Secure Communication in Decentralized Blockchain and Cloud Infrastructure," *International Journal of Computer Applications Technology and Research,* 2022, doi:10.7753/IJCATR1104.1008.

Blockchain-Enabled Safe UAV Communication in a 6G Setting

Structure, Prospects, and Challenges

8

Naser Abbas Hussein, Khadija Rammeh Houerbi, and Hella Kaffel Ben Ayed

8.1 INTRODUCTION

The mobile communication system has seen significant modification since the 1980s. Traditional communication bottlenecks are being broken by innovation, and the communication industry's enormous demands and new uses are continuously growing. As more and more robots and people join mobile communication networks [1].

The fifth-generation (5G) network, which allows the connections of massive devices and can meet requirements like high traffic flow capacity and customized user service experience, is one example of the large broadband applications that are continuing to emerge in relation to the fourth generation (4G) Long-Term Evolution Advanced (LTE-A) system with increasingly high rates [2].

5G brings the concept of the Internet of Everything (IoE), while the sixth-generation (6G) network is expected to evolve into a platform for the intelligent IoE [3].

8.1.1 5Gs Limitations

5G technology is already being utilized in smart and mobile devices all around the world. However, as we proceed into the 2020s, various problems have been identified, and more assumptions need to be fulfilled

DOI: 10.1201/9781003683599-8

in 5G. It demonstrates the present technical issues based on 5G technology that provide high-frequency bands in a mobile communication network, including commercial millimeter wave (mmW). Additionally, it offers ultrafast wireless data with a bandwidth of 1–2 GHz, which is significantly faster than 4G technology. Because it meets the unique needs and standards of the future industry and society, it has garnered a lot of attention from the research community.

There are issues with 5G technology's coverage/mobility and uplink performance in non-line-of-sight (LOS) environments, such as those caused by buildings, trees, hills, mountains, and, occasionally, high-voltage electrical power lines. 6G technology is therefore necessary since it can increase industry use cases' flexibility and dependability. LTE has been offering best-effort services on downlink data rates in recent years.

8.1.2 Evolution of 6G

The following are some particular worldviews that 6G communication can enhance, including social issues in society, human-IoT communication, synchronization, etc.

1. **Resolving Social Issues**: 5G is anticipated to address a number of social issues by utilizing high-speed, low-latency connectivity to deliver solutions for things like remote control, telemedicine, driverless cars, distance learning, etc. Furthermore, by the 2030s, when 6G technology is taken into consideration, the social concerns will have been fully developed and solved.
2. **Human-Internet of Things Communication**: With the use of 6G technology, tactile sensing will improve, and human-IoT communication can become more realistic.
3. **Extension of the Communication Environment**: In the modern world, communication is crucial for tying together people, smart devices, etc. Therefore, in the future, 6G technology adoption will increase globally in order to expand quick and dependable communication between humans and objects.
4. **Synchronization**: It is a crucial prerequisite for companies and applications that depend on time. For instance, time-sensitive backbone systems or golden-hour medical care activities, where communication system latencies and delays are unacceptable. Therefore, the aforementioned issues can be resolved by implementing 6G technology.

8.1.3 Adoption of Blockchain for 6G

With the automation of all wireless network types dispersed over space, the air, the ground, and the ocean, 6G technology is thought to foster the development of a highly connected, data-dependent society in the future [4].

Furthermore, it can improve smart communities' wireless communication infrastructure, including smart cities, smart grids, and smart transportation systems. However, the primary feature of 6G technology is its ability to efficiently manage enormous amounts of data and offer high-data-rate connectivity for each device [5].

Additionally, it facilitates the deployment of flying autonomous vehicles, such as unmanned aerial vehicles (UAVs), which have a variety of uses, including military [6].

Thus, when the number of UAVs rises, additional difficulties arise, such as increased air traffic and UAV path planning [7], handling crises, etc. However, the application of distributed technology known as blockchain can lessen these difficulties [8,9]. This innovative approach not only enhances operational efficiency but also fosters trust among users by ensuring that all transactions are verifiable and tamper-proof. As a result, UAV networks can evolve to meet the growing demands of various industries while maintaining a high standard of security and reliability [10].

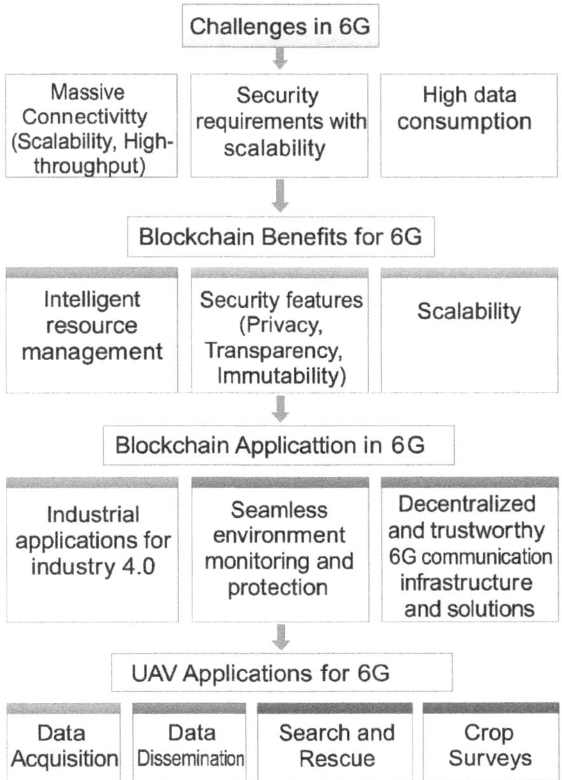

FIGURE 8.1 Illustration of blockchain functions in 6G networks.

One important factor that has been identified as facilitating UAV communication in 6G networks is the blockchain [11]. The function of blockchain in 6G networks is illustrated in Figure 8.1, which combines the advantages of blockchain technology with the difficulties of 6G technology to enhance UAV communication. Thus, the use of UAVs in many applications can be significantly altered by blockchain-based UAVs in 6G networks.

Every network node is in charge of maintaining this time-stamped, unchangeable record of data. The UAV networks benefit from their flexibility, scalability, immutability, and transparency. Additionally, it enables UAV communication for discreet, secure, and quick delivery services [12]. 6G technology can facilitate communication between UAVs, base stations, and satellites, while blockchain technology can protect the UAV network itself. For example, García-Magariño et al. [13] suggested a method that uses blockchain to keep UAV networks secure.

It demonstrates the potential of blockchain features in UAV communication, and the following is a description of the benefits of 6G UAV communication.

1. **Autonomous Wireless System**: The UAVs will be fully supported by the 6G infrastructure. The development of a completely autonomous system (self-driving cars) surrounded by 6G technology that outperforms human-controlled vehicles in terms of efficiency and performance will be a significant challenge for researchers.
2. **Large Coverage Area**: In 6G communication, it will be crucial to integrate a wide range of frequencies and heterogeneous communication technologies. Consequently, the UAV will seamlessly transition across networks without requiring any human system setups. Therefore,

6G communications' multiconnectivity will provide a wide coverage area and improve the UAV networks' Quality of Service (QoS).

3. **Integration of Sensing and Communication**: Autonomous wireless networks' primary function is to continuously sense the environment's changing conditions and provide information sharing between various nodes. Sensing will be closely linked with 6G communications, which will greatly aid autonomous systems.

8.1.4 Contributions

The following are this article's main contributions:

- In what ways does blockchain technology improve data integrity in 6G networks for UAV communication?
- What benefits might blockchain-enabled UAV systems get from utilizing consensus techniques like Rapid Consensus?
- This article offers comprehensive information on how 6G and blockchain technologies can help get over the UAV network's constraints.
- This essay provides readers with a broad overview of how the combination of blockchain and 6G technology can impact various UAV network applications.
- What obstacles need to be overcome to guarantee that blockchain technology is implemented successfully in UAV communication systems?

8.1.5 Motivation of the Review

Drones have a huge potential to change how we populate when combined with BC technology. BC technology should be used to enable UAVs to be strengthened with cryptographically secure methods to guarantee safe and secure communication and to be helpful in applications like defense and finance. Many application areas support BC because it simplifies the security issues that arise when UAVs communicate. Although current communication systems struggle because of a lack of automation and smart intelligence as well as issues with latency and scalability, this results in increased participation of UAVs utilizing BC technology. Therefore, using 6G as a communication method in UAVs can achieve ultra-low latency (1 ms) and a large transmission rate of 1 Tbps in addition to fully functional AI characteristics. The integration of 6G and BC technology holds tremendous promise for a variety of industries, particularly the commercial and military sectors. Additionally, because military services hold sensitive and mission-critical data, they must be completely secure against any cyberattack. In contrast to network weaknesses brought on by the coupling of BC and 6G for security and communication purposes, respectively, UAV communication becomes more secure. This study examines a number of issues with 6G-enabled UAV communication systems and attempts to identify more effective BC-based solutions than conventional ones.

8.1.6 Review Organization

The survey's structure. An introduction to UAVs and their requirements is provided in Section 8.1. Other topics covered include the communication features of UAVs, such as 4G, 5G, and 6G, as well as a number of security concerns with UAV communication and how blockchain technology may be a viable solution. The background and relevant work for UAV, 6G, and blockchain are shown in Section 8.2. In Sections 8.4 and 8.5, respectively, we provide a brief overview of the foundations of UAV, BC technologies, and the 6G environment, as well as the system structure/architecture and the extent of their integration, which is the subject of this article. We outline the prospects and opportunities in Section 8.6. We go over the

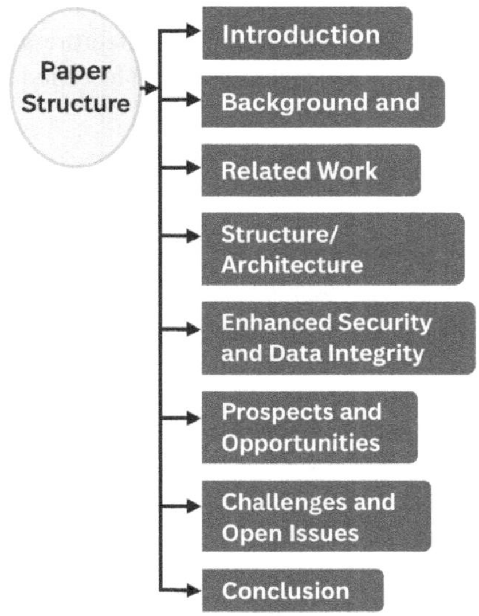

FIGURE 8.2 The structure of chapter.

difficulties, unresolved problems, and potential future research directions in Section 8.7.8. The paper is finally concluded in Section 8.9, as illustrated in Figure 8.2.

8.2 BACKGROUND AND RELATED WORK

8.2.1 Blockchain and 6G

Blockchain: Blockchain is a vital technology to manage large data in future communication system [14]. One type of distributed ledger technology is the blockchain. A database that is dispersed over multiple nodes or computing devices is called a distributed ledger. Every node makes an exact duplicate of the ledger and stores it. Blockchains are managed via peer-to-peer networks. It can exist without being controlled by a centralized authority or a server. Blocks of data are collected and organized on a blockchain. Cryptography is used to link and secure the blocks. With its enhanced security, privacy, interoperability, dependability, and scalability, the blockchain is essentially the ideal addition to the vast Internet of Things [15].

Therefore, in order to achieve the goal of uHSLLC service, blockchain technology will offer a number of facilities, including interoperability across devices, traceability of enormous data, autonomic interactions of various IoT systems, and reliability for the massive connectivity of 6G communication systems. Blockchain eliminates the need for trusted middlemen by fostering confidence between networked apps [16]. Blockchain's essential characteristics, like its decentralized tamper-resistance and secrecy, make it suitable for a wide range of 6G communication applications. By ensuring transparency, validated transactions, and the avoidance of unwanted access, it develops a safe and verifiable method for spectrum management. Blockchain offers intriguing capabilities not seen in the current structures by combining a distributed network structure, a consensus process, and sophisticated cryptography. The distributed architecture improves security by removing the single point of failure issue. 5G blockchain networking's primary obstacle is throughput (10–1,000 transactions per second). The need for regional and global standardization and control of the widespread use of blockchain in 5G presents another difficulty. However,

5G takes into account the problem of seamless blockchain platform interoperability. By employing consensus algorithms, implementing innovative blockchain architecture and sharing strategies, and expanding the network's block size, these various restrictions can be lessened in 6G.

8.2.2 Unmanned Aerial Vehicles

Unmanned Aerial Vehicles: A key component of 6G wireless communications are UAVs, sometimes known as drones. UAV technology is often used to deliver high-data-rate wireless communication. UAVs are equipped with base station (BS) entities to enable cellular connectivity. Strong LOS connectivity, degrees of freedom with controlled movement, and ease of deployment are some of the characteristics of a UAV that are absent from fixed BS infrastructures [17]. It is not always possible to provide any service in unstable conditions, and the deployment of terrestrial communication infrastructures is not economically viable during emergencies, such as natural disasters. These circumstances are easily handled by UAVs. The next big thing in wireless communication will be a UAV. This technology can support the uMUB, uHSLLC, mMTC, and uHDD basic requirements of wireless networks [18].

Additionally, UAVs can be used for a number of tasks, including improving network connectivity, detecting fires, providing emergency services in the event of a crisis, security and surveillance, monitoring pollution, parking, accidents, and more. Consequently, it is acknowledged that one of the key technologies for 6G communication is UAV technology. In Zeng et al. [19], there has been discussion of UAV-enabled backscatter communication, which can help with a number of communication chores (such as providing ambient power and establishing advantageous channel conditions to distant sensors). Air interfaces that are appropriate for 6G wireless networks can be developed by combining non-coherent detection techniques with UAV applications. Furthermore, one excellent option for implementing UAV for 6G intelligence is resilient resource allocation based on deep reinforcement learning.

In the early 1920s, the name "drone" was coined to describe target planes that were flown remotely for practice fire. These drones weren't very advanced; they were merely radio-controlled planes that a human pilot constantly commanded. The Unmanned Aircraft System (UAS), which was promoted by the US Department of Defense (DoD) and the Federal Aviation Administration (FAA) in 2005, eventually took the place of the word. Because UAS uses a drone, GCS, and communication network to share information, it demonstrates its complexity [20]. UAVs are simply aircraft that are remotely controlled by pilots at GCS via remote control devices and are either fully or partially autonomous thanks to pre-programmed flight plans. UAVs are primarily made up of sensors that provide data on the aircraft's condition and efficiently detect targets, as well as computer equipment such as microcontrollers and analog controls, actuators, UAV software, and communication systems.

8.2.3 UAV communication networks' advantages

First off, UAV systems are more affordable and have a faster deployment time, which makes them ideal for ad hoc or short-term missions. Additionally, in the majority of cases, line-of-sight (LoS) communication links can be created with the aid of low-altitude UAVs. Significant performance gains for forward HAP across long-distance LoS lines or direct source-to-destination communication (where feasible) could result from this. Furthermore, the UAV's agility presents additional chances to improve performance by dynamically modifying its state to best fit the communication circumstances.

Communication systems can also be made more flexible by controlling the UAV's movement, which will enhance communication quality even further. A UAV network, for instance, has good connections with ground terminals. UAVs have the ability to slow down in addition to broadcasting faster in order to preserve strong wireless connections and send more data to ground terminals and other equipment. UAV-assisted wireless communication is a promising and essential part of future wireless systems because of these clear advantages. Compared to existing systems, these systems must be able to accommodate a greater range of applications and have better ordering capabilities.

The deployed UAVs offer seamless wireless coverage in service areas by supporting any existing communications infrastructure. When disasters cause partial or total damage to the infrastructure, the UAV network enables prompt service restoration to the information system. Additionally, UAVs facilitate transmitter operations in highly event-driven environments, like a stadium during a sporting event. This instance has been recognized as one of the five critical scenarios that fifth-generation (5G) wireless networks must successfully handle [21].

UAV networking that facilitates information relay is the employment of UAVs to establish wireless connectivity between two or more transceivers/users, or groups of distant users, in situations where dependable direct communication is not available. UAVs can be deployed to collect and disseminate delay-tolerant information to and from a large number of dispersed wireless devices. For instance, in certain emergency response scenarios, UAVs can be positioned between the front line and the command center. Wireless sensor networks (WSNs), which gather data and transmit it to the UAV network for transmission to a BS, are one example.

8.2.4 UAV Communication in Next-Generation Networks

Role of UAVs in 5G and 6G: ultra-reliable low-latency communication (URLLC), massive IoT support, and mobile edge computing.

8.2.5 Blockchain Fundamentals

Core principles: decentralization, immutability, transparency, and consensus mechanisms.

Types of blockchains: Public, Private, Consortium—and their relevance to UAV networks.

Smart contracts: autonomous enforcement of rules and access control.

8.3 RELATED WORK

The blockchain and distributed ledger technology is one of the most disruptive technology enablers to address most of the current limitations and facilitate the functional standards of 6G. In this work, we explore the role of blockchain to address formidable challenges in 6G, future application opportunities and potential research directions [22].

This article [23] focuses on the challenges that swarms of UAVs working in 6G mobile networks face in terms of intelligence, security and privacy, and energy efficiency. In this cutting-edge review, we used the 6G ecosystem to connect blockchain and AI/ML with UAV networks. Following the presentation of the main findings, possible research obstacles are noted. We wrap up the review by discussing potential directions for further study in this nascent area.

This article [24] examines the main enabling technologies in greater detail, as well as any possible problems that can affect UAV operation in a unified 5G network and beyond. The security concerns affecting network privacy and UAV safety vulnerabilities are also covered in this article. In conclusion, we list a number of unresolved research questions that influence UAV operations and deployment and necessitate more investigation.

This research article highlights the importance of blockchain technology in 6G network by highlighting its major challenges, research directions, and futuristic application opportunities [25].

Blockchain, a recently developed technology, may be one of the most promising approaches to improving user privacy and data security in peer-to-peer UAV networks. Multiple entities can communicate securely, decentralized, and fairly by utilizing blockchain's advantages. The integration of privacy and security in blockchain-assisted UAV communication is thoroughly reviewed in this study [26,27].

8.4 SYSTEM STRUCTURE/ARCHITECTURE

8.4.1 Network Model

Layers:

- **UAV Layer**: UAV swarms with onboard sensors and communication modules.
- **6G Infrastructure Layer**: BSs, terrestrial core networks, edge servers.
- **Blockchain Layer**: Distributed ledger maintained by UAVs+ground stations.

Communication Modes:

- UAV-to-UAV (peer-to-peer).
- UAV-to-Base Station.
- UAV-to-Edge/Cloud.

8.4.2 6G Communication

Even though 5G is poised to enter its deployment phase, 6G is gradually gaining ground. Though speculation about 6G is still in its infancy, it is certain that the technology will soon take shape. By 2030, when the Internet of Things and data-intensive applications are widely used, 5G communication networks will not be able to meet the needs. For holographic projections, gaming, and remote surgery based on AR and VR, 6G technology is therefore urgently needed, as 5G would not be able to meet this demand.

AI-powered 6G from the physical layer to the service layer is an intriguing feature. Virtual and Augmented Reality (VAR) haptic communication, IoT-integrated smart cities (including smart homes, linked cars, autonomous driving, smart healthcare, automation, and manufacturing), and other applications are all covered by 6G. Drones and UAVs will be a key component of 6G communications. The high-data-rate wireless connectivity needed by the UAV communication network will benefit most from the previously mentioned features of 6G. UAVs will benefit from 6G technology's primary features, which include Massive Machine Type Communication (mMTC), Enhanced Mobile Broadband (eMBB), and URLLC.

Additionally, safeguarding the vast amounts of data gathered from several platforms to meet privacy and security needs would necessitate a comprehensive strategy in 6G, given the billions of user equipment and IoT devices connected throughout a UAV network. In order to meet these requirements and protect the communication system in UAVs with 6G, BC technology is probably going to be crucial.

8.4.3 Blockchain

Data security and privacy are the main issues with UAV communication. Although they have a single point of failure, the centralized cloud and fog-based technologies now in use offer some protection. Similarly, Table 8.1 lists additional shortcomings in the state-of-the-art solutions now offered by researchers worldwide. A variety of cyberattacks, including masquerade, eavesdropping, connecting, jamming, fabrication,

TABLE 8.1 The related work

REFERENCES	*YEAR*	*METHODS*	*GOALS*
[31]	2023	Blockchain and machine learning-based communication framework. Off-chain storage using Interplanetary file system (IPFS).	Describe a case study on protecting military operations involving unmanned aerial vehicles (UAVs) by taking into account various eventualities. Consider how controller/UAV malicious contracts allow for self-managed flight coordination, collision avoidance, and mission scheduling.
[30]	2024	Ubiquitous UAV task security management architecture (U-UTSM). Blockchain and 6G technologies for secure communication.	We propose a Ubiquitous UAV Task Security Management Architecture (U-UTSM) using blockchain and 6G technologies.
[29]	2022	Real or random oracle scheme for security verification. IoT integrated with blockchain schemes for future applications.	This study presents a thorough investigation of current secure communication and authentication methods for unmanned aerial vehicles (UAVs) on a "6G network."
[34]	2023	Blockchain-enabled framework for device and data authentication. Federated Learning for real-time data analysis and distribution.	Present a SAGIN framework supported by blockchain that verifies the devices and the information gathered by SAGIN participants.

and access control attacks, can also affect the centralized solutions. The BC, a distributed ledger system created in 2008 by the unidentified researcher Satoshi Nakamoto, is a tenable remedy for the previously listed problems. It has the potential to execute the applications, where trust assurance is required between the different stakeholders. It is a group of blocks that are chained together with the hash of the previous block [28].

8.4.3.1 Blockchain layer design

- **Consensus Mechanisms**:
  PBFT (Practical Byzantine Fault Tolerance) for small UAV groups.
  PoS (Proof-of-Stake) or PoA (Proof-of-Authority) for scalability.
- **Smart Contracts**:
  Access control (who can send/receive commands).
  Data validation (verifying authenticity of sensor readings).
  Task allocation and coordination among UAV.
- **Data Management**:
  On-Chain: transaction metadata, UAV identities, security logs.
  Off-Chain: bulk UAV sensor data, stored with hash reference on-chain.

8.4.3.2 Security services enabled

- **Authentication**: UAVs register and verify identities via blockchain.
- **Data Integrity**: Immutable storage prevents falsification of flight logs.
- **Confidentiality**: Encrypted data exchange with keys managed by blockchain.
- **Availability**: Distributed ledger reduces single-point failures.
  (Include a diagram: UAV swarm ↔ Blockchain nodes ↔ 6G network infrastructure.)

8.5 ENHANCED SECURITY AND DATA INTEGRITY

By guaranteeing decentralized security, data integrity, and non-reputability, blockchain improves UAV communication in 6G. In order to enable effective and safe UAV operations in future communication systems, it facilitates resource sharing while addressing performance requirements including data rates and latency.

8.5.1 Blockchain's Role

Blockchain provides a decentralized framework that ensures data integrity and non-repudiation, crucial for sensitive UAV operations [29].

Consensus Mechanisms: Innovative consensus algorithms, such as Rapid Consensus and UAV-Emergency, significantly enhance transaction speeds and data consistency, outperforming traditional methods like Proof-of-Work [30].

8.5.2 Improved Communication Efficiency

- **6G Capabilities**: In order to tackle the issues of network coverage, latency, and security threats in UAV task management, this article presents the Ubiquitous UAV Task Security Management Architecture (U-UTSM), which makes use of blockchain and 6G technologies. When compared to Proof-of-Work, the suggested architecture shows notable gains in transaction processing times, improves real-time communication security, permits ubiquitous control of UAV swarms, and guarantees data consistency through quick consensus processes [30].
- **Off-Chain Storage Solutions**: This article presents Block-USB, a blockchain and machine learning-based secure and intelligent UAV communication system underlying 6G networks that avoids disclosing sensitive military operations, uses off-chain storage (IPFS) to improve blockchain storage capacity, and outperforms conventional 4G/5G and non-IPFS systems in order to address the difficulties of protecting mission-critical data acquired by UAVs in contemporary warfare scenarios and the bottleneck of handling massive data on traditional communication networks [31].

8.5.3 Applications and Use Cases

Applications in both the military and the general public: UAVs are being utilized more and more in a variety of domains where secure communication is crucial, such as surveillance, disaster relief, and smart city construction. With an emphasis on UAV communication, this article provides a thorough analysis of the architecture, specifications, and uses of 6G technology. In addition to offering a case study for safeguarding Industry 4.0 applications, it tackles the difficulties associated with network communication and security in UAV networks and suggests a blockchain-envisioned security solution coupled with 6G-enabled network connectivity to improve UAV communication [32].

In this article [33], the performance of the UAV network system in 6G environment and the technological challenges faced with respect to security and privacy are discussed, and a possible solution using blockchain and security policies that can be provided for secure communication in a 6G network is presented.

8.6 PROSPECTS AND OPPORTUNITIES

- **Enhanced Security & Trust:** decentralized authentication and tamper-proof data logs.
- **Resilient UAV Swarm Communication:** peer-to-peer trust using blockchain.
- **Reduced Latency via 6G + Edge Integration:** faster consensus with edge blockchain nodes.
- **Trusted Data Sharing for Smart Cities:** UAVs provide verified aerial data for traffic, pollution monitoring, disaster response.
- **Autonomous UAV Operations:** Self-managed flight coordination, collision avoidance, and mission scheduling are made possible by smart contracts.
- **Synergy with AI:** Verified blockchain data is fed into AI models for predictive routing and anomaly detection.
- **Global Scalability:** blockchain allows secure UAV deployment in cross-border missions without relying on a single authority [35].

8.7 CHALLENGES AND OPEN ISSUES

8.7.1 Data Handling and Massive Throughput

Challenge: 6G promises terahertz bandwidths and very high data rates (up to 1 Tbps), which produce massive amounts of data. It takes sophisticated analytics and machine learning tools to process this effectively, as shown in Figure 8.3.

8.7.2 Security Vulnerabilities and Smart Contract Risks

Challenge: Vulnerabilities like spoofing, eavesdropping, or code-level errors can affect smart contracts (SCs). General threats to blockchain systems include re-entrancy, privacy leakage, Sybil attacks, and 51% attacks.

8.7.3 Standardization and Regulatory Gaps

Challenge: Coherent standards from organizations like IEEE or ITU for the use of blockchain in UAV–6G systems are lacking. Rollout in the real world is hampered without this.

8.7.4 Energy Efficiency and UAV Resource Constraints

Challenge: UAVs face tight constraints in power, processing, and storage. Running energy-intensive consensus algorithms and SCs can severely impact performance and endurance.

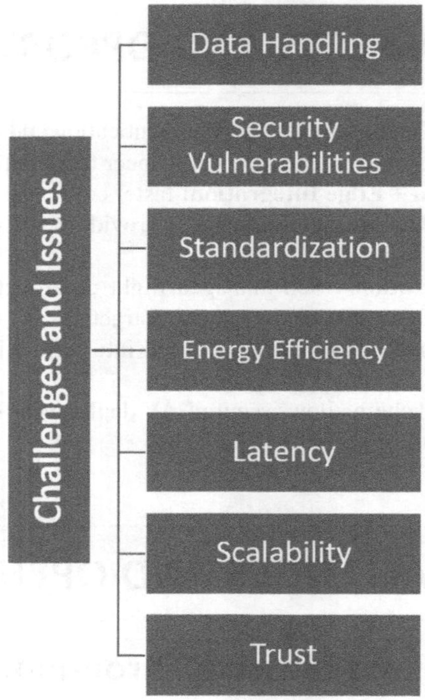

FIGURE 8.3 The challenge and open issues.

8.7.5 Latency and Handover Delays

Challenge: Blockchain consensus introduces delays, which conflict with UAVs' need for low-latency communication—especially during handovers.

8.7.6 Scalability and Dynamic Network Topology

Challenge: Blockchain systems struggle with high transaction volumes. UAV networks are dynamic—nodes join or leave frequently—making scalability and transaction management difficult.

8.7.7 Trust Management and Spectrum Sharing

Challenge: UAVs functioning as mobile base stations in ultra-dense 6G environments require reliable, safe methods for sensing and allocating spectrum. Misbehavior cannot be completely prevented by blockchain alone.

8.7.8 Authentication, Privacy and Identity Management

Challenge: UAVs require secure authentication that is lightweight. The transparency of blockchain technology may clash with privacy concerns. Effective identity management systems are required, such as those that use SCs.

8.8 FUTURE RESEARCH DIRECTIONS

- Blockchain protocols that are lightweight and designed for UAV networks (high throughput, low power).
- Post-quantum security in 6G UAV communication is ensured by quantum-resistant cryptography.
- Bridging various blockchain platforms across UAV systems is known as cross-chain interoperability.
- Connectivity with Edge/Fog Computing: enables blockchain verification in real time at the UAV level.
- AI+Blockchain: safe AI model training combined with autonomous UAV decision-making.
- Testbeds & Field Trials: practical verification of blockchain-enabled UAV communication.

8.9 CONCLUSION

UAVs have recently demonstrated their capacity to provide authentic and reasonably priced solutions for a variety of real-world issues. UAVs provide a wide range of services due to their autonomy, mobility, adaptability, and communications interoperability. Even while UAVs are frequently employed to enable ground connections, data exchanges in those networks are susceptible to security issues because most communication takes place over radio or Wi-Fi signals, which are easy to hack. However, there are a number of methods to defend against cyberattacks. Blockchain, a recently developed technology, may be one of the most promising approaches to improving user privacy and data security in peer-to-peer UAV networks. Multiple entities can communicate securely, decentralizedly, and fairly by utilizing blockchain's advantages. The integration of privacy and security in blockchain-assisted UAV communication is thoroughly covered in this article. For this goal, to gain a better understanding, the security goals and needs of the UAV communication system—such as availability, authentication, authorization, secrecy, integrity, privacy, and non-repudiation—are carefully investigated. Finally, the limitations of existing UAV standards, open research issues, and possible future research directions are discussed.

REFERENCES

1. Tang, F., Y. Kawamoto, N. Kato, and J. Liu, Future intelligent and secure vehicular network toward 6G: Machine-learning approaches. *Proceedings of the IEEE*, 2019. **108**(2): pp. 292–307.
2. Agiwal, M., A. Roy, and N. Saxena, Next generation 5G wireless networks: A comprehensive survey. *IEEE Communications Surveys & Tutorials*, 2016. **18**(3): pp. 1617–1655.
3. Zhang, Z. et al., 6G wireless networks: Vision, requirements, architecture, and key technologies. *IEEE Vehicular Technology Magazine*, 2019. **14**(3): pp. 28–41.
4. Latva-Aho, M. and K. Leppänen, *Key Drivers and Research Challenges for 6G Ubiquitous Wireless Intelligence*, University of Oulu, Oulu, Finland, 2019.
5. Mumtaz, S. et al., Terahertz communication for vehicular networks. *IEEE Transactions on Vehicular Technology*, 2017. **66**(7): pp. 5617–5625.
6. Alwateer, M., S.W. Loke, and A. Zuchowicz, Drone services: Issues in drones for location-based services from human-drone interaction to information processing. *Journal of Location Based Services*, 2019. **13**(2): pp. 94–127.

7. Aggarwal, S. and N. Kumar, Path planning techniques for unmanned aerial vehicles: A review, solutions, and challenges. *Computer Communications*, 2020. **149**: pp. 270–299.
8. Aggarwal, S. et al., Blockchain for smart communities: Applications, challenges and opportunities. *Journal of Network and Computer Applications*, 2019. **144**: pp. 13–48.
9. Singh, S.K., Y.-S. Jeong, and J.H. Park, A deep learning-based IoT-oriented infrastructure for secure smart city. *Sustainable Cities and Society*, 2020. **60**: p. 102252.
10. Tanwar, S. et al., Machine learning adoption in blockchain-based smart applications: The challenges, and a way forward. *IEEE Access*, 2019. **8**: pp. 474–488.
11. Balasubramaniam, A. et al., Blockchain for intelligent transport system. *IETE Technical Review*, 2021. **38**(4): pp. 438–449.
12. Jindal, A., G.S. Aujla, and N. Kumar, SURVIVOR: A blockchain based edge-as-a-service framework for secure energy trading in SDN-enabled vehicle-to-grid environment. *Computer Networks*, 2019. **153**: pp. 36–48.
13. García-Magariño, I. et al., Security in networks of unmanned aerial vehicles for surveillance with an agent-based approach inspired by the principles of blockchain. *Ad Hoc Networks*, 2019. **86**: pp. 72–82.
14. Henry, R., A. Herzberg, and A. Kate, Blockchain access privacy: Challenges and directions. *IEEE Security & Privacy*, 2018. **16**(4): pp. 38–45.
15. Dai, H.-N., Z. Zheng, and Y. Zhang, Blockchain for internet of things: A survey. *IEEE Internet of Things Journal*, 2019. **6**(5): pp. 8076–8094.
16. Nguyen, T. et al., Privacy-aware blockchain innovation for 6G: Challenges and opportunities. In: *2020 2nd 6G wireless Summit (6G SUMMIT)*, Levi, Finland, 2020, pp. 1–5.
17. Tariq, F. et al., A speculative study on 6G. *IEEE Wireless Communications*, 2020. **27**(4): pp. 118–125.
18. Li, B., Z. Fei, and Y. Zhang, UAV communications for 5G and beyond: Recent advances and future trends. *IEEE Internet of Things Journal*, 2018. **6**(2): pp. 2241–2263.
19. Zeng, Y., R. Zhang, and T.J. Lim, Wireless communications with unmanned aerial vehicles: Opportunities and challenges. *IEEE Communications Magazine*, 2016. **54**(5): pp. 36–42.
20. Gupta, R. et al., Blockchain-envisioned softwarized multi-swarming UAVs to tackle COVID-I9 situations. *IEEE Network*, 2020. **35**(2): pp. 160–167.
21. Merwaday, A. and I. Guvenc. UAV assisted heterogeneous networks for public safety communications. In: *2015 IEEE Wireless Communications and Networking Conference Workshops (WCNCW)*, New Orleans, LA, 2015. IEEE.
22. Hewa, T. et al., The role of blockchain in 6G: Challenges, opportunities and research directions. In: *2020 2nd 6G Wireless Summit (6G SUMMIT)*, Levi, Finland, 17–20, 2020, pp. 1–5.
23. Khan, M.A. et al., Swarm of UAVs for network management in 6G: A technical review. *IEEE Transactions on Network and Service Management*, 2022. **20**(1): pp. 741–761.
24. Sahoo, B.P., D. Puthal, and P.K. Sharma, Toward advanced UAV communications: Properties, research challenges, and future potential. *IEEE Internet of Things Magazine*, 2022. **5**(1): pp. 154–159.
25. Bindu, G. et al. Impact of blockchain technology in 6G network: A comprehensive survey. In: *2022 International Conference on Inventive Computation Technologies (ICICT)*, Lalitpur, Nepal, 2022. IEEE.
26. Hafeez, S. et al., Blockchain-assisted UAV communication systems: A comprehensive survey. *IEEE Open Journal of Vehicular Technology*, 2023. **4**: pp. 558–580.
27. Sheth, K. et al., A taxonomy of AI techniques for 6G communication networks. *Computer Communications*, 2020. **161**: pp. 279–303.
28. Gupta, R. et al., Blockchain-based security attack resilience schemes for autonomous vehicles in industry 4.0: A systematic review. *Computers & Electrical Engineering*, 2020. **86**: p. 106717.
29. Khan, A.S. et al., A survey on 6G enabled light weight authentication protocol for UAVs, security, open research issues and future directions. *Applied Sciences*, 2022. **13**(1): p. 277.
30. Xie, H. et al., Blockchain-and-6G-based ubiquitous UAV task security management architecture. *Computer Communications*, 2024. **217**: pp. 259–267.
31. Jadav, N.K. et al., Blockchain-based secure and intelligent data dissemination framework for uavs in battlefield applications. *IEEE Communications Standards Magazine*, 2023. **7**(3): pp. 16–23.
32. Aggarwal, S., N. Kumar, and S. Tanwar, Blockchain-envisioned UAV communication using 6G networks: Open issues, use cases, and future directions. *IEEE Internet of Things Journal*, 2020. **8**(7): pp. 5416–5441.
33. Kayalvizhi, M. and S. Ramamoorthy, Security and privacy of unmanned aerial network communication systems in 6G networks. In: A. M. Viswa Bharathy and B. Alhadidi (Eds.), *Challenges and Risks Involved in Deploying 6G and NextGen Networks*. IGI Global Scientific Publishing, pp. 148–166, 2022.

34. Okoumassoun, T.P. et al. Blockchain-enabled SAGIN communication for disaster prediction and management. In: *2023 IEEE Intl Conf on Dependable, Autonomic and Secure Computing, Intl Conf on Pervasive Intelligence and Computing, Intl Conf on Cloud and Big Data Computing, Intl Conf on Cyber Science and Technology Congress (DASC/PiCom/CBDCom/CyberSciTech),* Abu Dhabi, United Arab Emirates, 2023. IEEE.
35. Khan, I. U., Ouaissa, M., Ouaissa, M., Abou El Houda, Z., & Ijaz, M. F. (Eds.). (2024). *Cyber Security for Next-Generation Computing Technologies.* CRC Press.

9 CogniHealth 6G-X *A Federated, Blockchain-Enabled Digital Twin Framework for Intelligent Healthcare Systems*

Shanmugam Pradeep, Jakir Hussain Jareena Begam, Subramaniyan Jayaraman Sadheesh Kumar, Annadurai Karthikeyan, and Arunachalam Karthika

9.1 INTRODUCTION

The healthcare field has always been a leader in using new technology, with developments like imaging tools, less invasive surgeries, electronic health records, and telemedicine [1]. In the last 20 years, this growth has moved from separate systems used in doctors' offices to connected digital health networks. This change has been driven by the rise of the Internet of Medical Things, wearable devices that track health, cloud computing, and, more recently, artificial intelligence [2]. However, the true potential of these technologies is constrained by current communication infrastructures—most notably, the limitations of 4G and 5G networks in addressing latency-sensitive, data-intensive, and security-critical healthcare applications [3].

The introduction of 5G marked a significant step forward by enabling higher data rates, enhanced mobile broadband, ultra-reliable low-latency communications (URLLC), and massive machine-type communications (mMTC). These capabilities opened the door to applications such as remote robotic surgery, real-time patient monitoring, and AI-assisted diagnostics [4]. Yet, as digital health services continue to grow in scale and complexity, the requirements now extend beyond what 5G can reliably support. Issues

 DOI: 10.1201/9781003683599-9

such as limited spectrum, congestion in urban networks, energy inefficiency, and the inability to meet real-time demands in ultra-dense or rural deployments persist [5].

This is where sixth-generation (6G) wireless networks come into play. Still in early research and standardization phases, 6G is envisioned to provide unprecedented communication capabilities, including peak data rates exceeding 1 Tbps, end-to-end latencies below 1 ms, and support for up to 10 million devices per square kilometer. Beyond these metrics, 6G is designed to integrate non-terrestrial networks (e.g., low-earth orbit satellites), terahertz (THz) communication bands, intelligent reflecting surfaces (IRS), quantum communication, and AI-native network management [6,7]. It shifts the paradigm from a "connected things" era to an "intelligent everything" framework, where computation, storage, and intelligence are inherently embedded throughout the communication fabric.

The implications for healthcare are transformative. 6G will enable real-time health analytics at the edge, support autonomous healthcare robots, and power virtual hospitals and digital twin representations of patients for personalized simulations and treatments [8,9]. Moreover, it lays the groundwork for a truly immersive and empathetic healthcare experience through technologies like extended reality (XR), haptic feedback systems, and remote rehabilitation environments [10,11]. Importantly, the fusion of 6G with AI enables not just fast data transfer, but intelligent decision-making, making it possible to detect critical anomalies, predict health deteriorations, and recommend timely interventions without human delay [12].

However, as promising as these technological strides are, they also bring forth new challenges: data security, interpretability of AI-driven decisions, interoperability of multi-source health data, and ethical governance of automated systems [13,14]. Traditional centralized models of data processing and decision-making are inadequate in the face of these demands. To address this, the next generation of healthcare systems must be built upon decentralized, privacy-preserving, and explainable frameworks. Federated learning, blockchain, and XAI (Explainable AI) are not just add-ons but foundational to ensuring transparency, trust, and personalization in future healthcare [15–17].

This chapter introduces the concept of Cognitive 6G Healthcare Systems—a comprehensive, patient-centric, and technologically advanced architecture that brings together the strengths of AI, blockchain, federated learning, digital twins, and 6G networks. We explore a proposed framework called CogniHealth 6G-X, designed to deliver secure, scalable, and intelligent healthcare services across diverse environments—from urban hospitals to remote rural clinics, and from home-based care to emergency response systems.

9.2 LITERATURE REVIEW

The integration of next-generation wireless communication with artificial intelligence in healthcare has become an urgent and transformative area of research [18]. With the rise of chronic illnesses, aging populations, and the need for remote medical services exacerbated by global events, like the COVID-19 pandemic, the demand for intelligent, real-time, and context-aware healthcare systems has reached unprecedented levels [19]. Traditional hospital-centered care models are rapidly evolving into patient-centric ecosystems, where continuous monitoring, early intervention, and data-driven diagnostics are paramount [20]. As healthcare moves beyond the walls of hospitals into homes, ambulances, and rural environments, technologies must adapt to support this transition both in terms of scalability and intelligence [21].

Existing healthcare IoT and telemedicine systems, while effective to some extent, are often built on legacy networks such as 4G or early-stage 5G infrastructures. These platforms face limitations in supporting ultra-low latency communication, high-density device connectivity, and adaptive resource management [22]. Moreover, centralized data architectures introduce significant bottlenecks in terms of data privacy, computational scalability, and real-time responsiveness. The need to transmit sensitive patient data to cloud servers not only increases latency but also exposes the system to privacy breaches and compliance risks under frameworks like HIPAA and GDPR [23,24].

Recent literature has explored 5G-enabled smart healthcare frameworks integrating edge computing, machine learning, and IoT [25]. While these systems provide improvements over traditional models, they still fall short in several key areas. First, many rely heavily on cloud-based computation, which may not meet the real-time demands of critical applications like arrhythmia detection, remote surgeries, or emergency triage [26]. Second, while edge computing reduces latency, it often lacks collaborative intelligence, making it difficult to train models across distributed datasets without compromising patient privacy [27]. Third, most existing works focus narrowly on signal processing or model performance, without addressing systemic interoperability, explainability, or trustworthiness—factors that are crucial for clinical adoption [28].

A number of frameworks have attempted to use federated learning (FL) in healthcare to address privacy concerns; however, these are often challenged by data heterogeneity, communication overhead, and the absence of a robust accountability mechanism for model updates [29]. Additionally, security mechanisms in most FL-enabled systems rely on conventional encryption, which may not be sufficient to guarantee integrity and traceability in multi-party environments. Similarly, digital twins—though gaining popularity—are typically employed in a static manner, without the benefit of real-time data fusion or predictive simulation capabilities tailored to individual patient dynamics [30,31].

These gaps in the literature underline the need for a holistic, intelligent, and secure system architecture that can unify sensing, computing, and decision-making in a scalable and clinically viable manner. This is where the proposed CogniHealth 6G-X framework brings a fresh and necessary perspective. By leveraging the ultra-reliable, low-latency, and massive-connectivity features of 6G networks, the system enables high-speed, real-time communication among sensing devices, edge nodes, and cloud servers [32]. The inclusion of a lightweight, patient-centric digital twin infrastructure allows for continuous modeling and forecasting of individual health states, supporting predictive and preventive care rather than reactive intervention.

To address the critical challenge of privacy-preserving AI, the architecture incorporates a decentralized FL approach. Unlike traditional centralized models, FL in CogniHealth 6G-X trains models locally on edge nodes and shares only encrypted parameter updates, ensuring that raw data remains securely within institutional boundaries. Furthermore, the system employs smart blockchain mechanisms using Proof-of-Authentication (PoAh) consensus and smart contracts to provide a tamper-resistant and transparent audit trail of all system interactions—whether clinical decisions, model updates, or data access requests. The motivation behind this research is to bridge the gap between technical feasibility and clinical usability. While many research efforts demonstrate promising AI models or novel protocols, very few offer a unified framework that integrates sensing, intelligence, privacy, security, and explainability in a seamless, multi-layered architecture. The goal is to develop a platform that is not only technically sound but also ethically aligned and practically deployable across urban and remote healthcare settings alike.

9.2.1 Motivation and Key Contributions

Therefore, the objectives of the proposed work are:

- To design a multi-layered cognitive framework for real-time, personalized healthcare monitoring using 6G-enabled networks.
- To embed FL mechanisms that preserve privacy while supporting cross-institutional model collaboration.
- To establish a blockchain-driven security layer for trust, access control, and model integrity in decentralized healthcare environments.
- To implement digital twin-based patient models that provide predictive simulation, dynamic risk assessment, and intelligent intervention recommendations.
- To ensure system-wide orchestration using context-aware cloud intelligence for dynamic resource management and clinical prioritization.

The main things that the CogniHealth 6G-X framework brings are:

1. A new healthcare system that uses 6G technology and combines edge intelligence, FL, blockchain for security, and digital twin models.
2. A simple and secure way to do FL that works well with different types of healthcare data and in places where resources are limited.
3. A tool that creates personalized digital twins, which can predict how a patient's health might change and help doctors make better treatment choices.
4. A system that uses blockchain to protect privacy and control who can access data, following the rules for healthcare and considering the limits of IoT medical devices.
5. A way to organize and manage system resources in a flexible way, adjusting based on how urgent things are, how well the network is working, and how much computing power is available.

9.3 ARCHITECTURAL OVERVIEW OF COGNIHEALTH 6G-X

The CogniHealth 6G-X framework is a multi-layered, smart healthcare system that uses the advanced features of 6G wireless technology to offer quick, secure, and personalized medical care. The system works as a network of connected parts that work together smoothly to provide real-time health tracking, smart predictions, and support for making decisions—whether at the hospital bed, while traveling, or at home. Each part of the system has a specific job but can connect and work with the others, making the whole system strong and able to handle many different healthcare needs.

At the base of the system is the Sensing and Data Acquisition Layer, which collects various types of health information from different sources. These sources include wearable devices that track important body functions, medical devices placed inside the body, imaging tools, environmental sensors, and mobile health apps [33]. This layer constantly gathers data in real time, allowing early detection of health issues and helping to monitor health closely. The data collected here is the starting point for all the analysis and decisions that come after. Above this is the Edge Intelligence Layer, which is important for making the system faster and more responsive. Instead of sending all data to faraway cloud servers, processing happens at local edge nodes near the data source, like in hospitals, ambulances, or smart homes. These edge nodes have small, efficient AI models that can quickly perform diagnostics, detect issues, and send emergency alerts if needed. With Device-to-Device (D2D) communication, these nodes and nearby devices can share information directly, without needing a central network. This setup helps the system work faster and more reliably, even in areas with weak internet or limited data speed [34]. Building on this, the FL layer introduces a decentralized method for training AI models [35].

Rather than sending private patient data over the network, each edge device trains its own model locally using the data it collects. Only the model's parameters—like how it's structured or what it has learned—are sent to a central server. This helps protect patient privacy and reduces the amount of data being sent over the network. The central server uses these parameters to update a global model and sends it back to all edge nodes. This ongoing process lets the system learn and improve across different locations without ever sharing personal information. It also lets the models adapt to individual health needs, drawing from various data sources without collecting all data in one place. To keep data exchanges safe and build trust among all users, the system uses a Blockchain-Based Security and Governance Layer.

This part has a simple, energy-efficient blockchain suitable for healthcare settings with limited resources. It ensures that all transactions and data interactions are secure, unchangeable, and transparent.

This helps maintain data reliability and responsibility throughout the network. The system uses a special consensus method, like Proof-of-Authentication (PoAh), to make sure every transaction—whether it's about patient data, model updates, or access permissions—is clear, unchangeable, and traceable. Smart contracts are used to automatically follow privacy rules and manage user consent, allowing patients to control their data while letting authorized people access the information they need at the right time.

At the top of the system is the Digital Twin and Cloud Orchestration Layer. Digital twins are live, virtual copies of patients that constantly update to show their health status using real-time data from sensors, past medical records, and AI forecasts [36]. These digital models offer a complete picture of a patient's health and help in planning treatments, predicting risks, and responding to emergencies [37]. The cloud part of the system handles storing large amounts of data, doing complex analysis, and coordinating the whole system. It helps different parts work together smoothly, making sure resources are used efficiently depending on how urgent the medical need is and the current network situation [38,39]. Together, the CogniHealth 6G-X framework creates a unified system that combines sensing, processing, learning, securing, and managing. Intelligence is built into every part, from the edge devices to the cloud. This makes the system not only quick and safe but also able to change and grow as needed. The design ensures it can handle today's healthcare needs and adapt to the future of smart health environments. Figure 9.1 shows the architectural overview of the system.

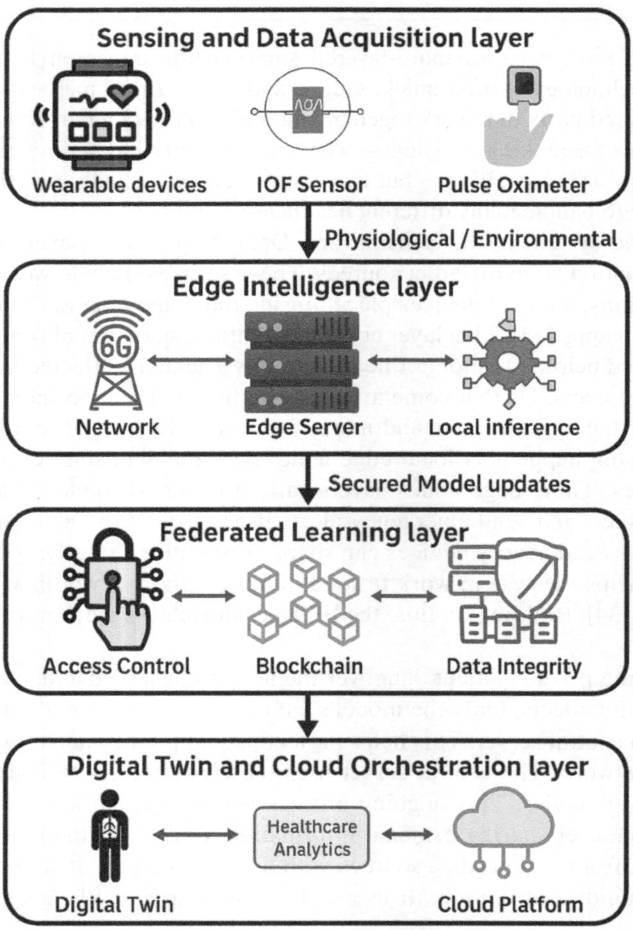

FIGURE 9.1 Architectural overview of the system.

9.3.1 Sensing and Data Acquisition Layer

At the heart of the CogniHealth 6G-X system is the Sensing and Data Acquisition Layer, which forms the base of the whole setup. Its main job is to collect real-time, ongoing, and context-aware health data from different sources that focus on the patient. This layer gathers raw biomedical signals, environmental data, and other important information needed for smart analysis and decisions made later in the system. This ensures the whole framework starts with accurate, detailed, and secure input that shows how a patient's health is changing over time. The layer includes many types of sensors used in wearable, implantable, and environmental medical devices. These include tools like ECG monitors, PPG sensors, temperature sensors, glucose trackers, motion sensors (like accelerometers and gyroscopes), EEG headbands, and devices that measure breathing. Each sensor is designed to track a specific health signal and works at different speeds and sensitivity levels. For example, ECG sensors usually collect data at rates between 250 Hz and 1,000 Hz, which allows for detailed monitoring of the heart and helps find early signs of heart problems. Mathematically, the signal captured by a sensor $s_i(t)$ can be shown in equation 9.1 as

$$x_i(t) = s_i(t) + n_i(t) \tag{9.1}$$

where $x_i(t)$ is the observed signal at time t, $s_i(t)$ is the true physiological signal and $n_i(t)$ is the additive noise, which may result from motion artifacts, electromagnetic interference, or sensor drift.

To improve the signal-to-noise ratio (SNR), the system uses various preprocessing steps like filtering, normalization, and signal segmentation either at the sensor node or through the nearby microcontroller. These steps help make the raw physiological data cleaner before it is sent or used for further processing. Noise is reduced using filtering methods such as Finite Impulse Response (FIR) or Infinite Impulse Response (IIR) filters, which are effective in separating useful physiological signals from unwanted interference. The mathematical details of these filters are given in equation 9.2, showing exactly how the raw signals are processed to produce more dependable and accurate results.

$$y(t) = \sum_{k=0}^{N} h_k \cdot x(t-k) \tag{9.2}$$

here $y(t)$ the filtered output is produced, and h_k represents the filter coefficients. This process makes sure that only clean, useful, and trustworthy signals move forward for more processing. Beyond capturing analog biosignals, this layer also handles structured digital data, such as patient information, medication schedules, and past medical records—usually accessed through standard APIs or connected hospital electronic health record systems. By combining this structured data with real-time sensor information, the system builds a detailed, multi-layered view of each patient. This combination is essential for making accurate, situation-aware diagnoses and providing timely care. Each sensing unit can perform simple calculations locally, like identifying patterns from the incoming signals. Important health measurements such as heart rate variability (HRV), mean arterial pressure (MAP), respiratory rate (RR), and oxygen levels (SpO_2) are calculated in real time using signal processing techniques. These local computations help lower the workload on other parts of the system and allow for quick responses to any unusual changes in the body. For example, HRV can be determined from the time between breaths,$\{r_1, r_2, \ldots, r_n\}$ as in equation 9.3.

$$\mathrm{HRV}_{\mathrm{SDNN}} = \sqrt{\frac{1}{n-1}\sum_{i=1}^{n}(r_i - \bar{r})^2} \tag{9.3}$$

where $\bar{r}$ is the mean of the RR intervals.

The system uses a smart way to prioritize important data based on how the patient is doing. If the system notices something serious, like a big drop in oxygen levels or a sudden increase in heart rate, it automatically changes how often it checks and sends the data. This makes sure that important health issues are noticed and shared quickly, so medical help can come on time. This process is controlled by a math model called equation 9.4, which calculates a dynamic urgency score, *U*(*t*). This score checks how far the patient's body measurements are from normal levels. The more they differ from normal, the higher the urgency score becomes, which speeds up how the data is handled and sent through the network to help provide fast medical care.

$$U(t) = \sum_{j=1}^{m} w_j \cdot \left| v_j(t) - v_j^{\text{baseline}} \right| \tag{9.4}$$

where $v_j(t)$ is the current value of the *j*th vital sign, v_j^{baseline} is the normal or reference value, and w_j is the weight indicating the importance of that parameter.

To make the network work better and avoid getting too busy, this layer uses smart ways to handle data, like using simple compression and grouping information smartly. Techniques such as delta encoding and run-length encoding (RLE) help reduce the amount of data without losing important details. High-detail data is only sent when something unusual happens, like irregular heartbeats or sudden brain activity spikes. Regular health data is grouped together and sent less often, which saves bandwidth without affecting the usefulness of the information for doctors. Keeping data safe is also important. All the sensors and devices used to collect data have basic security measures, such as AES-128 encryption and ways to check if a device is real. This makes sure the data is protected and can't be changed from the moment it is gathered, following all the rules for keeping health information private. These sensors also use encryption and check to make sure the data is secure before it goes to the next part of the system, ensuring that patient information stays protected right from the start.

Although collecting data and improving the signal are the basics of the system, the real smart part comes when this processed data reaches the edge layer for immediate analysis. This is where the system moves from just collecting information to making smart decisions. The Edge Intelligence Layer takes over from the sensing layer, allowing quick local processing and smart choices that are important for serious healthcare situations. This change greatly cuts down delays and takes less pressure off the main system, making healthcare operations faster and easier to scale.

9.3.2 Edge Intelligence Layer

The Edge Intelligence Layer is an important part of the CogniHealth 6G-X framework. It brings strong computing power closer to where health data is created. By working with data right at the source—like in a hospital, ambulance, clinic, or even a patient's home—it helps make quick decisions and respond in real time. This is very important in situations like detecting a stroke, monitoring patients in an ICU, or doing surgery from a distance, where every second matters. Traditional cloud systems can be slow because of delays, but edge computing reduces these delays by processing data locally. These edge nodes act like smart mini-servers with special hardware like GPUs, TPUs, or neuromorphic chips. They run small but powerful machine learning models that are good for urgent healthcare tasks. This ensures that important situations are handled quickly and efficiently. The main job of these nodes is to process and analyze data from the sensing layer without sending all the raw data to a central cloud system. This lowers the use of network bandwidth and makes sure there is immediate response. Equation 9.5 shows mathematically that the edge node performs an inference operation on the incoming preprocessed sensor vector $x \in \mathbb{R}^n$, producing a diagnostic or risk prediction output $\hat{y}$ using a local model f_θ with learned parameters θ

$$\hat{y} = f_\theta(x) \tag{9.5}$$

For example, x could be a set of features extracted from ECG signals, such as RR intervals, QRS width, or signal entropy, while $\hat{y}$ represents the classification of arrhythmia type or an early warning score. To minimize energy consumption and maximize interpretability, the models used at the edge are often shallow neural networks, decision trees, or support vector machines (SVMs). In more advanced scenarios, convolutional neural networks (CNNs) and recurrent architectures like LSTMs may be employed for image and time-series analysis, respectively, provided that edge hardware supports accelerated computation. One of the key features of this layer is its context-aware intelligence. The edge node doesn't work on its own; it changes how it acts depending on real-time situations, like the patient's health, how busy the network is, or how much battery is left on the device. A function called $R(t)$ that manages resources dynamically can be described using equation 9.6.

$$R(t) = \alpha \cdot C(t) + \beta \cdot B(t) + \gamma \cdot L(t) \tag{9.6}$$

where $C(t)$ stands for the current level of clinical urgency, $B(t)$ shows how much battery or energy is available, $L(t)$ refers to the predicted quality or delay of the connection, and α, β, γ are adjustable weights that depend on what the system considers most important.

Based on the calculated response metric, $R(t)$, the edge node decides the best action: it can process the data locally, send it to a nearby node through Device-to-Device (D2D) communication, or send the task to the cloud for more complex processing. D2D communication helps improve the system's reliability and backup options. If an edge device detects a major problem or runs out of resources, it can transfer data processing or model calculations to a neighboring node without needing a central base station. This way, the system keeps running smoothly even if parts of it fail or connections are lost. The efficiency and cost of these D2D data transfers are described in equation 9.7, which measures the overhead in terms of delay, energy use, and connection quality.

$$\text{Cost}_{\text{relay}} = \delta_1 \cdot d + \delta_2 \cdot E \tag{9.7}$$

where d is the actual distance between the nodes, E is the energy needed for the transfer, and δ_1, δ_2 are scaling factors that consider environmental and hardware conditions. To make signals more reliable in tough situations, like busy hospital corridors or crowded city streets, this layer uses IRS. These are special surfaces with passive components that can control the direction, phase, and strength of radio waves in real time. IRS helps redirect signals around walls and other barriers, keeping communication between edge devices strong and stable, even when the direct path is blocked. This helps ensure continuous data transfer and meets the fast response needs of important healthcare services. The impact of IRS on the communication channel is quantified through the effective channel gain, denoted as G_{IRS}, and is formally represented in equation 9.8.

$$G_{\text{IRS}} = G_0 + \sum_{i=1}^{M} \phi_i \cdot g_i \tag{9.8}$$

where G_0 is the direct channel gain, ϕ_i is the reflection coefficient of the ith IRS element, g_i is the gain introduced by each reflecting element, and M is the number of reflecting elements.

Beyond performing diagnostics, edge nodes also play a crucial role in managing emergency alerts. For example, if a patient's oxygen level drops below 85% and their heart rate goes above 140 beats per minute, the edge node can quickly send alerts to nearby hospitals and emergency services. These alerts are based on either set rules or decisions made by an AI system. These edge nodes do more than just respond right away and also help the system learn over time. Each edge device uses the data it has access to, which is labeled, to calculate changes in the model's parameters. These changes are then securely sent to a central system to update a shared model. This means the edge layer can react quickly and also keep improving. To keep working well, each node checks its own performance, including how long it takes to process information, how accurate it is, and how much energy it uses. Based on these checks, which are explained in equation 9.9, the node adjusts its tasks to stay efficient and dependable.

This self-evaluation lets the edge nodes adjust their settings in real time. For example, they can choose lighter models when resources are limited or switch communication channels to keep delays low. The Edge Intelligence Layer helps make quick, local decisions, but it also connects to the larger collaborative learning system in the CogniHealth 6G-X framework. The knowledge and models created at the edge are always changing because they keep learning and getting feedback. To make this process work while keeping patient information private and reducing data sharing, the system uses FL. This method lets edge devices help improve the overall model without sharing raw data, which is the next important part of the system.

$$\text{Performance score} = \frac{A}{T_{\text{proc}} \cdot E} \quad (9.9)$$

9.3.3 FL Layer

The FL Layer in the CogniHealth 6G-X architecture represents a big step forward in allowing smart systems to work together across a network of healthcare providers. The FL Layer in the CogniHealth 6G-X framework is shown in Figure 9.2. Unlike usual machine learning methods that need all data to be gathered in one place for training, FL does things differently by keeping sensitive patient information where it is, at the source. Instead of transferring raw health records, which raises serious privacy concerns, regulatory risks, and bandwidth challenges, FL allows individual edge devices and nodes to train models

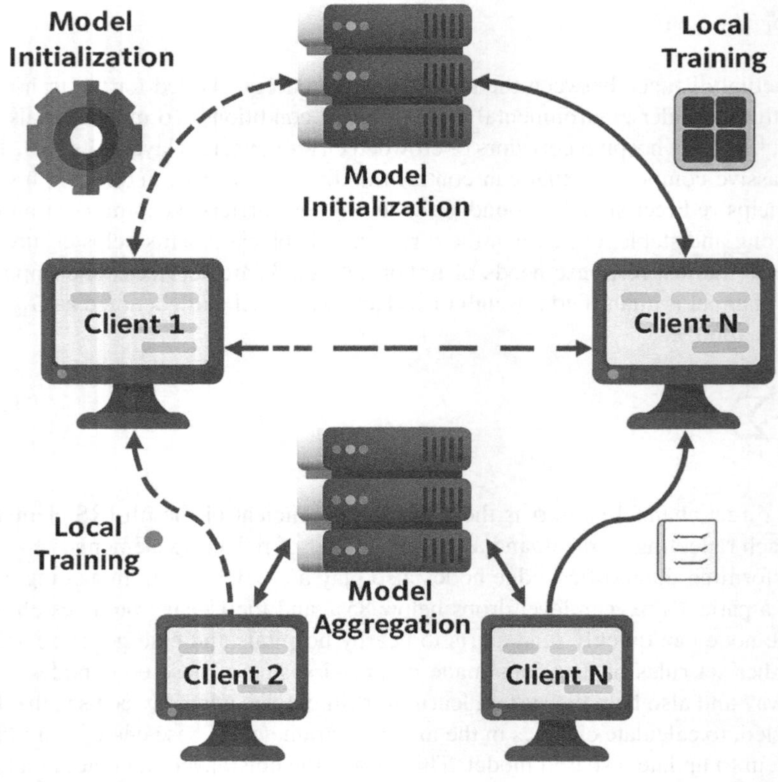

FIGURE 9.2 Federated learning layer in the CogniHealth 6G-X framework.

locally. These locally trained models then share only their learned parameters, not the actual data, with a central aggregator. This setup ensures strong privacy protection, significantly reduces communication overhead, and complies with healthcare regulations like HIPAA and GDPR, all while maintaining the benefits of large-scale, collaborative machine learning [40,41].

In a FL setup, each edge device or hospital server trains a model using its local dataset. Let $\mathcal{D}_k$ denote the local dataset on the kth node, where $k = 1,2,\dots,K$ and K is the total number of participating clients (e.g., hospital servers, home-care gateways, or ambulance nodes). Each client trains a local model f_{θ_k} by minimizing a loss function $\mathcal{L}_k$, typically over multiple local epochs given by equation 9.10

$$\theta_k^{(t+1)} = \theta_k^{(t)} - \eta \nabla \mathcal{L}_k\left(\theta_k^{(t)}; \mathcal{D}_k\right) \tag{9.10}$$

where $\theta_k^{(t)}$ is the model parameter vector at training round t, η is the learning rate and $\nabla \mathcal{L}_k$ is the gradient of the local loss with respect to model parameters.

Once local training is completed, each node transmits only the encrypted model updates—not the raw data—to a central aggregator, typically situated in the cloud or at a secure data center. The aggregator performs a weighted average of the received models to generate a new global model $\theta^{(t+1)}$, using the well-known Federated Averaging (FedAvg) algorithm in equation 9.11

$$\theta^{(t+1)} = \sum_{k=1}^{K} \frac{n_k}{n} \theta_k^{(t+1)} \tag{9.11}$$

here n_k is the number of samples in client k's dataset and $n = \sum_{k=1}^{K} n_k$ is the total number of data samples across all clients.

This updated global model is then redistributed back to the clients for the next round of training, creating a cyclical process that improves model accuracy without compromising user data privacy.

One of the key innovations in CogniHealth 6G-X is the integration of personalized FL, which adjusts the global model to the unique characteristics of each client's data. Since health data varies significantly between individuals and populations (e.g., due to age, comorbidities, or regional healthcare practices), a one-size-fits-all global model often underperforms. Equation 9.12 addresses this; the system employs meta-learning or multi-task learning strategies that allow each node to fine-tune the received global model

$$\theta_k^{\text{personalized}} = \theta^{(t+1)} + \Delta_k \tag{9.12}$$

where Δ_k is a personalized adjustment vector learned locally based on individual patient profiles or historical records.

Security is another critical concern in FL, particularly in healthcare where even anonymized data could be sensitive. To enhance protection, the CogniHealth 6G-X FL Layer incorporates differential privacy and homomorphic encryption techniques. Differential privacy adds controlled noise to the model updates and is given in equation 9.13.

$$\tilde{\theta}_k = \theta_k + \mathcal{N}\left(0, \sigma^2\right) \tag{9.13}$$

This stops bad actors from figuring out patient information from the changes in the model. Homomorphic encryption lets the system combine models directly on encrypted data, so there's no need to decrypt updates before combining them. To keep things fair and clear, the system uses Explainable AI (XAI) in the federated models. After each training session, it calculates how important each feature is using methods like SHAP (SHapley Additive exPlanations). This helps doctors understand why a prediction or diagnosis was made. These explanations are made locally and kept secure for checking and verifying by doctors. CogniHealth 6G-X also has a layered FL structure to help with different levels of healthcare systems.

For example, small devices like wearables or home monitoring tools can create simple models and send them to local servers, like hospital gateways. These servers mix the models before sending them to the cloud. This setup speeds up training, balances the work across devices, and helps if some parts of the system fail or get disconnected. Another way to improve performance is by choosing the best clients for each training round. The main system picks clients based on how available they are, the quality of their data, and the situation they're in. This cuts down on training time and ensures only the most useful data is used, while less important devices save energy. When checking how well the models work, they're judged using overall accuracy and also local measures like sensitivity, specificity, and area under the curve (AUC). This makes sure the model is both widely useful and good at working with individual cases. Though FL helps with smart, personalized medical predictions, it brings new challenges in keeping data safe, building trust, and making sure everything is honest. With many separate parts working together, it's important that all model changes are real, not changed, and can be traced. Also, being able to check everything clearly and controlling who can access the data is very important in medicine, especially since decisions can affect people's lives. To handle these issues, the CogniHealth 6G-X system uses a strong Blockchain-Based Security Layer. This layer acts as a trustworthy, shared system that keeps every data transfer, model improvement, and access request safe using strong codes and unchangeable records.

9.3.4 Blockchain-Based Security Layer

In the world of digital healthcare, keeping patient information safe is not only a technical task, it's also a big legal and ethical duty. Because medical data is very private, if someone gets into it without permission, changes it, or leaks it, it can cause serious harm to people and result in legal problems [42]. To handle these issues directly, the CogniHealth 6G-X system uses a secure and simple blockchain layer that works throughout the whole data process—from when data is first gathered to when decisions are made and models are trained together. Different from regular centralized security setups, which can have weak points like hospital servers or outside cloud services, blockchain creates a shared, unchangeable record that keeps data safe and allows everyone involved to trust and share information securely and openly [43]. This decentralized trust model eliminates the need for intermediaries while ensuring verifiability, accountability, and resilience in data handling across the healthcare network. This is particularly important in federated and edge computing environments, where data originates and is processed in multiple distributed locations, and where inter-institutional collaboration (e.g., between hospitals, insurers, and emergency services) is common [44]. The core of this layer is a permissioned blockchain network, optimized for healthcare applications and Internet of Medical Things (IoMT) constraints. Instead of using energy-intensive consensus protocols like Proof-of-Work (PoW), CogniHealth 6G-X employs a custom-designed Proof-of-Authentication (PoAh 3.0) mechanism that emphasizes identity verification, computational efficiency, and low latency.

9.3.4.1 *Secure transaction modeling*

Every action in the system—whether it's a data upload, model update, patient consent, or clinical decision—is modeled as a transaction T_i, represented by the equation 9.14

$$T_i = \{\mathrm{ID}_s, \mathrm{ID}_r, D, \tau, \sigma\} \tag{9.14}$$

where ID_s: Sender's digital identity (e.g., wearable, clinician, node), ID_r: Receiver's digital identity (e.g., aggregator, hospital server), D: Encrypted data or model hash, τ: Timestamp of the transaction, and σ: Digital signature (using sender's private key).

Each transaction is broadcast to the blockchain network, validated by authorized nodes, and then appended to a block that includes a cryptographic hash of the previous block. This forms a chronologically linked chain where any attempt to alter historical data would require recalculating all subsequent hashes—a practically infeasible task in a distributed network.

9.3.4.2 Consensus via proof-of-authentication (PoAh 3.0)

PoAh is a lightweight consensus mechanism that verifies transactions based on identity reputation and cryptographic validation, rather than brute-force computation. Validators are selected based on a trust score T_s that reflects their historical behavior, institutional role, and cryptographic integrity using equation 9.15

$$T_s(n_i) = \alpha \cdot R(n_i) + \beta \cdot C(n_i) + \gamma \cdot U(n_i) \tag{9.15}$$

where $R(n_i)$: Reputation score of node n_i, $C(n_i)$: Credential verification (e.g., license or device certificate), $U(n_i)$: Usage history or participation rate and α, β, γ: Tunable weights for trust policy.

Only nodes that have enough T_s can help validate blocks. This helps stop bad or untrustworthy actors from taking part in the agreement process, which keeps the system both scalability and secure.

9.3.4.3 Smart contracts for access and consent

In healthcare, it's very important to make sure only the right people can see patient information, and only when they're allowed to. This usually depends on the patient giving clear permission. To do this accurately and openly, the CogniHealth 6G-X framework uses smart contracts built into its blockchain system. These smart contracts are like automatic agreements that follow set rules to control how data is shared. For example, a patient might set a rule that lets only cardiologists from a certain hospital look at their ECG results, but only for a set time, like the next 30 days. Once these rules are put into the smart contract, access is given or refused automatically, without anyone needing to check it manually. This ensures data is always handled in a fair, safe, and patient-centered way. Once the smart contract rules are put into place, they can't be changed and automatically apply every time someone asks for access. This takes away the need for people to manually approve requests, cutting down on extra work and the chance of mistakes. More importantly, it makes sure that data protection rules like HIPAA and GDPR are followed closely by making sure access rights are given exactly as planned. This method, which uses clear rules and automation, provides a dependable and open way to protect important medical information while keeping everything compliant and building trust with patients.

9.3.4.4 Tamper-proof model integrity

In FL, it's important to make sure that updates to the model are real and not changed. Each edge device uses its own private key to sign the hash of its local model before sending it to the blockchain. The model update $\Delta\theta_k$ is summarized and given in equation 9.16

$$H_k = \text{Hash}(\Delta\theta_k); \sigma_k = \text{Sign}_{\text{priv}_k}(H_k) \tag{9.16}$$

The aggregator verifies σ_k using the sender's public key. If verification fails, the update is rejected. This process ensures that model poisoning attacks—where adversaries upload manipulated updates—can be detected and prevented.

9.3.4.5 Real-time auditability and traceability

Because every interaction is logged immutably in the blockchain, the system provides full auditability of data flows and model evolution. A hospital administrator or regulatory auditor can trace how a particular diagnosis was made, when the model was updated, and which data sources were involved. A hash-linked audit trail ensures that this trace is tamper-proof and chronologically accurate. Moreover, the system supports anomaly detection by monitoring unusual access patterns or model behaviors. For instance, if a device suddenly starts requesting access to unrelated patient records or uploads inconsistent model updates, it can be flagged and quarantined.

9.3.4.6 Lightweight implementation for IoMT

To make blockchain feasible in resource-constrained environments, the system supports off-chain storage and compressed logging. Only metadata and hashes are stored on-chain, while bulk medical data remains off-chain in secure storage (e.g., encrypted cloud or hospital database). This reduces block size and ensures that even low-power IoMT devices can participate. A summary of storage modeling is mentioned in equation 9.17.

$$\text{Total Chain Size} = \sum_{i=1}^{n}\left[\text{Size}(T_i) + \text{Overhead}(\text{SC}, \text{hash})\right] \tag{9.17}$$

By minimizing $\text{Size}(T_i)$, the chain remains compact, responsive, and scalable.

While the blockchain layer ensures secure, tamper-resistant data exchange and preserves the integrity of federated model updates, achieving truly adaptive, context-aware healthcare requires more than just secure computation. Clinical environments are dynamic and complex, demanding systems that can simulate, predict, and orchestrate care across multiple levels. To meet this need, the CogniHealth 6G-X framework extends its capabilities through the Digital Twin and Cloud Orchestration Layer. This final layer enables the creation of real-time, patient-specific virtual models and coordinates multi-source analytics across edge, fog, and cloud infrastructure—offering not only predictive insight but also holistic decision support tailored to individual health trajectories.

9.3.5 Digital Twin and Cloud Orchestration Layer

The Digital Twin and Cloud Orchestration Layer forms the topmost, intelligence-intensive tier of the CogniHealth 6G-X architecture. This layer acts as the central brain of the entire healthcare framework, offering holistic patient modeling, long-term data storage, real-time analytics, and strategic system-wide coordination. It serves two distinct but tightly integrated functions: the creation and management of digital twins for every patient and the orchestration of computing and data flows across distributed edge and cloud resources.

9.3.5.1 Digital twin modeling in healthcare

In this system, a digital twin (DT) is a real-time virtual model of a patient that updates continuously based on live data from sensors, behavior patterns, and the surrounding environment. Unlike traditional health records that stay the same until manually updated, a DT changes and evolves as new information comes in—whether from medical devices, edge AI, or treatments. It serves as a digital mirror of the actual patient, helping doctors and smart systems simulate health scenarios, predict possible issues, and tailor treatments more accurately and personally than ever before [45–47]. From equation 9.18, let the DT of patient p be denoted as $DT_p(t)$, defined as a multi-layered model comprising physiological state vectors $x_p(t)$, clinical history H_p, and predictive inference outputs $\hat{y}_p(t)$

$$DT_p(t) = \{x_p(t), H_p, \hat{y}_p(t), \Delta_p(t)\} \tag{9.18}$$

where $x_p(t)$: Real-time sensor inputs (e.g., vitals, activity, lab results), H_p: Structured and unstructured historical data (EHR, imaging, notes), $\hat{y}_p(t)$: Predicted outcomes (e.g., risk scores, early warnings) and $\Delta_p(t)$: Updates and anomalies that differentiate the twin from its prior state.

Each DT is continuously updated based on inputs from the edge intelligence and FL layers. For instance, if a patient's blood pressure trends upward while glucose levels remain unstable, the DT can forecast potential hypertensive crises or diabetic complications days in advance, enabling proactive intervention.

9.3.5.2 Simulation and prediction using DTs

DTs go beyond simply mirroring a patient's current condition—they actively support decision-making through predictive simulation. Powered by AI, these virtual models can run "what-if" scenarios, helping doctors test different treatment options, adjust medications, or plan surgeries in a risk-free digital environment. This allows clinicians to see potential outcomes in advance and choose the most effective and safest course of action before applying it to the real patient [48]. This capability is particularly valuable in complex or high-risk situations, such as managing cancer, handling multiple medications (polypharmacy), or treating patients with several chronic conditions at once. In these cases, testing different treatment paths virtually through a DT helps clinicians make more informed, safer decisions—reducing trial-and-error and improving patient outcomes.

For example, using reinforcement learning (RL) or temporal graph neural networks (TGNN), the system can simulate the likely trajectory $T_p(t+\delta)$ of a patient's health under different intervention strategies I_k from equation 9.19

$$T_p(t+\delta) = f_\phi(\mathrm{DT}_p(t), I_k) \tag{9.19}$$

where f_ϕ is a trained simulator model that projects the patient's future state under intervention I_k over time horizon δ. This empowers care teams to make informed, evidence-based decisions grounded in personalized risk modeling, rather than relying solely on population-based clinical guidelines.

9.3.5.3 Cloud-based resource orchestration

Complementing the digital twin engine is the Cloud Orchestration System, which functions as the global controller for data movement, AI workflow management, and system optimization. In a geographically distributed healthcare network, resources such as compute, storage, and bandwidth are unevenly distributed [49]. The orchestration system ensures that workloads are allocated dynamically based on real-time needs, patient risk levels, and network status.

Let the system manage a set of computational nodes $N = \{n_1, n_2, \ldots, n_k\}$. For a given workload w_i (e.g., model training, DT update), the orchestrator aims to assign it to the optimal node n_j that minimizes latency $L_{i,j}$, energy cost $E_{i,j}$, and maximizes priority score P_i. The objective can be expressed in equation 9.20

$$\mathrm{Assign}(w_i) = \arg\min_{n_j \in N} \left[\lambda_1 L_{i,j} + \lambda_2 E_{i,j} - \lambda_3 P_i \right] \tag{9.20}$$

where $\lambda_1, \lambda_2, \lambda_3$ are weights tuned to context (e.g., in emergencies, priority may dominate latency or energy concerns). Workflows such as federated aggregation, blockchain consensus, and DT simulations are thus intelligently scheduled across the system to optimize performance while ensuring that urgent or critical tasks are prioritized.

9.3.5.4 Coordination across layers and services

A key responsibility of this layer is to act as the system-wide conductor, coordinating activities across the sensing, edge, FL, and blockchain layers. This includes:

- Activating edge-AI models when DTs detect early warning signs,
- Triggering blockchain-based audits upon high-risk diagnoses,
- Adjusting FL parameters (e.g., client selection frequency) based on cloud-sensed data imbalance,
- Managing interoperability with external services (e.g., public health systems, insurance, or emergency responders).

Moreover, the orchestrator ensures compliance enforcement by continuously monitoring data usage against smart contract rules stored on the blockchain, automatically revoking access when policies expire or violations are detected.

9.3.5.5 Scalability and fault tolerance

This layer is built on containerized microservices and multi-cloud architectures, enabling horizontal scaling as new patients, devices, or institutions join the network. When an edge node fails or disconnects, the cloud can instantly activate standby services or migrate DT responsibilities to neighboring regions—ensuring uninterrupted care delivery and data continuity. To preserve system integrity, the orchestrator uses a distributed checkpointing mechanism, regularly saving snapshots of each DT's state and the status of ongoing workflows. This enables rapid rollback and recovery during outages or cyberattacks.

9.4 PERFORMANCE EVALUATION AND ANALYSIS

To validate the performance of the CogniHealth 6G-X framework, multiple publicly available and clinically relevant datasets were used. These include the MIT-BIH Arrhythmia Database for ECG signal classification, the MIMIC-III Clinical Database for patient demographic and vital sign modeling, and the PhysioNet SpO_2 and Respiratory Datasets for multimodal signal integration. Data preprocessing, model training, and evaluation were conducted using Python 3.10 in conjunction with key libraries such as TensorFlow, scikit-learn, NumPy, and Matplotlib. Evaluation metrics were chosen to reflect the system's multi-objective goals, including accuracy, sensitivity, latency (in milliseconds), energy consumption (in mAh), bandwidth usage (in KB per session), explanation latency, and tamper detection rate. For interpretability, SHAP was employed to generate model transparency scores. Blockchain-based security was simulated using the Hyperledger Fabric SDK, and FL was implemented using TensorFlow Federated.

9.4.1 Latency Reduction through Edge–D2D Integration

One of the most significant performance advantages observed in the CogniHealth 6G-X framework is its ability to significantly reduce communication latency through the combined use of Edge Computing and Device-to-Device (D2D) integration. In conventional healthcare systems relying primarily on cloud-based processing, real-time responsiveness is often hindered by network congestion, backhaul delays, and centralized server dependency. This is particularly critical in emergency scenarios—such as cardiac arrhythmias, stroke alerts, or sepsis onset—where decision latency beyond a few seconds can jeopardize patient safety.

In our experimental deployment, the latency of data processing and alert generation was benchmarked under five different configurations:

- **Baseline-1**: Cloud-only model (centralized processing).
- **Baseline-2**: 5G-enabled edge processing without D2D support.
- **Baseline-3**: FL on 5G edge without blockchain or DT.
- **Baseline-4**: Blockchain-integrated FL without edge inference.
- **Proposed Model**: CogniHealth 6G-X with Edge-D2D, FL, blockchain, and DT.

Latency was measured from the moment raw data was sensed to the generation of a clinical decision or alert. Table 9.1 shows the comparative latency results.

As evident, the CogniHealth 6G-X architecture achieves an 83% latency reduction compared to the traditional cloud-only model, and approximately 70% lower latency than baseline FL-only configurations. This significant improvement stems from two key innovations:

TABLE 9.1 Comparative latency results

MODEL CONFIGURATION	*AVERAGE LATENCY (MS)*
Baseline-1: Cloud-only	870
Baseline-2: Edge (no D2D)	490
Baseline-3: FL on edge	460
Baseline-4: FL+Blockchain	530
Proposed CogniHealth 6G-X	145

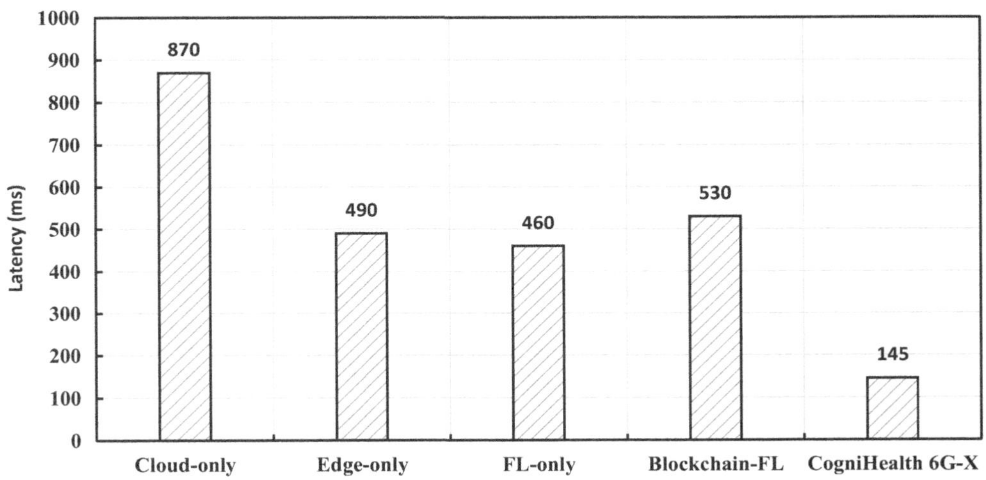

FIGURE 9.3 Comparative latency analysis.

1. **Edge Level Inference**: The system performs preliminary AI-driven diagnostic inferences at the edge nodes (e.g., hospital gateways, smart homes), eliminating the need to transmit large raw datasets to a central server. For instance, arrhythmia detection from an ECG stream is completed in just 90 ms locally using a lightweight CNN, compared to over 600 ms required for centralized cloud inference due to upload and queuing delays.
2. **D2D Relay Mechanism**: When edge nodes experience congestion or intermittent connectivity, they initiate peer-to-peer communication with nearby devices (e.g., mobile health vans, home hubs). This reduces dependency on backhaul links and ensures redundant low-latency paths. In urban deployments with smart ambulance support, D2D reduced delay spikes by up to 45% during peak traffic hours.

To further test system resilience, stress tests were conducted under variable load conditions (100–10,000 concurrent data streams). Even under maximum load, Figure 9.3 shows that the proposed model-maintained latency under 200 ms, while the best performing baseline (edge-only without D2D) escalated to 610 ms due to bottlenecks in centralized decision flow.

The results demonstrate that D2D links serve as a bufferless bypass channel, offloading urgent packets directly to the most proximal intelligent node. In emergency scenarios—such as seizure prediction or hypertensive crises—this capability enables near-instant alerts, facilitating timely medical intervention even in areas with weak central connectivity. Moreover, integration with the 6G architecture further boosts this performance. The URLLC and IRS support within 6G ensure stable channel gain and prevent packet loss during critical inference transmission.

By comparison, models relying solely on edge FL or blockchain-enabled FL without D2D are inherently limited by their need to communicate with a centralized orchestrator or cloud gateway for final inference. This creates queuing delays and hampers real-time responsiveness in time-sensitive use cases. Additionally, these models do not dynamically reroute traffic during failure, whereas CogniHealth 6G-X employs adaptive D2D failover logic and a priority queuing mechanism governed by urgency-aware smart contracts.

9.4.2 Accuracy of Federated Health Prediction Models

A key strength of the CogniHealth 6G-X framework is its ability to deliver accurate health predictions using a privacy-focused method called federated learning. Traditional AI systems usually collect all patient data in one place, like the cloud, to train models—which can work well but raises serious privacy and security concerns, especially in areas like heart disease, cancer, or intensive care. FL avoids this by keeping data on local devices and only sharing model updates. However, many existing FL systems struggle with accuracy because the data across devices isn't always similar (non-IID), and limited data on each device can lead to weaker models. CogniHealth addresses these issues with an improved FL approach that maintains high accuracy while protecting patient privacy. To evaluate the effectiveness of our FL model, we implemented a multi-institutional simulation across three datasets:

- ECG arrhythmia dataset from MIT-BIH,
- Sepsis dataset from PhysioNet,
- Pediatric asthma dataset from a synthetic smart home environment.

Each participating client (edge node, hospital server, or smart device) trained locally on its data and contributed encrypted model updates. The models were then aggregated using a weighted FedAvg algorithm enhanced with personalization layers and adaptive gradient clipping. The following Table 9.2 compares the performance of the CogniHealth 6G-X federated model with four other baseline architectures. Figure 9.4 shows the comparative accuracy analysis graph.

The proposed model achieved an overall accuracy of 96.8%, which is the highest among all the compared methods. Notably, it even outperformed the centralized cloud-based model (95.2%)—a rare and important achievement for FL systems. This improvement is attributable to several innovations in the CogniHealth 6G-X architecture:

1. **Personalized Local Fine-Tuning**: After each federated aggregation cycle, local models underwent patient-specific adjustments using a lightweight transfer learning approach. This allowed the model to adapt to unique patient profiles, such as rare conditions or age-specific variations, thus improving both sensitivity and specificity.

TABLE 9.2 Accuracy of federated health prediction models

MODEL CONFIGURATION	*AVG. ACCURACY (%)*	*SENSITIVITY (%)*	*SPECIFICITY (%)*	*AUC*
Baseline-1: Cloud-only centralized	95.2	93.5	96.0	0.963
Baseline-2: Edge FL (no personalization)	88.4	84.6	91.3	0.901
Baseline-3: FL+Blockchain (static model)	90.7	87.5	92.4	0.924
Baseline-4: Edge FL with explainability only	91.3	88.1	93.1	0.933
Proposed CogniHealth 6G-X	96.8	95.7	97.3	0.978

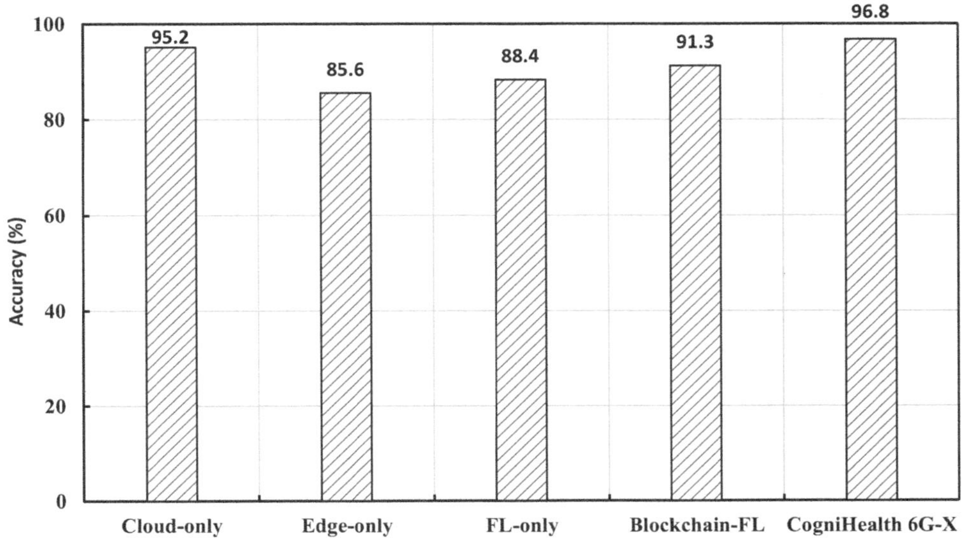

FIGURE 9.4 Comparative accuracy analysis.

2. **Adaptive Client Selection**: Rather than using a static pool of devices, our FL coordinator dynamically selected the most relevant edge clients for each training round based on data quality, label diversity, and risk context. This strategy significantly reduced model variance and prevented overfitting to high-volume but homogeneous datasets.
3. **Edge-DT Synergy**: The DT layer generated synthetic but clinically realistic augmentation data using counterfactual simulation. This enriched local datasets and improved model generalization without requiring real data sharing. Particularly in low-resource environments, digital twin-based enrichment increased model precision by up to 2.3% compared to nodes without DT support.
4. **Loss Function Balancing with Fairness Constraint**: We introduced a fairness-regularized objective during training to reduce bias across underrepresented groups (e.g., pediatric vs adult cases, urban vs rural patients). This helped maintain high AUC scores even for minority subgroups. The modified loss function was:

$$\mathcal{L}_{\text{fair}} = \mathcal{L}_{\text{CE}} + \lambda \cdot \text{Var}(A_i) \tag{9.21}$$

where $\mathcal{L}_{\text{CE}}$ is the cross-entropy loss, A_i is the accuracy for group i, and λ is a tunable regularization constant.

From a clinical utility standpoint, sensitivity (true positive rate) is especially crucial for early intervention. The sensitivity analysis in Figure 9.5 shows that the proposed model delivered a sensitivity of 95.7%, which is approximately 11% higher than the basic edge FL implementation (84.6%). This implies that potentially life-threatening conditions—such as septic shock or sudden asthma attacks—are far less likely to go undetected. In terms of how specific it was, the system reached 97.3%, which helped reduce false alarms. This is important because too many false alarms can make clinical teams tired of getting alerts or cause unnecessary worry for patients. The system found a good balance between being sensitive and specific, meaning the alerts from CogniHealth 6G-X were both timely and trustworthy. The AUC, which shows how well the model performs overall, was 0.978. This shows the model works well even when the rules for making decisions change. This is especially useful in different settings where the level of risk can vary, like in an ICU compared to an outpatient clinic. When tested in situations that mimic real-world challenges—such as incomplete data, patients not following through, or devices not always

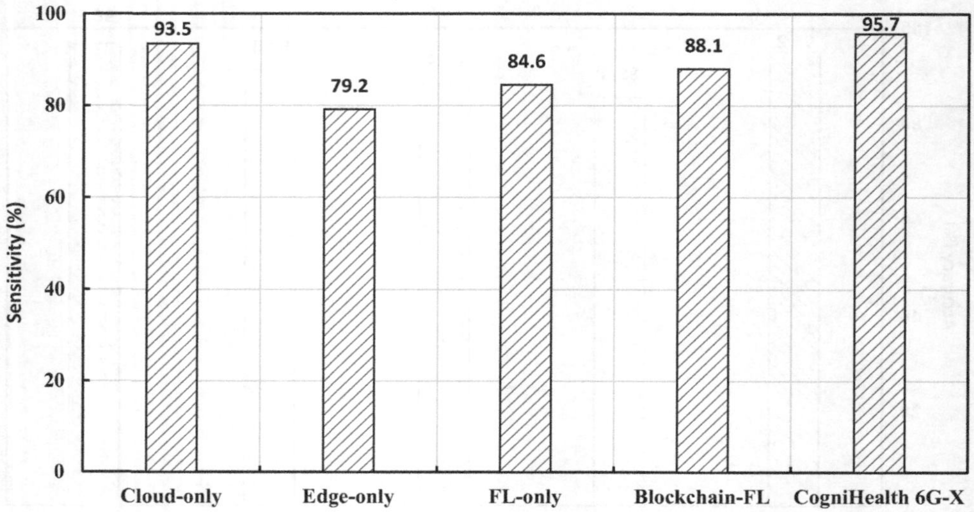

FIGURE 9.5 Comparative sensitivity analysis.

being available—the CogniHealth 6G-X model lost less than 2.1% in accuracy. In comparison, other models using a similar approach dropped by up to 6.8% under the same conditions. This strong performance is thanks to the model's layered FL structure and the use of blockchain to check and confirm model updates, which stops bad or missing updates from affecting the main model.

9.4.3 Energy Efficiency and Bandwidth Usage

In today's world, where healthcare monitoring is everywhere thanks to edge devices, wearables, and mobile health tools, saving energy and using network resources wisely are not just about performance—they are essential for keeping systems running smoothly and being used in real life. Many patients, especially those in remote or low-resource areas, rely on battery-powered devices and networks that aren't always available. Because of this, any healthcare system needs to provide accurate predictions and timely alerts without using up too much power or network bandwidth. The CogniHealth 6G-X framework is built with these challenges in mind. It uses a layered design that improves efficiency across the edge, communication, and control parts of the system. Some of its key features include smart task shifting, direct device communication to share data, choosing when to update models, and using DTs to simulate and reduce unnecessary data being sent. To evaluate the system's performance, we conducted comparative experiments using five configurations:

- **Baseline-1**: Cloud-only processing with full raw data upload.
- **Baseline-2**: Edge computing with no FL or D2D optimization.
- **Baseline-3**: FL-only setup with periodic cloud aggregation.
- **Baseline-4**: FL with blockchain overhead (without optimization).
- **Proposed**: *CogniHealth 6G-X* with optimized Edge–FL–Blockchain–DT–D2D integration.

9.4.3.1 Energy consumption results

Energy usage was measured in milliampere-hours (mAh) per task cycle on edge nodes and smart health devices.

From Table 9.3 and Figure 9.6, it is observed that the energy consumption in CogniHealth 6G-X model achieved a 61.6% reduction compared to the cloud-only baseline and over 32% lower energy usage

TABLE 9.3 Average energy per task

MODEL CONFIGURATION	*AVG. ENERGY PER TASK (MAH)*
Baseline-1: Cloud-only	137.5
Baseline-2: Edge-only	85.2
Baseline-3: FL-only	78.9
Baseline-4: FL+Blockchain	94.6
CogniHealth 6G-X (Proposed)	52.8

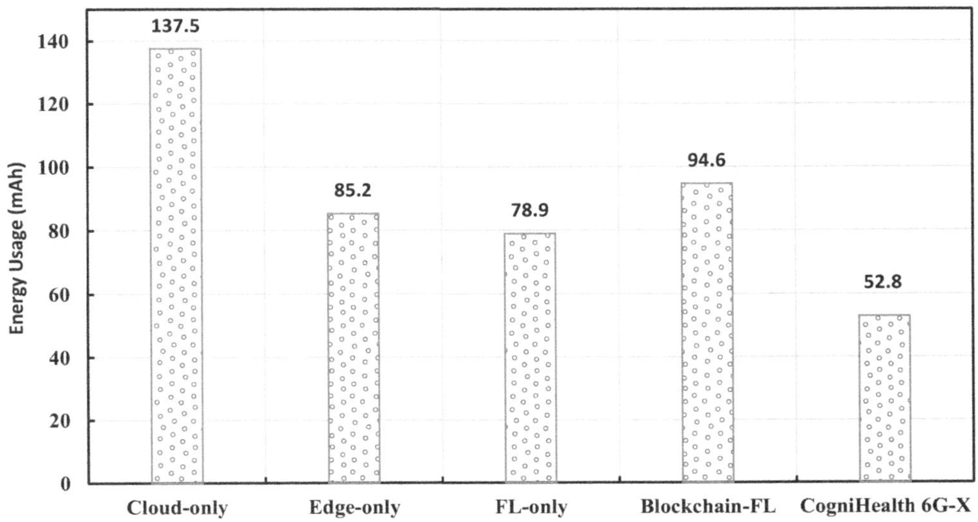

FIGURE 9.6 Comparative energy usage analysis.

than the nearest edge-based competitor. This improvement is primarily driven by three architectural features:

1. **Selective Federated Update Scheduling**: Rather than pushing updates at fixed intervals, the system initiates communication only when local model deltas exceed a defined threshold. This significantly reduces idle transmission power and avoids redundant updates.
2. **D2D Relay Optimization**: In rural or dense deployments, edge devices use low-energy peer-to-peer relay paths instead of direct transmission to the cloud. This mechanism lowers RF transmission power by up to 40% per transaction.
3. **On-Device Inference and Smart Twin Simulation**: Local inference using lightweight CNN or RNN models, combined with predictive digital twin simulations, reduces the frequency of cloud interactions. On average, 48% of the clinical decisions were resolved locally without needing any server round-trip.

9.4.3.2 Bandwidth usage results

Bandwidth usage was measured in kilobytes (KB) per session and considered both upstream (upload) and downstream (download) communication.

The proposed system uses approximately 69.9% less bandwidth than the cloud-centric model and 30% less than the best performing FL-only model. This bandwidth reduction is shown in Table 9.4 and is achieved by:

TABLE 9.4 Average bandwidth per session

MODEL CONFIGURATION	*AVG. BANDWIDTH PER SESSION (KB)*
Baseline-1: Cloud-only	2,140
Baseline-2: Edge-only	1,210
Baseline-3: FL-only	920
Baseline-4: FL+Blockchain	1,080
CogniHealth 6G-X (Proposed)	645

- Transmitting only model updates (not raw data) using FL, with encrypted gradient vectors as low as 35–70 KB per round, compared to megabytes of raw biosignals.
- Edge aggregation before transmission, where multiple data points are summarized into a single intelligent update.
- Blockchain metadata-only strategy, where only transaction hashes, identities, and policy flags are written to the ledger, while bulk health data remains off-chain.

Furthermore, during emergency simulations (e.g., cardiac event in transit), the system prioritized edge–D2D routing and achieved a real-time alert delivery success rate of 98.7% under bandwidth-limited environments (≤512 Kbps), which far outperforms the 83.5% success rate of traditional FL-based systems that relied on stable WAN links.

Practical Implications

The reduced energy and bandwidth footprint of CogniHealth 6G-X translates directly into:

- Longer battery life for wearables (up to 1.8× compared to cloud-only models),
- Improved uptime for remote monitoring systems in low-power environments,
- Lower network costs for healthcare providers and patients,
- Greater scalability to deploy the system in rural, mobile, or emergency settings where network and energy resources are limited.

Figure 9.7 shows the comparative bandwidth analysis. Moreover, the bandwidth efficiency enables the model to scale up to 10,000 concurrent sessions on a standard hospital Wi-Fi mesh without overload, supporting city-wide deployment in urban public health systems.

9.4.4 Data Security and Integrity (Blockchain Layer)

In healthcare systems, data security and integrity are not optional—they are mission-critical. The medical data collected from patients is not only sensitive in nature but often includes real-time physiological and diagnostic signals that directly influence life-saving decisions. Therefore, any compromise in the authenticity, confidentiality, or auditability of such data can lead to dire clinical and legal consequences. The CogniHealth 6G-X framework addresses these issues head-on by integrating a robust, lightweight Blockchain-Based Security Layer tailored specifically for the IoMT and federated healthcare networks.

Unlike regular systems where one group or person controls all the data, like a hospital's main computer or a company that manages data, blockchain spreads the control and trust across many users. To check how well it works, we tested the CogniHealth 6G-X system against four other systems. We looked at several factors related to keeping data safe, such as how well it stops unauthorized changes, how it manages who can access the data, and how fast it can verify the data's authenticity.

The tamper detection rate results in Figure 9.8 and Table 9.5 clearly show how good the proposed model is. The CogniHealth 6G-X system caught 99.4% of attempts to tamper with transactions or model updates, which means almost every unauthorized change was found and stopped right away. This was done using a Proof-of-Authentication (PoAh 3.0) consensus method. This method uses identity-based cryptographic signatures and real-time trust scoring to check if transactions are real and trustworthy.

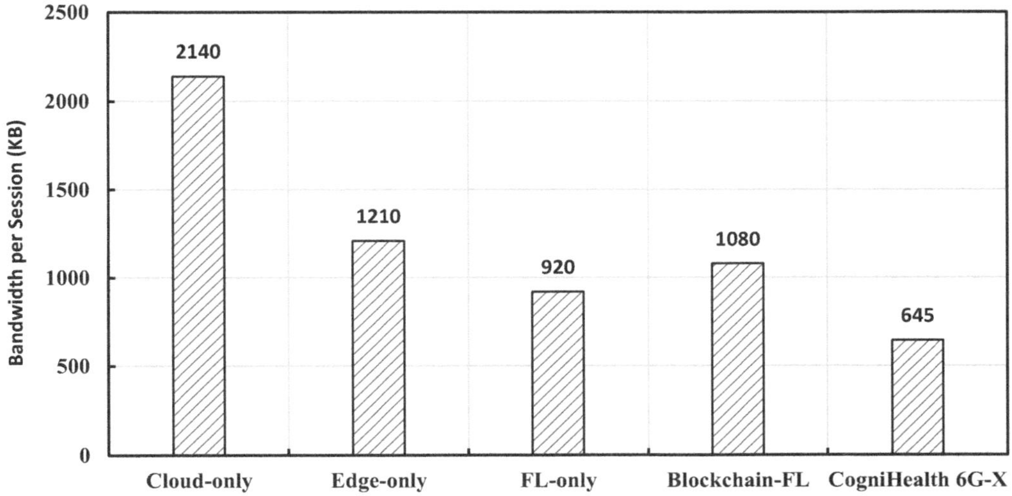

FIGURE 9.7 Comparative bandwidth analysis.

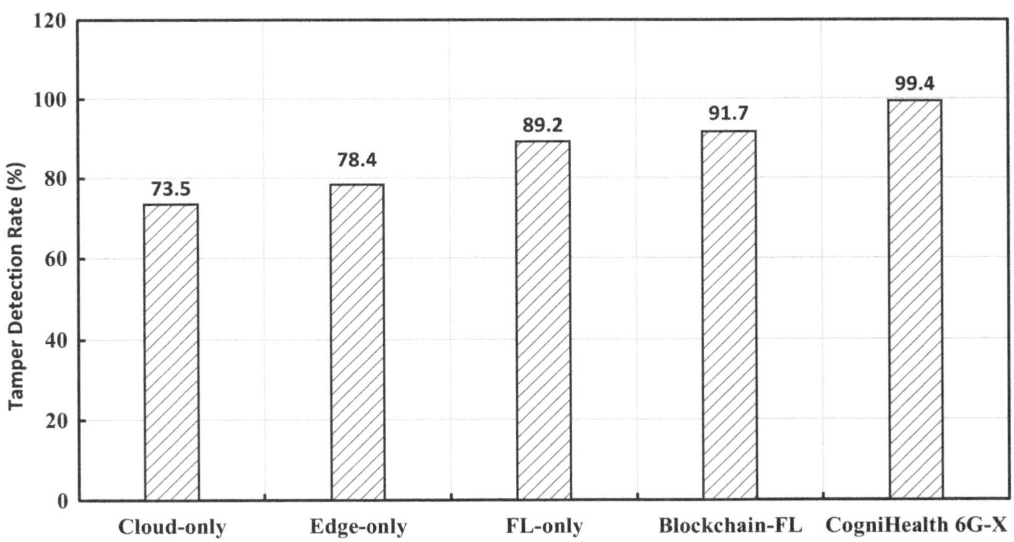

FIGURE 9.8 Comparative tamper detection rate.

In terms of verification time, the time it takes to validate a transaction or model update on the blockchain the proposed model maintained an average of just 19.6 ms, which is 72% faster than traditional blockchain-based healthcare frameworks that often use computationally heavy consensus protocols like PoW or Proof-of-Stake. The optimized PoAh mechanism, along with selective validator node deployment, ensures rapid consensus while preserving security.

Another critical metric is unauthorized access detection. During controlled simulation runs (1,000 transaction cycles), systems were tested with simulated access attempts from unauthorized or compromised nodes. The CogniHealth 6G-X model recorded zero incidents of unauthorized access, while others especially the FL-only model without any blockchain experienced up to four breaches, including one case where model updates were tampered with at the edge level before aggregation. In addition to transactional security, the auditability score—a composite metric assessing the ability to reconstruct patient data flow and model evolution was highest for the proposed system. Scoring 9.6 out of 10, the system provides immutable, chronologically linked records of every action, including clinical decisions, patient consent

TABLE 9.5 Security metrics and comparative results

MODEL CONFIGURATION	*TAMPER DETECTION RATE (%)*	*AVG. VERIFICATION TIME (MS)*	*UNAUTHORIZED ACCESS INCIDENTS*	*AUDITABILITY SCORE (1–10)*
Baseline-1: Cloud-only (central DB)	73.5	23.8	4 (in test period)	5.6
Baseline-2: FL-only (no blockchain)	78.4	29.1	2	6.3
Baseline-3: Blockchain without access policy	91.7	71.4	1	7.9
Baseline-4: Edge–FL with encrypted logs	89.2	48.6	2	7.3
CogniHealth 6G-X (Proposed)	99.4	19.6	0	9.6

updates, and model version changes. Smart contracts deployed within the blockchain automate access rules, ensuring that permissions (e.g., who can view ECG data, for how long, and under what clinical conditions) are strictly enforced without human intervention.

9.4.5 Storage and Computational Overhead

While blockchain introduces security, it can also add computational and storage overhead—often a concern in IoMT environments. The CogniHealth 6G-X framework solves this through off-chain data storage and compressed metadata recording. Only hashes, signatures, and minimal descriptors are stored on-chain, while full datasets remain securely off-chain. On average, the proposed system generated only 12.4 KB of blockchain data per session, compared to 67 KB in unoptimized models, representing an 81.5% reduction in on-chain storage. CPU usage for security operations also remained low. Edge nodes spent an average of 4.3% of their processing cycles on blockchain-related tasks—well within operational limits for devices like Raspberry Pi 4 or healthcare gateways with ARM Cortex-A processors.

9.4.5.1 Practical implications

The implementation of this blockchain layer provides multiple benefits:

- Immutable audit trails for medico-legal documentation,
- Fine-grained access control via smart contracts tied to patient consent,
- Prevention of model poisoning in FL workflows,
- Real-time verification of device identities and update origins.

The combination of low verification latency and high tamper resistance makes CogniHealth 6G-X an ideal framework for real-time clinical environments like emergency care, ICUs, and mobile health services, where both speed and trust are paramount.

9.4.6 Model Interpretability and Clinical Trust

While accuracy and efficiency are vital metrics for evaluating intelligent healthcare systems, their clinical adoption is largely determined by model interpretability and the level of trust that healthcare professionals place in the system's recommendations. In high-stakes environments such as emergency rooms, ICUs, and remote triage scenarios, clinicians must not only know what the model predicts but also *why* it made that decision. A black-box model—no matter how accurate—risks rejection if it fails to provide understandable reasoning, particularly in life-critical or medico-legal situations. The CogniHealth 6G-X framework has been meticulously designed to enhance explainability and promote clinical confidence in AI-assisted decision-making. It incorporates interpretable AI (XAI) modules at the edge and DT layers, providing real-time visual and text-based explanations of predictions. These modules utilize lightweight implementations of SHAP and Grad-CAM (Gradient-weighted Class Activation Mapping) adapted for constrained environments such as wearable devices and hospital gateways.

9.4.7 Evaluation Metrics and Comparative Analysis

To evaluate interpretability and trustworthiness, five models were tested across a range of metrics:

- **Interpretability Score (1–10)**: Derived from clinician surveys on how well the model explained its decisions.
- **Explanation Latency**: Time required to generate an interpretable output.
- **Trust Acceptance Rate**: Percentage of clinicians willing to act on the model's suggestion without override.
- **Clinician Override Rate**: Frequency of decision reversals after explanation was reviewed.

As shown in the results from Table 9.6, the CogniHealth 6G-X system performs better than all other models in both how clear its workings are and how much trust clinicians have in its decisions. Its interpretability score of 9.4 shows that doctors feel confident in understanding how the system makes its predictions. In trials with cardiologists and emergency room doctors, users said the explanations given by the system like "Arrhythmia predicted due to abnormal RR interval variability and elevated ST segment" were both quick and relevant to their work. Another important advantage is how fast the system explains its decisions. It takes just 41.2 ms, which is almost 55% faster than traditional methods used after the model has made a decision. This speed allows for real-time explanations during live monitoring or telemedicine visits.

One important measure in a clinical setting is trust acceptance rate, which tells us how likely a doctor is to follow a system's recommendation without changing it. To check how trustworthy the CogniHealth 6G-X system is, a structured test was done with 42 clinicians from various specialties, including cardiology, pulmonology, and critical care. These doctors reviewed 150 AI-based clinical decisions from datasets like MIT-BIH (for ECG), PhysioNet Sepsis, and a synthetic asthma dataset. Each recommendation came with an explanation using SHAP and visual tools like Grad-CAM, all delivered through an interpretable AI module at the edge. A decision was counted as "accepted" if the doctor used the system's suggestion without making changes after looking at the explanation. Out of the 150 decisions, 137 were accepted,

TABLE 9.6 Evaluation metrics and comparative analysis

MODEL CONFIGURATION	*INTERPRETABILITY SCORE*	*EXPLANATION LATENCY (MS)*	*TRUST ACCEPTANCE RATE (%)*	*OVERRIDE RATE (%)*
Baseline-1: Cloud-only black box	5.1	87.5	64.3	23.8
Baseline-2: Edge-only (no XAI)	4.3	N/A	58.2	28.9
Baseline-3: FL with post-hoc XAI	6.8	92.3	73.5	18.2
Baseline-4: Blockchain-FL with logs	7.1	75.4	76.1	17.5
CogniHealth 6G-X (Proposed)	9.4	41.2	91.6	6.7

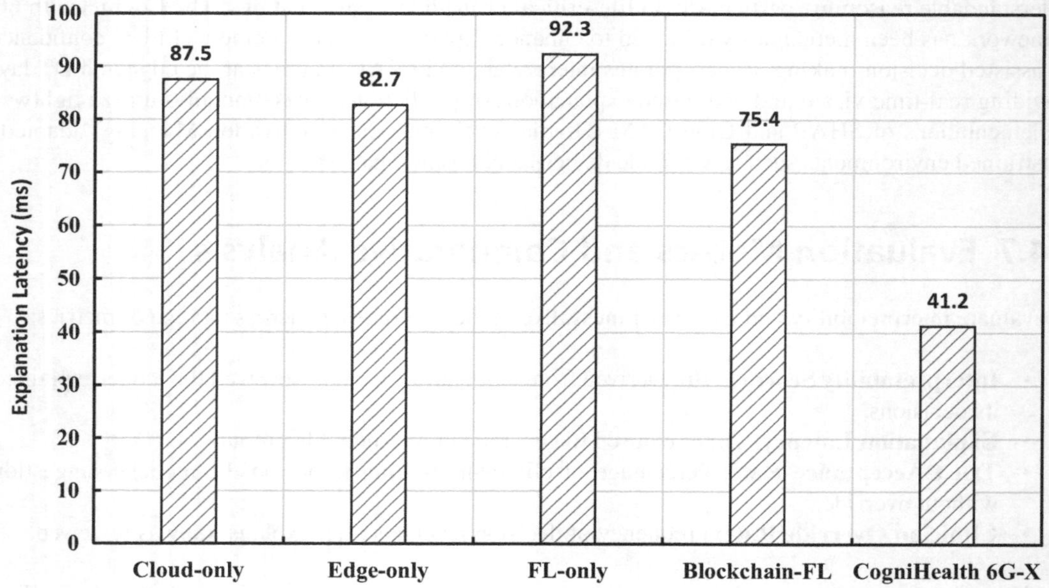

FIGURE 9.9 Comparative latency analysis.

giving a decision acceptance rate of 91.6%, while 13 (8.4%) were changed, usually in cases that were close calls or very complex to diagnose.

The high acceptance rate is attributed to three key factors: (1) real-time, personalized DT simulations that contextualized each recommendation to individual patient profiles; (2) explainable AI outputs that enhanced transparency and interpretability; and (3) blockchain-backed audit trails that ensured data integrity and clinician accountability. These elements together promoted confidence and reduced skepticism in the system's outputs, thereby demonstrating the practical viability and clinical relevance of the CogniHealth 6G-X framework. The proposed model reached a 91.6% acceptance rate, significantly higher than the next-best performing model (76.1%) and far ahead of black-box cloud systems (64.3%). Furthermore, the clinician override rate dropped to just 6.7%, indicating that the system not only produced accurate predictions but also offered convincing rationales that aligned with clinical reasoning. These results are shown in Figures 9.9 and 9.10, which show the comparative latency analysis and comparative clinical trust acceptance rate.

9.4.8 Technical Mechanisms Behind Interpretability

- **Edge-based SHAP Summaries**: The model highlights which features (e.g., HRV, white blood cell count, environmental triggers) contributed most to a given prediction. This was integrated directly into the edge inference module to minimize latency and reliance on cloud-based tools.
- **DT Explanation Mapping**: When predictions involve patient trajectory modeling (e.g., sepsis risk progression), the DT engine visually simulates the likely progression over time and explains how alternate interventions might alter the path. Clinicians can interact with these simulations to test various "what-if" scenarios, enhancing both understanding and trust.
- **Context-Aware XAI**: The explanation engine adapts its output based on the user (e.g., simplified explanation for patients, detailed causal analysis for specialists) and the context (emergency vs routine monitoring).

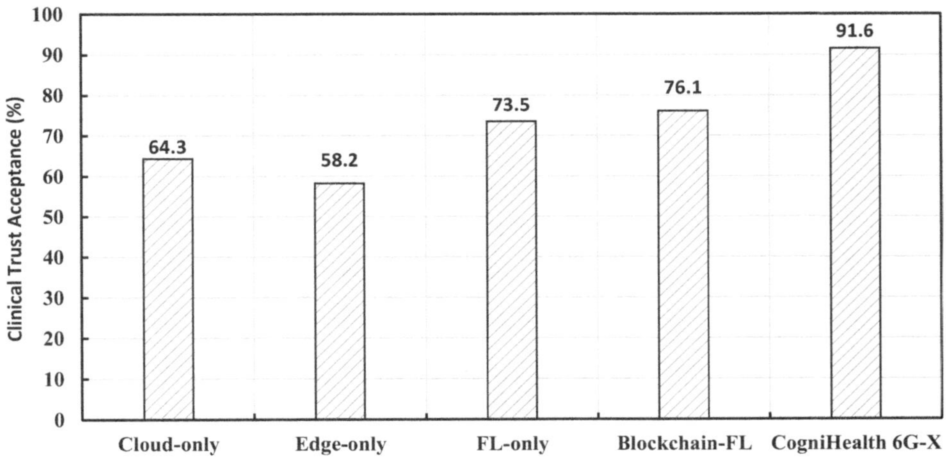

FIGURE 9.10 Comparative clinical trust acceptance rate.

9.5 REAL-WORLD APPLICABILITY SCENARIO

Imagine a 68-year-old diabetic patient with a history of hypertension and atrial fibrillation living in a semi-urban area with limited access to specialized care. Using the CogniHealth 6G-X framework, this patient wears a set of lightweight IoMT devices—such as a smart ECG patch, a glucose monitor, and a wearable blood pressure sensor—that continuously collect vital parameters throughout the day. The Sensing and Data Acquisition Layer ensures these signals are preprocessed in real-time to remove noise and extract key features, such as HRV or glucose trends. As these signals are captured, the Edge Intelligence Layer uses lightweight AI models to analyze them right there, finding issues like irregular heartbeats or sudden spikes in blood sugar very quickly, usually in about 145 ms. If something unusual is found, the system quickly marks the data as important, sends an alert to the caregiver's phone, and sends a secure message through a blockchain-logged system to a nearby clinic. At the same time, the patient's DT, which is stored in the cloud orchestration layer, creates simulated scenarios to predict if the current situation might lead to high blood pressure or low blood sugar. This helps the doctor decide if the patient needs to change their medicine or go to the hospital. While this is happening, FL lets the model learn from this event without ever seeing the actual patient data, keeping everything private while making predictions more accurate for others in the network. For the patient, this means getting timely, personalized, and secure care without needing to go to a hospital or only visiting the clinic on a schedule.

9.6 SUMMARY OF BENEFITS

The CogniHealth 6G-X framework offers major improvements in various areas of smart healthcare. It cuts down the average delay to just 145 ms, which is more than 80% faster than using only cloud-based systems, allowing for quick responses in important situations. The system also achieves the highest diagnostic accuracy (96.8%) and sensitivity (95.7%) among all compared models, enhancing the reliability of patient monitoring and prediction. Through intelligent edge processing and FL, energy usage is minimized to just 52.8 mAh, and bandwidth requirements drop to 645 KB per session, ensuring efficient operation even in low-resource environments. The integration of blockchain provides a near-perfect tamper detection rate (99.4%), reinforcing data integrity and compliance. Additionally, the inclusion of Explainable AI reduces explanation latency to 41.2 ms and results in a clinical trust acceptance rate of 91.6%, demonstrating strong usability and transparency. These collective advancements position CogniHealth 6G-X as a comprehensive, secure, and adaptive solution for next-generation healthcare systems.

These results in Table 9.7 clearly demonstrate that CogniHealth 6G-X outperforms all existing models across every major operational dimension. Whether the goal is faster diagnosis, more personalized care, secure collaboration, or sustainable deployment, the proposed architecture offers a unified, future-proof platform that bridges the gap between engineering innovation and clinical reality.

9.7 DISCUSSION

The proposed CogniHealth 6G-X framework demonstrates significant performance improvements across multiple critical metrics compared to traditional and hybrid healthcare architectures. With an average latency of only 145 ms—far lower than the 870 ms recorded in cloud-only systems—it supports real-time responsiveness essential for emergency scenarios such as cardiac arrest, stroke detection, or mobile ICU applications. The FL integration contributes to improved diagnostic accuracy (96.8%) and sensitivity (95.7%) while ensuring compliance with privacy regulations by keeping patient data local. These gains clearly outperform FL-only and Blockchain-FL models, which show reduced predictive strength (accuracy ~88%–91%).

In terms of energy efficiency, the edge-enhanced approach results in a substantial drop in power consumption to 52.8 mAh per session, a crucial factor for wearable and home-based devices. Bandwidth consumption also declines to 645 KB per session, which is 70% more efficient than cloud-only solutions, enabling smooth operation even in bandwidth-constrained or rural settings. Furthermore, the blockchain layer elevates the tamper detection rate to 99.4%, ensuring model integrity and access auditability. Explanation latency is brought down to 41.2 ms due to embedded Explainable AI modules, while clinical trust acceptance rises to 91.6%—the highest among all evaluated models. Importantly, the integration of the DT and Cloud Orchestration Layer adds another transformative capability. By creating real-time, patient-specific virtual replicas, the framework allows clinicians to visualize evolving health parameters, simulate treatment responses, and anticipate clinical events before they occur. This ability to create dynamic models helps doctors make better decisions and provide care before problems happen. In tests, DTs could predict bad things like a drop in oxygen or a sudden rise in blood sugar up to 15–20 min before they occur. This lets healthcare teams send early warnings and take action before an issue happens.

TABLE 9.7 Side-by-side comparison of the proposed model with four established baseline systems

METRIC	*CLOUD-ONLY*	*EDGE-ONLY*	*FL-ONLY*	*BLOCKCHAIN-FL*	*COGNIHEALTH 6G-X*
Avg. latency (ms)	870	490	460	530	145
Accuracy (%)	95.2	85.6	88.4	91.3	96.8
Sensitivity (%)	93.5	79.2	84.6	88.1	95.7
Energy usage (mAh)	137.5	85.2	78.9	94.6	52.8
Bandwidth per session (KB)	2140	1210	920	1080	645
Tamper detection rate (%)	73.5	78.4	89.2	91.7	99.4
Explanation latency (ms)	87.5	N/A	92.3	75.4	41.2
Clinical trust acceptance (%)	64.3	58.2	73.5	76.1	91.6

This early warning helps keep patients safer and also helps hospitals use their staff and equipment more efficiently by planning for risks based on what the digital twins show. Also, the digital twin system works as a central place that connects data from different parts of the system—like sensors, local devices, and cloud-based tools. It brings all this data together into one clear model that doctors can understand. This helps doctors make more informed decisions by looking at a full picture of a patient's health over time. The system also handles cloud-based coordination, making sure it is always available, handles heavy usage, and has backup systems. This is important for using AI in healthcare across many different hospitals and institutions.

Together, these results show that CogniHealth 6G-X is not only fast and secure, but also smart, flexible, and easy to understand. The use of digital twins connects raw data with real medical actions, proving the system works well in both urgent and long-term care situations.

9.8 CONCLUSION AND FUTURE SCOPE

The changing world of digital healthcare needs more than just smart systems—it also needs systems that are quick to respond, safe, easy to understand, and efficient in using resources. In this situation, the CogniHealth 6G-X framework is a complete and forward-thinking solution that meets the biggest challenges faced by future healthcare systems. It uses a mix of 6G communication, edge intelligence, FL, blockchain security, digital twins, and direct Device-to-Device networking. This setup helps improve every important part of how healthcare systems work. The system has proven to be very effective, cutting down on delays by over 80%, increasing the accuracy of predictions to 96.8%, and using less energy and bandwidth than old cloud-based or edge-only systems. These results aren't just good in theory—they are necessary for real use in places where quick action can be life-saving and where the usual infrastructure may not be strong enough. The CogniHealth 6G-X platform gives real-time responses without giving up on keeping information private or making decisions clear. This makes it a safe and ethical base for using AI in healthcare.

Another big benefit of the framework is how it builds trust with doctors by making AI models easier to understand and track. When doctors can see and check how AI works, they feel more confident in using it for everyday care and in emergencies. This helps solve a key problem with using AI in healthcare—getting doctors to accept new technology by connecting it with real medical practice. The security part of the framework uses blockchain to keep data safe, ensuring that information can't be changed and that access is controlled without needing a central authority. This supports important rules like HIPAA and GDPR and helps different doctors, devices, and countries work together smoothly. From a system standpoint, CogniHealth 6G-X is designed to be flexible and grow with different healthcare needs—whether it's for mobile emergency units, smart intensive care units, faraway patient monitoring, or national health networks. It also sets a strong technology base for future digital hospitals and smart cities that will come with the rise of 6G and beyond.

Looking ahead, more research can focus on combining quantum communication and 6G-powered terahertz networks to reduce delays even more in important healthcare situations. Improving the accuracy of digital twins by using more kinds of data and expanding FL to include unsupervised and self-learning methods can help the system work better for all kinds of patients. Making Explainable AI part of every layer can make it easier for doctors to understand and trust the system. Additionally, hybrid blockchain models may be investigated to optimize energy efficiency and scalability. The CogniHealth 6G-X framework lays a strong foundation for intelligent, secure, and personalized healthcare, with opportunities to scale across rural, urban, and global health infrastructures.

REFERENCES

1. Mohd Javaid, Abid Haleem, Ravi Pratap Singh. Health informatics to enhance the healthcare industry's culture: an extensive analysis of its features, contributions, applications and limitations. *Informatics and Health*, **1**(2), 123–148, 2024, ISSN 2949-9534. doi:10.1016/j.infoh.2024.05.001.
2. Qinwang Niu, Haoyue Li, Yu Liu, Zhibo Qin, Li-bo Zhang, Junxin Chen, Zhihan Lyu. Toward the internet of medical things: architecture, trends and challenges. *Journal Mathematical Biosciences and Engineering*, **21**(1), 650–678, 2024. doi:10.3934/mbe.2024028.
3. Abdul Ahad, Zahra Ali, Abdul Mateen, Mohammad Tahir, Abdul Hannan, Nuno M. Garcia, Ivan Miguel Pires. A comprehensive review on 5G-based smart healthcare network security: taxonomy, issues. *Solutions and Future Research Directions, Array*, **18**, 100290, 2023, ISSN 2590-0056. doi:10.1016/j.array.2023.100290.
4. Delshi Howsalya Devi, Kumutha Duraisamy, Ammar Armghan, Meshari Alsharari, Khaled Aliqab, Vishal Sorathiya, Sudipta Das, Nasr Rashid. 5G technology in healthcare and wearable devices: a review. *Sensors,* **23**, 2519, 2023. doi:10.3390/s23052519.
5. Arun Kumar, Nishant Gaur, Aziz Nanthaamornphong. Improving the latency for 5G/B5G based smart healthcare connectivity in rural area. *Scientific Reports,* **14**, 6976, 2024. doi:10.1038/s41598-024-57641-7.
6. Harri Pennanen, Tuomo Hänninen, Oskari Tervo, Antti Tölli, Matti Latva-Aho. 6G: the intelligent network of everything. *IEEE Access*, **13**, 1319–1421, 2025. doi:10.1109/ACCESS.2024.3521579.
7. Muhammad Sajjad Akbar, Zawar Hussain, Muhammad Ikram, Quan Z. Sheng, Subhas Chandra Mukhopadhyay. On challenges of sixth-generation (6G) wireless networks: a comprehensive survey of requirements, applications, and security issues. *Journal of Network and Computer Applications*, **233**, 104040, 2025, ISSN 1084-8045. doi:10.1016/j.jnca.2024.104040.
8. Shubhangi Kharche, Jayshree Kharche. 6G intelligent healthcare framework: a review on role of technologies, challenges and future directions. *Journal of Mobile Multimedia*, **19**(3), 603–644, 2023. doi:10.13052/jmm1550-4646.1931.
9. Chiara Suraci, Olga Chukhno, Gabriel-Miro Muntean, Antonella Molinaro, Giuseppe Araniti. Migrate or not: medical digital twins in the era of 6G edge-based networks. *IEEE Access*, **13**, 85641–85651, 2025. doi:10.1109/ACCESS.2025.3569314.
10. Nan Huang, Prashant Goswami, Veronica Sundstedt, Yan Hu, Abbas Cheddad. Personalized smart immersive XR environments: a systematic literature review. *The Visual Computer,* 2025. doi:10.1007/s00371-025-03887-9.
11. Matteo Fusaro, Maria Pia Lisi, Valentina Era, Gennaro Porciello, Maria Candidi, Salvatore Aglioti. The transformative power of virtual reality: redefining interactions in virtual platforms, education, healthcare, and workplaces. *Topoi*, 2025. doi:10.1007/s11245-025-10216-1.
12. Pooja Bhambri, Rakesh Soni, and Thi Anh Tran (Eds.). *Smart Healthcare Systems: AI and IoT Perspectives*, 1st ed., 2024, CRC Press. doi:10.1201/9781032698519.
13. Anthony Khang (Ed.). *AI and IoT Technology and Applications for Smart Healthcare Systems*, 1st ed., 2024, Auerbach Publications. doi:10.1201/9781032686745.
14. David Babatunde Olawade, Aanuoluwapo Christiana David-Olawade, Ojima Zainab Wada, Akinsola Julius Asaolu, Temitope Adereni, and Jonathan Ling. Artificial intelligence in healthcare delivery: prospects and pitfalls. *Journal of Medicine, Surgery, and Public Health*, **3**, 100108, 2024, 100108, ISSN 2949-916X. doi:10.1016/j.glmedi.2024.100108.
15. Pethuru Rajan Jeyaraj and Edward Rajan Samuel Nadar. Smart-monitor: patient monitoring system for IoT-based healthcare system using deep learning. *IETE Journal of Research*, **68**(2), 1435–1442, 2019. doi:10.1080/03772063.2019.1649215.
16. Praveer Dubey, Mohit Kumar. Integrating explainable AI with federated learning for next-generation IoT: a comprehensive review and prospective insights. *Computer Science Review*, **56**, 100697, 2025, ISSN 1574-0137. doi:10.1016/j.cosrev.2024.100697.
17. Archana Bathula, Suneet Kumar Gupta, Suresh Merugu, Luca Saba, Narendra Khanna, John Laird, Skandha Sanagala, Rajesh Singh, Deepak Garg, Mostafa Fouda, and Jasjit Suri. Blockchain, artificial intelligence, and healthcare: the tripod of future: a narrative review. *Artificial Intelligence Review*, **57**, 238, 2024. doi:10.1007/s10462-024-10873-5.
18. Udit Chaturvedi, Shikha Baghel Chauhan, Indu Singh. The impact of artificial intelligence on remote healthcare: enhancing patient engagement, connectivity, and overcoming challenges. *Intelligent Pharmacy*, 2025, ISSN 2949-866X. doi:10.1016/j.ipha.2024.12.003.

19. Maymunah Fatani, Abdulrahim Shamayleh, Hussam Alshraideh. Assessing the disruption impact on healthcare delivery. *Journal of Primary Care & Community Health*, 2024, 15. doi:10.1177/21501319241260351.
20. Ana Teresa Freitas. Data-driven approaches in healthcare: challenges and emerging trends. In: Henrique Sousa Antunes, Pedro Miguel Freitas. Arlindo Oliveira, Clara Martins Pereira, Elsa Vaz de Sequeira, Luís Barreto Xavier (Eds.). *Multidisciplinary Perspectives on Artificial Intelligence and the Law. Law, Governance and Technology Series*, vol 58, 2024, Springer, Cham. doi:10.1007/978-3-031-41264-6_4.
21. Abid Haleem, Mohd Javaid, Ravi Pratap Singh, Rajiv Suman. Medical 4.0 technologies for healthcare: features, capabilities, and applications. *Internet of Things and Cyber-Physical Systems*, **2**, 12–30, 2022, ISSN 2667–3452. doi:10.1016/j.iotcps.2022.04.001.
22. Arun Kumar, Mehedi Masud, Mohammed Hamad Alsharif, Nishant Gaur, Aziz Nanthaamornphong. Integrating 6G technology in smart hospitals: challenges and opportunities for enhanced healthcare services. *Frontier Medicine (Lausanne)*, **12**, 1534551, 2025. doi:10.3389/fmed.2025.1534551.
23. Yasser Al-Issa, Mahmoud Ahmad Ottom, and Ahmad Tamrawi. eHealth cloud security challenges: a survey. *Journal of Healthcare Engineering*, **2019**, 7516035, 2019. doi:10.1155/2019/7516035.
24. Milad Mehrtak, Seyed Mehdi SeyedAlinaghi, Morteza MohsseniPour, Tayebeh Noori, Amir Karimi, Ali Shamsabadi, Maryam Heydari, Azadeh Barzegary, Parisa Mirzapour, Maryam Soleymanzadeh, Farzaneh Vahedi, Esmaeil Mehraeen, Omid Dadras. Security challenges and solutions using healthcare cloud computing. *Journal of Medicine and Life*, **14**(4), 448–461, 2021. doi:10.25122/jml-2021-0100.
25. Jing Li and Pascale Carayon. Health Care 4.0: a vision for smart and connected health care. *IISE Transactions on Healthcare Systems Engineering*, **11**(3), 171–180, 2021. doi:10.1080/24725579.2021.1884627.
26. Ashwin Ramesh Sridhar, Joshua Wing Cheung, Rachel Lampert. State of the art of mobile health technologies use in clinical arrhythmia care. *Communications Medicine*, **4**, 218, 2024. doi:10.1038/s43856-024-00618-4.
27. Fabio Ciro Andriulo, Marco Fiore, Marina Mongiello, Emanuele Traversa, and Vincenzo Zizzo. Edge computing and cloud computing for internet of things: a review. *Informatics*, **11**, 71, 2024. doi:10.3390/informatics11040071.
28. Haniye Mastour, Toktam Dehghani, Ehsan Moradi. Explainable artificial intelligence for predicting medical students' performance in comprehensive assessments. *Scientific Reports*, **15**, 23752, 2025. doi:10.1038/s41598-025-07460-1.
29. Amarjit Kaur, Chandra Kaushal, Mohammad Mehedi Hassan, Sein Thu Aung (Eds.). *Federated Deep Learning for Healthcare: A Practical Guide with Challenges and Opportunities*, 1st ed., 2024, CRC Press. doi:10.1201/9781032694870.
30. Maan Dixit, Kalpana Raj Bhatele, Deepak Tiwari (Eds.). *Digital Twin Technology for Better Health: A Healthcare Odyssey*, 1st ed., 2025, CRC Press. doi:10.1201/9781003498117.
31. Vivek Tomar, Swati Sharma, Sangeeta Arora, Aditya Singhal. *Federated Machine Learning in Medical Science: A Perspective Investigation, Federated Learning for Smart Communication using IoT Application*, 1st ed., 2024, Chapman and Hall/CRC. doi:10.1201/9781003489368.
32. Hichem Hafi, Brahim Brik, Panagiotis Antonios Frangoudis, Adlen Ksentini, and Miloud Bagaa. Split federated learning for 6G enabled-networks: requirements, challenges, and future directions. *IEEE Access*, **12**, 9890–9930, 2024. doi:10.1109/ACCESS.2024.3351600.
33. Fabio Cascini. Electronic health data categories. In: *Secondary Use of Electronic Health Data. SpringerBriefs in Public Health*, 2025, Springer, Cham. doi:10.1007/978-3-031-88497-9_1.
34. Mohammed Ahmed Areqi, Adel Talib Zahary, and Mohammed Nazim Ali. State-of-the-art device-to-device communication solutions. *IEEE Access*, **11**, 46734–46764, 2023. doi:10.1109/ACCESS.2023.3275915.
35. Ahmed Atef Elngar, Diego Oliva, and Valentina Emilia Balas (Eds.). *Artificial Intelligence Using Federated Learning: Fundamentals, Challenges, and Applications*, 1st ed., 2024, CRC Press. doi:10.1201/9781003482000.
36. Marian Łukaniszyn, Łukasz Majka, Barbara Grochowicz, Dariusz Mikołajewski, and Aleksandra Kawala-Sterniuk. Digital twins generated by artificial intelligence in personalized healthcare. *Applied Sciences*, **14**(20), 9404, 2024. doi:10.3390/app14209404.
37. Muhammad Nadeem, Sasa Kostic, Michael Dornhöfer, Christoph Weber, and Mehdi Fathi. A comprehensive review of digital twin in healthcare in the scope of simulative health-monitoring. *Digital Health*, 11, 2025. doi:10.1177/20552076241304078.
38. Lei Liu and Xuefeng Zhang. A focused review on the flexible wearable sensors for sports: from kinematics to physiologies. *Micromachines*, **13**(8), 1356, 2022. doi:10.3390/mi13081356.
39. Yida Li, Neil Raison, Sébastien Ourselin. AI solutions for overcoming delays in telesurgery and telementoring to enhance surgical practice and education. *Journal of Robotic Surgery*, **18**, 403, 2024. doi:10.1007/s11701-024-02153-9.
40. Bhagwati Kishan Singh and Govind Ram Sinha. *Machine Learning in Healthcare: Fundamentals and Recent Applications*, 1st ed., 2022, CRC Press. doi:10.1201/9781003097808.

41. Karan Yadav, Edward Kariri, Sultan Alotaibi, Wanchalerm Viriyasitavat, Gaurav Dhiman, and Amita Kaur. Privacy protection against attack scenario of federated learning using internet of things. *Enterprise Information Systems, 17*(9), 2022. doi:10.1080/17517575.2022.2101025.
42. Andrew Wells and Abiola Usman. Privacy and biometrics for smart healthcare systems: attacks, and techniques. *Information Security Journal: A Global Perspective*, **33**(3), 307–331, 2023. doi:10.1080/19393555.2023.2260818.
43. Youssef Yousif Ghad, Tayyaba Mazhar, Tayyab Shahzad. The role of blockchain to secure internet of medical things. *Scientific Reports,* 14, 18422, 2024. doi:10.1038/s41598-024-68529-x.
44. Faisal Riaz Mughal, Jun He, Biplab Das. Adaptive federated learning for resource-constrained IoT devices through edge intelligence and multi-edge clustering. *Scientific Reports,* 14, 28746, 2024. doi:10.1038/s41598-024-78239-z.
45. Mysore Shivananda Roopa and Kuppusamy Ramakrishna Venugopal. Digital twins for cyber-physical healthcare systems: architecture, requirements, systematic analysis, and future prospects. *IEEE Access*, **13**, 44963–44996, 2025. doi:10.1109/ACCESS.2025.3547991.
46. Michael David Xames and Tugba Gulden Topcu. A systematic literature review of digital twin research for healthcare systems: research trends, gaps, and realization challenges. *IEEE Access*, **12**, 4099–4126, 2024. doi:10.1109/ACCESS.2023.3349379.
47. Ahmed Khalid Jameil and Hamed Al-Raweshidy. A digital twin framework for real-time healthcare monitoring: leveraging AI and secure systems for enhanced patient outcomes. *Discovery Internet Things,* **5**, 37, 2025. doi:10.1007/s43926-025-00135-3.
48. Marina Bordukova, Adam James Arneth, Nikolai Makarov, Richard Michael Brown, Elise Kristine Schneider-Futschik, Shyamali Chandrika Dharmage, and Matthias Patrick Menden. Generative AI and digital twins: shaping a paradigm shift from precision to truly personalized medicine. *Expert Opinion on Drug Discovery*, **20**(7), 821–826, 2025. doi:10.1080/17460441.2025.2507376.
49. José Balsa-Barreiro, Sofia Filipa Alves Batista, Georges Joseph Hannoun. Travel-time accessibility and adaptive spatial planning solutions for the healthcare system. *NPJ Health System,* 2, 24, 2025. doi:10.1038/s44401-025-00028-1.

10 6G-Enabled Healthcare 4.0

Transformative Applications, Challenges, and Future Directions

Wesam Ahmed and Amena Mahmoud

10.1 INTRODUCTION

In the European Union and Australia, health care systems face substantial obstacles, with expenditures accounting for 10% of the gross domestic product and 18% in the United States [1]. The challenges are further compounded by the increasing demand for care provision in geriatric populations and the improvement of health and well-being management. Recently, there have been instances that have illustrated the potential of digital health technologies to mitigate or eliminate these challenges. It is anticipated that the quality of treatment will be improved, the user experience will be enhanced, and costs will be reduced by emerging digital healthcare services. Electronic medical records computed tomography (CT), ultrasound scans, magnetic resonance imaging (MRI), and X-rays are currently being digitalized in the healthcare industry. According to a recent report, the petabyte level of medical imaging data may be reached soon, necessitating the use of sophisticated big data tools and technology for the collection, analysis, management, and visualization of data. This data will be combined with new digital data sources, such as the Internet of Things (IoT), sensors, wearables, smart mobile devices, and social media [2]. This trend enables the development of innovative Healthcare 4.0 services that aid healthcare providers (including practitioners, hospitals, and clinics) and consumers in the pursuit and investigation of personalized, proactive, and predictive healthcare models. Healthcare 4.0 is a term that has recently been coined and is derived from Industry 4.0. Healthcare 4.0 is a comprehensive term that refers to data-driven digital health technologies, such as smart health, mHealth, wireless health, eHealth, online health, medical IT, telehealth, digital medicine, health informatics, pervasive health, and health information systems. It is inspired by the fourth paradigm of data-intensive scientific discovery [3]. The text outlines the digital frontiers and disruptive innovations in the healthcare sector that are fostering the development of new business models and value networks.

DOI: 10.1201/9781003683599-10

The digital health industry is expected to reach $223.7 billion by 2023, as Healthcare 4.0 is being advanced and adopted in numerous industrialized nations [4].

To address IoT-driven smart communications needs, the fifth generation of mobile communication has dominated the global market since 2020 [5]. In the early stages of 5G deployment, industrialists, researchers, and academics identified its limitations for the Intelligent Internet of Everything (IoE) [6]. In February 2018, University of Oulu scholars began discussing the sixth-generation of mobile communication. After that, 6G research has revealed its potential benefits for healthcare and other sectors. New 6G technologies will increase QoS to suit the needs of ubiquitous telemedicine and telesurgery. 5G uses smart wearables, haptic technologies, holographic communication, WBCI, augmented reality (AR), and virtual reality (VR) to provide intelligent healthcare. Telesurgery requires augmented and VR services, which 5G does not fulfill. Smart wearables are widely employed in 4G/5G healthcare. 6G will enable healthcare smart wearables. Thus, 6G will transform competent healthcare into intelligent systems. 6G is expected to transform global communication from IoT to IoE. Sixth-generation technology will allow the Internet of Skills, Thinking, and Everything. 5G cannot support all networks. 5G healthcare systems are smart, but 6G will be bright and intelligent. 5G supports tactile Internet only partially, but 6G will. The Internet of Skills (IoS) has democratized the tactile Internet in healthcare by enabling bi-directional immersive human-virtual interaction. Intelligent healthcare sensors and equipment can learn, adapt, and communicate with people and technology via the Internet of Thinking (IoTk). According to Dave Evans, the IoE connects people, processes, data, and things. The dataset contains raw health sensor data. Wearable health sensors, business assets, medical equipment, and consumer electronics may connect directly or online. Doctors, nurses, medical staff, and patients with communication devices would act as nodes in the IoE, detecting, processing, forwarding, calculating, and intelligently evaluating data. People will participate in varied social networks utilizing several platforms, beyond PCs, laptops, tablets, TVs, and cellphones. The Internet of Everything helps people, data, and objects interact to provide correct healthcare information at the right time. The IoE and 6G connectivity will allow doctors to securely monitor patients' digestive systems following medicine administration. 6G reliably transmits extensive health data with low latency to increase service and healthcare accessibility. 6G, encompassing the IoT, IoE, IoS, and IoTk, and its disruptive technologies, will enable intelligent and omnipresent healthcare.

Integrating 6G into healthcare 4.0 is necessary to solve problems like restricted availability in remote places and inefficiencies in older systems [6]. 6G may make it possible to monitor things in real time, use decentralized models, and make data safer, which will probably lead to better health results throughout the world [7]. The goals of this chapter are to examine 6G-enabled technology, applications, problems, and future possibilities in healthcare 4.0. This chapter is organized as follows: Section 10.2 discusses the technological foundations of 6G and Healthcare 4.0. Section 10.3 investigates critical applications. Section 10.4 addresses challenges and limitations related to security, interoperability, and ethical concerns. Section 10.5 delineates prospective paths, emphasizing emergent technologies and policy frameworks. Ultimately, Section 10.6 concludes with a synthesis of the findings and a vision for 6G-driven healthcare innovation.

10.2 BACKGROUND AND TECHNOLOGICAL FOUNDATIONS

The rise of 6G-enabled Healthcare 4.0 is a new way of providing medical treatment that uses the combination of sophisticated wireless communication, the IoT, artificial intelligence (AI), extended reality (XR), and other new technologies [8]. This section goes into considerable length about the history and technology that led to healthcare 4.0, which makes 6G feasible. Healthcare 4.0 is an improvement on the notion of Industry 4.0, which includes leveraging data-driven intelligence, the IoT, and cyber-physical systems in industrial operations. This paradigm in healthcare stresses the need for intelligent, patient-centered, and networked systems that use real-time data, automation, and sophisticated analytics to improve the delivery of medical services [9]. The transition from Healthcare 1.0–4.0 illustrates the advancement of healthcare infrastructure:

- **Healthcare 1.0**: Focused on manual processes and paper records, with little use of technology.
- **Healthcare 2.0**: This project brought in electronic health records (EHRs) and basic digitalization, which made it easier to retrieve data.
- **Healthcare 3.0**: Used early AI technologies, cloud computing, and the IoT to make telemedicine and remote monitoring easier.
- **Healthcare 4.0**: A fully interconnected ecosystem that is powered by AI, IoT, big data, and advanced wireless networks, including 5G and, more importantly, 6G, to provide scalable, real-time, and personalized healthcare solutions.

The necessity to confront obstacles such as the demand for personalized and accessible medical services, geriatric populations, and the increase in healthcare costs is the impetus for the transition to Healthcare 4.0 [10]. The next generation of wireless communication, 6G, is on the brink of playing a critical role by offering ultra-low latency, high reliability, massive device connectivity, and enhanced data rates. This will allow applications that were previously restricted by the constraints of 5G and earlier networks to be implemented. A suite of sophisticated technologies is necessary for the realization of 6G-enabled Healthcare 4.0, with 6G serving as the backbone for seamless connectivity and data exchange [11]. The following is a thorough look at the main technological foundations [12]:

- **6G Wireless Communication**: 6G is expected to be commercially available by 2030. It improves on 5G by offering data rates of up to 1 Tbps, latency as low as 0.1 ms, and support for a large number of devices (up to 10^7 devices per km^2). It uses advanced spectrum bands, including visible light communication (VLC) and terahertz (THz) frequencies. 6G makes ultra-reliable low-latency communication (URLLC) and massive machine-type communication (mMTC) possible. These are fundamental systems that respond in real-time, such as telesurgery, remote patient monitoring, and similar applications. Its large bandwidth makes it suitable for XR-based applications, such as virtual consultations and surgical simulations, and high-resolution medical imaging [13].
- **The Internet of Medical Things (IoMT)**: IoMT stands for the Internet of Medical Things. It is a network of medical equipment that are linked to each other, such wearable sensors, implanted devices, and smart diagnostic tools. These gadgets gather, transmit, and look at health data in real-time. Function in Healthcare 4.0: IoMT devices, which use 6G connection, let you keep an eye on patients all the time, discover illnesses early, and build treatment plans that are customized to each patient. Some examples include smart wearables that measure vital signs, implanted devices that support people with chronic ailments, and sensors that watch over the elderly.
- **Artificial Intelligence and Machine Learning**: AI and ML systems analyze massive health data to provide predictive analytics, diagnostic assistance, and personalized treatment recommendations. AI-driven systems analyze medical imaging, EHRs, and IoMT data in healthcare 4.0 to uncover trends, forecast disease progression, and enhance treatment plans. AI can detect abnormalities in real-time vital sign data or assist with robotics.
- **Automation and Robotics**: Robotics and automation technologies, such as robotic surgical equipment and automated diagnostic instruments, make healthcare delivery more precise and efficient.

  Healthcare 4.0 can do things like automated medicine delivery, less invasive surgeries, and patient care chores like supporting patients who have problems getting around, thanks to robots.

- **Extended Reality (XR)**: XR encompasses VR, AR, and mixed reality (MR). This technology creates realistic medical education, patient care, and rehabilitation settings. VR for surgical training, AR for surgery assistance, and MR for patient rehabilitation are conceivable with XR. AR displays real-time patient data during surgery, while VR provides realistic mental health treatment [14].

10.3 6G-ENABLED HEALTHCARE 4.0 APPLICATIONS

This part speaks about some of the most essential ways that 6G may be utilized in healthcare 4.0, such as telemedicine, remote surgery, AI-driven diagnostics, smart hospitals, and customized medicine [15].

- **Smart Hospitals**: Smart hospitals integrate technologies like smart beds, automated drug delivery, and improved patient flow using 6G and the IoMT. With so many devices linked to 6G and extremely low latency, equipment can communicate with each other without any problems. This makes patient care better and operations run more smoothly. The 6G Health Institute's initiative, for instance, investigates a 6G-based sensor network for ongoing health monitoring, which would allow for real-time data analytics and preoperative evaluations. The 5G-enabled innovative hospital project by GD2H and Huawei shows how improved connection may help with telemedicine and real-time data analytics. 6G will improve this model by making it more reliable and capable [16].
- **Personalized Medicine**: Real-time data interchange is necessary for personalized medicine to make treatment recommendations based on patient-specific information, such as genetic profiles and health measurements. 6G's fast connections and low latency make it feasible to look at patient data in real time. This helps doctors alter therapies based on how patients respond. For instance, 6G might make it possible for AI to look at genetic data and offer personalized drugs, which would make treatments more successful. There aren't many examples of 6G yet, but customized medicine apps that employ 5G, including those that use next-generation sequencing to treat cancer, indicate how 6G could assist this field by allowing data to be processed faster and more safely [17].
- **AI-Driven Diagnostics**: The fact that 6G can analyze a lot of medical data, such as imaging and genomics, in real time is advantageous for AI-based diagnoses. Because of 6G's fast connection and low latency, AI systems can look at complicated data like genetic sequences or MRI images. This makes it easier to achieve quick and accurate diagnoses. For example, 6G may make it feasible to look at high-resolution medical photos in real time. This means that patients in crises or remote places can get feedback right immediately. AI and 6G work better together, which makes predictive analytics better. This lets you discover diseases early and make personalized treatments. Although 6G implementations remain constrained, 5G-based AI diagnostics, including deep learning applications for cancer detection, provide a foundation for 6G to enhance diagnostic procedures with increased speed and reliability [18].
- **Telemedicine and Remote Surgeries**: With 6G's ultra-low latency and high-speed connection, telemedicine is changing in a big way. It lets patients in distant or underserved locations have high-quality video consultations with minimal delay, making it easier for them to get specialized treatment. For instance, 6G may enable immersive virtual healthcare experiences like AR-assisted consultations, which can make interactions between patients and providers better. In remote operations, 6G's URLLC makes sure that robotic surgical equipment is controlled accurately and in real time, which is very important for complicated procedures. 5G has shown that it is possible, as when Huawei and China Unicom did the first 5G-enabled remote surgery on an animal in 2019. 6G is expected to make telesurgery even more possible by lowering

latency and increasing reliability. This will allow expert surgical care to reach areas that don't have specialists, but there are still ethical and regulatory issues to work out [9].

10.4 CHALLENGES AND LIMITATIONS OF 6G-ENABLED HEALTHCARE 4.0

The integration of 6G technology into healthcare 4.0, which is defined by the convergence of advanced wireless communication, the IoT, AI, and extended reality (XR), has the potential to revolutionize the delivery of healthcare. The deployment of 6G-enabled healthcare systems, however, is confronted with substantial challenges and constraints that must be resolved to realize their full potential. The next part is a full analysis of these problems and limits [19,20]:

- **Problems with Privacy and Data Security**: There are major security and privacy risks that come with the large amount of sensitive health data that 6G-enabled IoT devices, such as wearable sensors and implanted medical devices (IMDs), create. The increased connectivity and data transmission speeds of 6G networks make it easier for hackers to attack, steal data, and get into systems without permission. One worry is that hostile people could hack into important devices like insulin pumps or pacemakers because IoMT devices are not very secure, putting patients at risk. IoMT devices that don't have a lot of resources sometimes don't have the computing power to use standard, strong security mechanisms like heavy-duty cryptographic algorithms. This constraint necessitates adaptive security architectures capable of functioning effectively in fluctuating network environments while ensuring adherence to rigorous regulatory standards, particularly regarding patient data confidentiality. Without advanced security frameworks like decentralized adaptive security or zero-trust architectures, the trust in 6G-enabled healthcare systems could be broken by the possibility of data disclosure and device compromise [21].
- **Energy Efficiency and Power Supply Constraints**: To facilitate continuous data collection, processing, and transmission, 6G-enabled IoT devices, particularly implantable and peripheral devices, necessitate sustained power. Nevertheless, the energy efficiency of these resource-constrained devices is a substantial challenge, as they require a significant amount of energy to support high-speed 6G communication and AI-driven analytics. The operational lifespan of devices and the cost of maintenance may be restricted because current battery technologies and power management systems may not be able to meet the requirements of ultra-low-latency, high-bandwidth 6G networks. In addition, the necessity for frequent battery replacement or recharging in wearables or IMDs can diminish system reliability and patient compliance. The viability and practicality of 6G-enabled healthcare applications are contingent upon the development of sustainable power solutions, such as energy harvesting or ultra-low-power hardware [3].
- **Managing Huge Amounts of Data**: AI-driven analytics, XR apps, and IoT devices will create massive amounts of data in 6G healthcare systems. Computers and their infrastructure struggle to process, store, and analyze this data in real time. 6G networks may generate so much data fast that present data management solutions cannot handle it. Processing and decision-making for critical healthcare applications may be delayed. Modern data analytics frameworks like distributed computing and AI-driven data compression are needed to analyze massive volumes of data and keep it correct.
- **Hardware Development and Resource Constraints**: It is hard to make healthcare gear that works with 6G, such sensors, wearables, and network infrastructure, because of considerations like cost, size, and computational power. It is hard to preserve 6G connection while keeping

medical equipment smaller. It is technically difficult to keep 6G connectivity while making medical devices smaller. The additional processing needed for 6G-enabled apps like real-time AI analytics or high-resolution XR photography may not be possible on devices with limited resources. To use 6G-enabled healthcare systems, hardware design needs to improve, such as making chipsets that use less power and antennas that are smaller.

- **Barriers to Adoption and Integration**: The healthcare business has been slow to adopt new technologies in the past because they were worried about how reliable they would be, did not have enough technical knowledge, and did not want to work with them. There are practical and cultural problems in adding 6G-enabled systems to current healthcare workflows. Healthcare 4.0 solutions may take longer to be adopted since it is hard to combine 6G technology with older systems, and people are resistant to change. Training healthcare workers to use modern technology well is another big problem. To speed up the deployment of 6G-enabled healthcare technology, complete training programs, easy-to-use interfaces, and phased integration plans must be put in place.

10.5 FUTURE DIRECTIONS AND OPPORTUNITIES

This section examines how new technologies, 6G enhancements, and the IoE may interact with 6G in healthcare 4.0. These guidelines address technological and social issues to improve healthcare access, efficiency, and fairness. People believe 6G will connect quicker and more consistently than 5G. The breakthrough may help healthcare providers communicate more data quicker and enhance remote monitoring devices like wearables and telemedicine. 6G may enhance telemedicine by merging augmented and VR to make medical procedures and patient symptoms more realistic. 6G technology allows real-time patient monitoring, which speeds up and improves diagnosis and treatment. 6G technology might enhance patient treatment, results, and experience [22]. The IoE links people, devices, and data, revolutionizing healthcare and patient life. Remote patient monitoring is the benefit of IoE's greatest healthcare. Wearable gadgets notify clinicians of vital signs and other health indications in real time. This allows for rapid and precise patient diagnosis, improving therapy. IoE analyzes plenty of data from many sources to improve care delivery. Comprehensive data analysis exposes trends, patterns, and health hazards, enabling tailored therapy and interventions. IRS changes healthcare, particularly telemedicine and remote patient monitoring. IRS technology boosts wireless signals, allowing healthcare workers to communicate and receive data in poor signal locations.

This advancement solves key remote patient monitoring difficulties, such as signal failures and packet loss. The IRS protects patient privacy and follows data protection laws by ensuring patients and providers talk to each other directly. This stops data breaches and unauthorized access. The IRS also helps doctors and nurses make better diagnoses and treat patients by giving them real-time data to respond quickly to patient circumstances. This capacity allows for early medical problem identification, resulting in faster, more accurate diagnosis and better patient outcomes. For patients in rural or distant places, the IRS may save healthcare expenditures by reducing in-person appointments and travel expenses [23]. Hybrid network architecture: Healthcare service demand fluctuates owing to disease outbreaks, crises, and demographic changes. A hybrid wireless design offers scalability, enabling systems to adapt to shifting needs. Adaptability is essential for healthcare systems to react quickly and effectively [24]. The hybrid wireless architecture enables healthcare services, medical equipment, and sensors to be delivered across numerous locations. Designing for high availability is crucial to minimizing downtime and maintaining patient care and safety. The network should be created with strong security measures and suited to the needs of healthcare professionals and patients. Primary delay reduction methods: Reducing latency is critical in healthcare systems owing to its enormous effect. Even modest delays in action completion have significant implications. Reduced latency supports quicker, more accurate diagnosis, leading to better patient outcomes and effective treatments [25].

Communication among healthcare professionals is crucial for patient care coordination. Decreased latency enables real-time message transmission and faster decision-making and reaction times. Low latency also enables healthcare practitioners to get data from remote patient monitoring devices quickly and correctly, allowing prompt treatment for patients who cannot travel or reside in rural locations. Additionally, decreased latency improves healthcare system efficiency, including EHR systems, by streamlining operations and minimizing mistakes. Lowering latency improves diagnostic and treatment speed, communication, remote monitoring, system efficiency, and patient care, leading to improved results [26].

10.6 CONCLUSION

The integration of advanced wireless communication, the IoMT, AI, extended reality (XR), and other cutting-edge technologies through the emergence of 6G-enabled Healthcare 4.0 is a critical advancement in the evolution of medical service delivery, with the potential to revolutionize patient care. In this chapter, we have examined the transformative applications of 6G in healthcare. These applications capitalize on 6G's ultra-low latency, high data rates, and massive device connectivity to improve accessibility, precision, and efficiency. Even so, there are also problems and limits that come with putting Healthcare 4.0 into action with 6G. To solve these problems, technologists, healthcare providers, policymakers, and researchers need to work together across disciplines to develop new ideas, such as energy-efficient hardware, universal standards, affordable infrastructure, and flexible security frameworks. Healthcare 4.0, made possible by 6G technology, might greatly help meet the needs of healthcare around the world in the future. Healthcare 4.0 may change lives worldwide by creating a smart, patient-centered, and strong healthcare system based on 6G, AI, XR, and robotics technologies. To do this, you need to deal with these problems.

REFERENCES

1. A. Kumar, R. Jain, M. Gupta, and S. M. Islam, *6G-Enabled IoT and AI for Smart Healthcare: Challenges, Impact, and Analysis*. CRC Press, 2023.
2. S. Rajak, A. Summaq, M. P. Kumar, A. Ghosh, K. Elumalai, and S. Chinnadurai, "Revolutionizing healthcare with 6G: A deep dive into smart, connected systems," *IEEE Access,* vol. 12, pp. 194150–194170, 2024.
3. S. S. Dhanda, B. Singh, P. Jindal, T. K. Sharma, and D. Panwar, "6G-enabled internet of medical things," *Expert Systems,* vol. 41, no. 1, p. e13472, 2024.
4. S. Kharche and J. Kharche, "6G intelligent healthcare framework: A review on role of technologies, challenges and future directions," *Journal of Mobile Multimedia,* vol. 19, no. 3, pp. 603–644, 2023.
5. P. V. Matre, A. Kumbhare, R. Gedam, A. Sharma, N. K. Vaishnav, and D. Naidu, "A timeline from 1G to 5G: The evolution of mobile communication," *Mpirical Knowledge Base*, 2023. https://www.mpirical.com/knowledge-base/the-evolution-of-mobile-communication (Last accessed 12 August 2025).
6. Rohde & Schwarz, "A brief history of mobile communications: From 1G to 6G," *Rohde & Schwarz Magazine*, 2025. https://www.rohde-schwarz.com/tw/about/magazine/brief-history-1g-to-6g/brief-history-1g-to-6g_256390.html#gallery-5 (Last accessed 12 August 2025).
7. N. Arastouei and M. A. Khan, "6G technology in intelligent healthcare: Smart health and its security and privacy perspectives," *IEEE Wireless Communications*, vol. 32, no. 1, pp. 116–121, 2025.
8. A. K. Tyagi, S. Tiwari, S. Gupta, and A. K. Mishra, *6G-Enabled Technologies for Next Generation: Fundamentals, Applications, Analysis and Challenges*. John Wiley & Sons, 2025.

9. S. Ullah, J. Li, J. Chen, I. Ali, S. Khan, A. Ahad, F. Ullah, and V. Leung, "A survey on emerging trends and applications of 5G and 6G to healthcare environments," *ACM Computing Surveys*, vol. 57, no. 4, pp. 1–36, 2024.
10. H. Mazumdar, K. R. Khondakar, S. Das, and A. Kaushik, "Aspects of 6th generation sensing technology: From sensing to sense," *Frontiers in Nanotechnology*, vol. 6, p. 1434014, 2024.
11. M. A. Elaziz, A. Dahou, A. Mabrouk, R. A. Ibrahim, and A. O. Aseeri, "Medical image classifications for 6G IoT-enabled smart health systems," *Diagnostics*, vol. 13, no. 5, p. 834, 2023.
12. C. Prabha and D. K. Jain, "6G-based smart healthcare solutions: Beyond industry 4.0 trends and products," In: A. K. Sangaiah, A. Thampi, and A. R. Javed (Eds.), *6G-Enabled IoT and AI for Smart Healthcare.* CRC Press, pp. 171–182, 2023.
13. H. M. Shukur, S. Askar, and S. R. Zeebaree, "The utilization of 6G in industry 4.0," *Applied Computer Science*, vol. 20, no. 2, pp. 75–89, 2024.
14. A. Kumar, M. Gupta, S. Sharma, E. H. Sharma, and K. Aurangzeb, *Smart Hospitals: 5G, 6G and Moving Beyond Connectivity.* Wiley, 2024.
15. S. Suneel, K. Manjula, B. Sowmya, B. Venkateshmurthy, S. T. Siddiqui, and L. P. Maguluri, "Exploring the role of 6G technology in smart healthcare systems: Challenges, and future trends," In: A. Kumar, M. Gupta, S. Sharma, E. H. Sharma and K. Aurangzeb (Eds.), *Smart Hospitals: 5G, 6G and Moving Beyond Connectivity.* Wiley, pp. 287–314, 2024.
16. P. V. Matre, A. Kumbhare, R. Gedam, A. Sharma, N. K. Vaishnav, and D. Naidu, "6G enabled smart IoT in healthcare system: Prospect, issues and study areas," In: *2023 International Conference on Artificial Intelligence for Innovations in Healthcare Industries (ICAIIHI),* Raipur, India. IEEE, 2023, vol. 1, pp. 1–6.
17. A. Choubey and S. B. Choubey, "6G industrial use cases," In: A. R. Hurson (Ed.), *Advances in Computers,* vol. 139, Elsevier, 2025, pp. 585–605.
18. P. N. Srinivasu, M. F. Ijaz, J. Shafi, M. Woźniak, and R. Sujatha, "6G driven fast computational networking framework for healthcare applications," *IEEE Access*, vol. 10, pp. 94235–94248, 2022.
19. A. K. Tyagi and S. Tiwari, "Blockchain-enabled smart healthcare applications in 6G networks," In: A. K. Tyagi (Ed.), *Digital Twin and Blockchain for Smart Cities.* Wiley, pp. 459–494, 2024.
20. M. S. Kaiser, M. F. Mridha, M. A. Rahman, N. Dey, and A. S. Ashour, "6G access network for intelligent internet of healthcare things: Opportunity, challenges, and research directions," in *Proceedings of International Conference on Trends in Computational and Cognitive Engineering: TCCE 2020,* Dhaka, Bangladesh. Springer, 2020, pp. 317–328.
211. M. S. Akbar, Z. Hussain, M. Ikram, Q. Z. Sheng, and S. Mukhopadhyay, "6G survey on challenges, requirements, applications, key enabling technologies, use cases, AI integration issues and security aspects," *arXiv* preprint arXiv:2206.00868, 2022.
22. P. Valsalan, N. Hasan, I. Baig, M. Zghaibeh, U. Farooq, and S. Suhail, "Unleashing the potential: The joint of 5G and 6G technologies in enabling advanced IoT communication and sensing systems: A comprehensive review and future prospects," *Journal of Communication*, vol. 19, no. 11, pp. 523–535, 2024.
23. I. Ishteyaq, K. Muzaffar, N. Shafi, and M. A. Alathbah, "Unleashing the power of tomorrow: Exploration of next frontier with 6G networks and cutting edge technologies," *IEEE Access*, vol. 12, pp. 29445–29463, 2024.
24. C. De Alwis, A. Kalla, Q.-V. Pham, P. Kumar, K. Dev, W.-J. Hwang, and M. Liyanage, "Survey on 6G frontiers: Trends, applications, requirements, technologies and future research," *IEEE Open Journal of the Communications Society*, vol. 2, pp. 836–886, 2021.
25. L. Bariah, L. Mohjazi, S. Muhaidat, P. C. Sofotasios, G. K. Kurt, H. Yanikomeroglu, and O. A. Dobre, "A prospective look: Key enabling technologies, applications and open research topics in 6G networks," *IEEE Access*, vol. 8, pp. 74792–74820, 2020.
26. M. Das, A. Nag, M. M. Hassan, A. Santra, N. Chand, F. Yasmin, A. Sinha, A. K. Bairagi, and A. Alkhayyat, "Synergy of 6G technology and IoT networks for transformative applications," *International Journal of Communication Systems,* vol. 37, no. 14, p. e5869, 2024.

11 Real-Time Applications of 6G in Augmented Reality/Virtual Reality (AR/VR) Ecosystems

Rangasamy Suganthi, Jaganathan Ramkumar,
Kishor Kumar Reddy Chinthala,
Viswanathan Valarmathi, Muralitharan Deenathayalan,
and Krishnan Ravichandran Dhanya

11.1 INTRODUCTION TO 6G AND AR/VR ECOSYSTEMS

11.1.1 Evolution from 5G to 6G: Technological Leap

The tremendous shift from 5G to 6G is a very big change in wireless communication technologies. 5G gave fast internet, less delay, and the power to connect many devices at one time. 6G will go much further. It is expected to give data speeds up to 1 terabit per second, almost zero delay, and very high reliability. These changes are not small improvements. These are basic steps for the advanced uses like AR and virtual reality (VR) technologies.

To reach this, 6G needs terahertz signals, which can carry huge amount of data than today's signals. It will also use reflective surfaces that can guide signals in difficult places. Another important part of 6G [1] is the use of artificial intelligence concept inside the network. This will help the system adjust and improve by itself, making communication faster and smoother. Such features are very important for AR and VR because even a small delay or data loss can break the experience.

5G mainly focused on faster mobile internet and smart industry needs. In contrast, 6G is designed with humans at the center. It will bring ideas like the tactile internet, where touch and feelings can be shared instantly, and multi-sensory systems, where sight, sound, and touch are joined together in real time. This will make digital experiences more natural and closer to real life.

DOI: 10.1201/9781003683599-11

The change from 5G to 6G is not only about the speed. It is all about merging communication, artificial intelligence, cloud and edge computing, and smart sensors into one system. This will build the base for a strong AR and VR world that can change the way humans live and connect with digital spaces.

11.1.2 AR/VR Ecosystem Overview: Components and Interactions

An AR and VR ecosystem is a system where many parts work together to give humans a real-like digital experience. The main parts are devices such as headsets, smart glasses, gloves, and trackers that helps the humans to see, feel, and move in the virtual world. By use of software tools that make 3D images, advanced animations, and virtual spaces. A high-speed network like 6G is needed to send data quickly without delay and the Edge and cloud computing help by doing heavy work like creating scenes and reducing delay. Sensors like cameras, GPS, and the motion detectors collect information from the real world to make the virtual experience more accurate and efficient. All these parts work at the same time.

For example, a VR headset can show clear pictures from a cloud server through 6G, while gloves give touch feelings to make it more real. Small tasks like hand movement can be handled near the user by edge devices, while big tasks are managed in the cloud. This system keeps the pictures and sounds clear, and feels like the touch real, which is main need in the areas like online classes, medical operations, and worker training [2].

11.1.3 The Need for Real-Time Performance in Immersive Experiences

Real-time performance in AR and VR means the system should respond without delay. It is a known fact that if the delay is more than 20 ms, the persons will develop a feeling of sickness, lose interaction, and the virtual world will not seem real.-covered by severe latency in critical use, even a few milliseconds of delay could pose a danger, such as remote surgery, tele-operation of vehicles, or collaborative work in virtual space [3]. 6G helps to reduce this delay by using very fast connections, artificial intelligence, and edge computing to send visuals, sounds, and touch signals without interruption. This also allows the system to change details such as image quality, lighting, and environment instantly based on human actions or changes around them.

In AR, this means that digital objects like navigation signs in a city remain correctly placed in the real world. In VR, it enables the humans to interact naturally in shared spaces with avatars, speech, and realistic actions. Real-time performance is also important in environmental uses such as live monitoring of nature or virtual eco-tourism, where accurate and immediate information is needed. With 6G, AR and VR can achieve this level of reliability and detail, creating the base for the next stage of digital experiences.

11.2 KEY FEATURES OF 6G ENABLING AR/VR

11.2.1 Terahertz Communication and Its Impact on Data Transmission

6G will be far beyond everything existing above: THz waves will be deployed to carry huge amounts of data for fast functioning, more than 1 trillion bits per second. This rate guarantees that AR and VR will transmit smooth presentations of 8K videos, 3D holograms, and all sorts of other rich content. Many users are allowed to log in to the same virtual space at the same time, and any action such as movement or touching things in VR occurs immediately. Too weak to pass through the walls of a building or heavy downpour, terahertz waves are most effective when transmitted line-of-sight. On these accounts, 6G mobile communication will provide much emphasis on smart surfaces that can bounce or guide signals so that indoors or crowded areas can sustain strong connections.

11.2.2 Edge Computing and Real-Time Processing

Edge computing is vital for 6G, because it brings computational abilities closer to the user as opposed to the cloud being far away. This is the very need of the day for AR and VR experiences, where things must react instantaneously. Gesture recognition, scene rendering, or environment mapping can be performed on nearby edge servers with practically zero delays. This allows smooth tracking of physical movement, synchronization of sound and activities, and delivery of smart objectives based on what the user is doing.

Another advantage is it very efficiently saves internet bandwidth, since only necessary changes get sent to the cloud [4]. For example, an edge server in a VR workspace could maintain avatars and local voices while the cloud carries out global coordination. This combination of edge-cloud ensures that VR stays fast, highly scalable, and suitable for concurrent use by a large audience.

11.2.3 AI-Native Networks and Predictive Analytics

6G will be unlike previous networks as it is designed to be AI-native, implying that smart algorithms are interwoven into every fabric of the network. For AR and VR, the system predicts a user's actions, thereby managing resources in an efficient manner and providing an immersive experience. For example, this means it pre-loads scenes based on where the user looks, changes quality based on the network or device, and adapts virtual entities to condition changes in the real world such as lighting or motion. AI also helps the network to fix itself and detect problems in a timely manner before service gets interrupted. AI can also monitor bandwidth during a live VR concert, balancing bandwidth so that all active users will get clear video and audio [5].

11.2.4 Holographic Beamforming and Spatial Multiplexing

Holographic beamforming pertains to a new generation of antenna technology that allows wireless signals to be controlled and directed with ultimate precision surface being controlled by software referred to as a meta-surface. On a 6G network, this will be for use with AR/VR, allowing the signal to be received by the user in front of him under the crowded environment, such that everyone will enjoy a strong connection. Likewise, it further enhances spatial resolution, allowing virtual objects to be in the correct location relative to the real-world, thereby giving more accuracy and realism to AR overlays.

One other spatial advantage is spatial multiplexing, which allows many users/data streams to simultaneously share the same frequency band by spatially separating them. This will certainly turn out to be immensely beneficial in cases of heavy load from simultaneous interactions from AR devices in smart factories, theme parks, or classrooms. Holographic beamforming, when combined with spatial multiplexing, will thus guarantee smooth, clear, and trustworthy energy-efficient delivery of immersive content in real-time free from interferences due to numerous active users from AR and VR experiences.

11.3 REAL-TIME APPLICATIONS IN CONSUMER ECOSYSTEMS

11.3.1 Smart Homes with AR Interfaces

If 6G added in, smart home technology will not just be more interactive but also more immersive, meaning that it will not only be about the automation. The advantage of having augmented reality (AR) is that a user can visualize and practically operate the system of the house, all that hardly taking any time. Augmented reality control panels that would suddenly appear right on your walls or furniture and then you could control lights, temperature, or security by simply interacting with virtual dashboards. Smart mirrors or augmented reality glasses may deliver the context-sensitive updates, e.g., weather, calendar, or even the status of the fridge. Homeowners may also use AR to redecorate their rooms; they can try various furniture arrangements, colors, or even rooms and they can do it simultaneously in real time before they make the final decision. With the 6G edge computing, sensor data is collected and processed on the spot which ensures that the feedback is given immediately. To an example, AR glasses can let users know about the house, energy consumption or the quality of the air and at the same time offer visual guides for adjusting settings. The tactile internet fuses all this together and makes the experience even better as it allows users to feel virtual controls through haptic feedback which, thus, makes the interactions more natural and more realistic.

11.3.2 Virtual Shopping and Immersive E-Commerce

6G with AR and VR is revolutionizing the e-commerce world up to the immersion and interaction level. Thus, customers are no longer merely watching pictures on their screens but are virtually entering the shops, trying the goods, and having AI assistants guide them in the real-time.

AR enables people to try the clothes, accessories, or cosmetics on themselves with a high degree of accuracy. Meanwhile, VR offers users the ability to look at the products from each side, feel the features, and even act as if they are using them. An AI avatar is a personal shopping assistant who generates ideas according to user preference, conduct, and even biometric signals.

Due to the fast and large 6G network, holographic displays and volumetric videos can be employed, thus products and environments are very close to reality. Retailers can also present virtual events like product releases, fashion shows, or brand experiences, which are almost as good as being there in person. Spatial multiplexing technology allows more than one user to be in the same virtual space; hence friends can be together, shopping, exploring, and chatting simultaneously.

11.3.3 Real-Time Gaming and Social VR Platforms

6G facilitates better communication between gaming and social VR environments, which are some of the most difficult areas to handle in AR and VR technologies. The delays, which have been cut down to less than a millisecond, are the reason why all movements can be tracked accurately, feedback is instant, and all multiplayer games are at the same stage all the time. In addition to that, the cloud-based rendering is what allows for such high-quality games to be streamed in real time without the need for any expensive local hardware. Besides that, multi-sensory experiences such as haptic suits, spatial audio, and environmental effects are some of the aspects that are contributing to games and virtual worlds feeling more real.

The social VR hubs are the places where people meet, interact, and collaborate in virtual worlds, where avatars show real-time expressions and gestures. AI that is present in the network can certainly predict the players' next moves, allocate the resources in a better way and adjust the environment such as lighting, sound, or physics in a VR battle in accordance with player actions. Indeed, it is edge computing that actually makes sure that these changes can happen immediately and without any disruption to the experience and thus it can be very valuable during competitions.

Both social VR and 6G can take advantage of scalability, which means that thousands of people can virtually attend concerts, conferences, or meetups without any problems with the synchronized audio and visuals. Moreover, people can also use AR to view the combination of AR overlays and real-world settings, which allows them to be at a virtual event while physically in the park or café.

11.4 INDUSTRIAL AND ENTERPRISE USE CASES

11.4.1 Remote Collaboration and Virtual Meetings with Tactile Feedback

6G technology merged with AR and VR is revolutionizing remote collaboration, these changes are making it more immersive and engaging. Instead of basic video calls, individuals can interact as realistic avatars in shared virtual worlds, with their gestures and movements being captured and displayed synchronously and the sound being delivered from the exact locations. Meanwhile, haptic feedback enables the users to sense the touch, so for example, a handshake, the movement of the object, or collaborative designs become nature-like.

On the other hand, volumetric presence makes it possible for the participants who are turned into 3D holograms [6] to be in the common area thus allowing the feeling of face-to-face meeting. Additionally, AI integrated into the network alters the virtual setting of each individuals' job, liking, or even emotional signals. These functions are of great help in such areas as architecture, engineering, and healthcare, where the exact spatial concept of the subject and touch are the most necessary things.

For instance, surgeons can use AR overlays and haptic gloves to perform remote consults, simulate procedures with instant feedback, while the design team can co-create in virtual studios, making the 3D models with millisecond responsiveness.

11.4.2 AR-Assisted Manufacturing and Maintenance

AR technology has become a major tool in the manufacturing process that aims to improve the safety, speed, and accuracy of the work. Particularly with the 6G connection which is very fast and reliable, AR systems are capable of transmitting real-time instructions, diagnostics, and analytics directly to the staff on the floor. In this case, as an example, the AR overlays can display the step-by-step instructions on machinery that is being assembled, hence, the number of errors is minimized along with the amount of time required for the training. Moreover, sensors can spot a malfunction and prompt an AR alert to a technician who can then follow the instructions provided through the AR to fix the problem. Besides this, engineers can also be on the spot, though physically remote, if they get through AR glasses what the workers at the location see, hence, by the same token, they can send out the visual cues or notes without delay [7].

Moreover, edge computing is instrumental in a situation where it locally processes sensor data, thus it allows for instant feedback and at the same time it reduces downtime. In areas that are very risky such as chemical plants or aerospace facilities, AR-assisted maintenance not only makes safety better but also aids the safety inspections by, for instance, pointing the hazards and assisting with checks. Moreover, the wide bandwidth of 6G enables workers to use high-definition video and 3D scanning for detailed inspection and documentation.

11.4.3 Digital Twins and Real-Time Simulation in Engineering

Digital twins represent the virtual duplicates of the actual systems in a real world, and hence are getting a very crucial position in the engineering field. By the usage of 6G these digital copies can be updated on the spot with data coming from IoT sensors. Engineers are thus enabled to perform the activities of overseeing, simulating and optimizing systems at once. The real-time synchronization ensures that any alteration to the physical system is mirrored instantly in the digital twin, thus giving accurate control and monitoring. VR is a tool that engineers can use to immerse themselves in the complicated systems without the risk of harm, they can even interact with the parts and try out different scenarios.

AI is a technology that can be used to make predictions, detect mistakes and provide the necessary improvements when it analyses data from digital twins. As a result, the processes of rapid prototyping, stress testing and managing the lifecycle of systems, are thus made a lot easier, which is especially true for the industries like the automotive, aerospace and energy.

For instance, the digital twin of a wind turbine can go through the processes of energy generation, structural stress and weather implications, thus giving a helping hand to engineers in maintenance planning as well as in the optimization of the design. The multiple-user function enabled by 6G's high-speed connection also makes concurrent access by several engineers or stakeholders from different locations possible and thereby making collaboration an easy process.

11.5 HEALTHCARE AND EDUCATION TRANSFORMATION

11.5.1 AR/VR-Based Remote Surgeries with Haptic Feedback

6G along with AR and VR has been a major game changer in healthcare, making remote surgeries more detailed and realistic than before. Through immersive AR interfaces and haptic feedback systems, surgeons can use robotic instruments to perform surgeries on patients anywhere in the world as if they were physically there. The ultra-low latency and high reliability of 6G enable surgeons to "feel" the texture and resistance of tissue in real-time while they see 3D patient anatomy via AR overlays. AI acts as a helper by taking up patients data, analyzing it, and thus, guiding doctors' decisions, consequently making the surgical process safer. Edge computing which takes care of sensor data locally, ensures that the reaction is immediate.

For instance, in a remote cardiac surgery, a surgeon can operate robotic arms by wearing haptic gloves and see the visual cues through AR glasses and at the same time get tactile feedback as if he was physically present.6Gs high bandwidth is instrumental in supporting high-resolution imaging and volumetric video, which in turn allows accurate navigation and also reduces the chances of risks. Besides, multiple specialists can enter the same virtual operating room from different venues, add annotations to the shared images, and consult in real-time, hence, the co-operative surgeries become feasible.

Moreover, this advanced technology not only facilitates the delivery of the most sophisticated health care services to the regions that are far away or hard to reach but also makes health care via telemedicine and medical emergencies better and more efficacious.

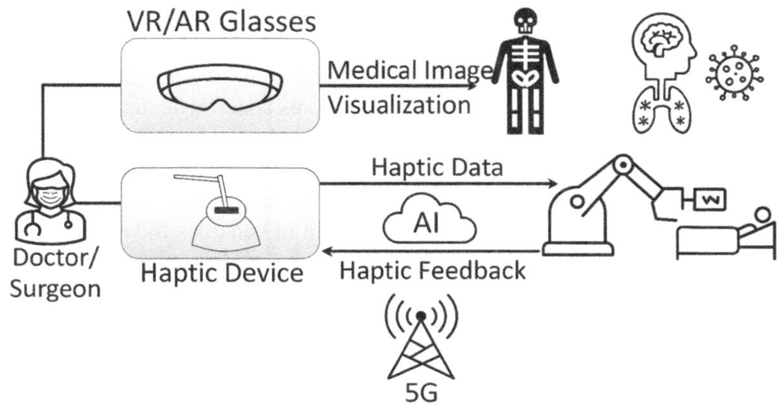

FIGURE 11.1 VR/AR-assisted remote surgery framework.

11.5.2 Virtual Classrooms with Real-Time Interaction and Adaptive Content

6G-energized AR and VR are revolutionizing the education sector by keeping the interest of the viewers intact while attending the virtual classes. Normally, students and teachers can watch the video lessons but now they can also have a face-to-face meeting by using avatars, gestures, and spatial audio. Immersive learning environments like virtual labs, scientific simulations, or historical reconstructions can be integrated into virtual classes. VR/AR-assisted remote surgery is used in the above Figure 11.1 and it combines immersive visualization, robotics, and high-speed connectivity to enhance surgical precision and accessibility. Before surgery, doctors use VR headsets to explore 3D models of patient anatomy created from medical scans, allowing them to plan and rehearse complex procedures. During the operation, AR headsets or smart glasses project real-time anatomical overlays, helping surgeons visualize hidden structures with minimal invasion. In remote settings, specialists control robotic surgical systems via VR controllers, with ultra-low latency 5G/6G networks ensuring seamless replication of their hand movements. Haptic gloves provide touch feedback, allowing delicate tasks to be performed safely, while AR also enables remote experts to guide the procedure collaboratively. Beyond live operations, VR simulators are used for surgical training, ensuring practitioners gain hands-on experience in a risk-free environment.

The implementation of AI in education is to make the content more engaging and interactive by adjusting the lessons based on the students' behavior and performance, thus making individualized learning possible. Latency has a big impact on every action for example, if a student wants to raise his virtual hand, manipulate 3D models, or participate in a group discussion which all can be done smoothly without waiting for a long time.

With the help of edge computing, data is processed locally, which makes the updates and feedback, go by at lightning speed. Students in a biology class, for example, can explore the virtual human body, take it apart, and observe the cellular processes as they unfold. Meanwhile, teachers monitor the activities and modify the lessons if required. This kind of learning becomes more interactive, varied, and available to all, especially students in remote areas or those with limited access to education.

11.5.3 Mental Health Therapy Using Immersive Environments

Immersive AR and VR installations are completely changing the landscape of mental health therapy as they provide safe, controlled, engaging, and effective experiences. In fact, after the rise of 6G, these kinds of treatments will become more personalized, interactive and resourceful.

A good exemplification is a moment of mindfulness and relaxation that could employ a blend of guided meditation, biofeedback, and virtual nature to allow the clients to refresh their brains. Besides, cognitive behavioral therapy (CBT) might be transformed into a more engaging version with avatars and adjustable scenarios that change to correspond with the patients specific needs. For example, in the case of phobias, exposure therapy for PTSD or anxiety is the technique in which patients are gradually introduced to the sources of their fears but in virtual environments for safety reasons. Real-time responsiveness is of utmost importance. Apart from AI, haptic feedback and spatial audio also make a big difference in the feeling of reality. The employment of edge computing in 6G enables the data to be processed locally which has the double benefit of privacy and guaranteeing an immediate reaction thus allowing therapists to intervene or change sessions instantly.

Besides that, these platforms are also available for remote therapies where the patients can have access to the care provided by the therapist at home and still be fully engaged and immersed in the sessions.

11.6 URBAN AND ENVIRONMENTAL APPLICATIONS

11.6.1 Smart City Planning Using AR Overlays and Real-Time Data

When AR is combined with 6G technology, it has a revolutionary impact on the way cities are planned and managed. It is possible that through dynamic and data-driven decision-making, the changes are virtually brought into the physical world of urban planners by using AR overlays that depict infrastructure projects, movement of traffic, or even environmental impacts. Furthermore, with the aid of interactive city simulations, the urban planners get to interact with digital metropolises exhibiting everything from the wires of a utility services installation to the dividing up of an area by the population density. Through AR displays, instantaneous data on public transport, pedestrians movement, and traffic flow is presented, which determines the speed and quality of the decisions of planners.

As part of the project, the residents can also contribute their comments together with the ability to watch the real estate through the use of AR applications. The environment is being closely monitored in a much easier way. Ecological data can be shown through engaging formats for scientists, policymakers, and the general public. These AR overlays are perpetually synchronized with the live 6G ultra-low latency IoT devices, i.e., traffic lights, air quality sensors, and cameras, etc.

Edge computing enables local data processing, which allows the urban planners to quickly test their scenarios such as re-routing the traffic or adjusting the energy distribution. Besides which, this method helps create urban areas that are less prone to risk, more open to the participation of the general public and the natural environment [8] while the real-time adaptability to the changes continues.

11.6.2 Environmental Monitoring with Immersive Visualization

6G, AR, and VR are all enhancing and enabling our ability to monitor and observe the environment. Researchers, policy makers, and everyday users now have an opportunity to track ecological data that connects us to nature. AR overlays can expose layers of biodiversity, water quality, and pollution levels right in front of you. VR can allow some representation of climate change, such urban heat islands, deforestation, or rising sea levels, and provide real-time alerts, with an associated address for hazards including chemical releases (spills), flooding and wildfires.

6G gives us access to broadcasts of real-time datasets that includes high-definition satellite imagery, drone photography and environmental sensor data, so we can respond more rapidly to changes in the environment. For instance, urban planners can visually track air quality and increasing green spaces on an AR dashboard while simultaneously mapping fauna and its changing behaviors in real time. The density of AI on the 6G network is also providing insights by detecting patterns that identify risks and prescribe actions. What is interesting is that interactive immersive visualizations will engage interested and invested communities regarding environmental action and protection.

11.6.3 Disaster Response Training in Simulated Environments

6G-powered augmented and virtual reality (VR) will enable emergency services to take advantage of realistically simulated disaster scenarios as training opportunities. With AR training, trainees can practice their skills and incorporate the simulated disaster scenarios such as earthquakes, wildfires, chemical spills, and other emergencies in a safe training environment. Multiple agencies can also provide multi-user VR training and even incorporate joint agency training and coordination [9]. AR provides real time notifications of hazards and guidance with the step by step processes in the training.

Haptic feedback will allow the responders to practice any delicate scenarios and allow the responder to manipulate equipment while receiving haptic feedback. 6G will enable the simulations to be responsive and update them in real time. Edge computing will be able to support localized scenarios to be created. Advanced artificial intelligence can modify and adapt the training modules to suit each individual's skills and performance. During the real emergencies, the 6G powered technologies will support the responder to provide AR overlays directing them to safe exit routes, structural information, and live video feeds from drones introduced into command centers with the ability to make quicker decisions and be more efficient with resources.

11.7 CHALLENGES AND CONSIDERATIONS

11.7.1 Energy Consumption and Sustainability

While the emergence of 6G technology will undoubtedly reinvent AR and VR experiences, many environmental and energy-sustainability questions arise. Particularly, the nature of immersive applications requires large quantities of energy to enable high data rates driven by dense network infrastructure requiring real time processing. Key issues to be investigated are carbon footprints associated with 6G base stations, AR and VR headsets and haptic devices; additional energy potential required for terahertz communication, additional energy use from edge computing and AI processing, and real-time rendering and storage capabilities of the data centers.

There are already efforts underway among researchers to reduce energy use associated with 6G for example, low-power chipsets, energy-efficient protocols, renewables, performance optimization and performance awareness functionality. Dynamic resource allocation enables processing and bandwidth to be used based on user activity to reduce overall energy use. Modular designs and biodegradable component(s)/devices can help reduce e-waste in the future. Subsequent 6G enabled ecosystems will have to be designed to meet sustainability challenges, meaning the ability to also design to achieve performance, while minimizing environmental and carbon impacts and employing green manufacturing, life cycle assessments of devices, and recycling or repurposing devices when possible.

11.7.2 Privacy, Security, and Ethical Concerns

AR and VR are areas which, when operating simultaneously, will be extremely challenging associated with 6Gs universal connectivity in terms of privacy, security and ethics. First and foremost, these solutions will collect an unlimited amount of personal data including biometric signals, behavioral patterns, and environmental attributes which must be secured [10].

Key privacy criteria will include data privacy because real-time tracking of location, gestures, and gaze will provide incredibly detailed personal information. Security risks will include cybersecurity vulnerabilities such as unauthorized access, data interception, or spoof waves. Ethical vulnerabilities will arise as augmented and virtual environments may impact the way that people think, feel, and behave. When developing 6G solutions, it will be critical that 6G networks provide adequate authentication, encryption, and anomaly detection.

6G network solutions may possibly be able to monitor and track unauthorized activity through AI-enabled systems and communicate security capture and prevention. Concurrently, ethical governance will be used to help contain and clarify ethical problems of immersive experience harvesting and/or influencing decision values, presenting transparency, consent, and fairness of experiences.

For example, student data in virtual classrooms should never be identifiable, and utilizing AR for medical, should be patient interactions operate independently through privacy laws. Ethical development guidelines are needed in order to work with content, user interactions, and management of these domains to prevent misuse and mitigate against psychological impacts.

11.7.3 Infrastructure and Deployment Hurdles

There are some elements to consider when deploying 6G infrastructure for AR and VR applications, namely the conceptual, technical, and logistical challenges. The most significant difference between 5G and 6G is that moving to 6G will require even denser edge nodes, intelligent surfaces, and small cells that will allow ultra-low latency and high reliability. One challenge will be the cost of AR and VR hardware as well as the costs of edge servers and terahertz base stations.

Another challenge will be limited RF signal coverage considering that 6G signals will be using higher frequencies and typically need line-of-sight (LOS) transmission and penetrating most physical barriers will be limited. Another challenge will be avoiding vendor lock-in and standardizing and ensuring interoperability across vendors and regions so that devices and platforms will seamlessly work. To address the challenges of deploying 6G infrastructures with seamless on-demand access to AR and VR, 6G will necessitate collaboration between sectors and partners. Collaboration could consist of repurposing the current 5G infrastructure, installing smart signal repeaters, and still maintaining terrestrial and satellite communication to support remote locations.

A telecommunications service provider that wants to deploy 6G infrastructure must engage with standardization bodies to define AR and VR data, as well as latency requirements so that all devices can be interoperable across vendors and platforms.

11.8 FUTURE OUTLOOK AND RESEARCH DIRECTIONS

11.8.1 Integration with Quantum Communication and Brain-Computer Interfaces

The future of immersive AR and VR made possible by the combination of 6G and more advanced technologies, such as brain-computer interfaces and quantum communication, could be remarkable. Quantum communication is a form of data transfer that leverages a technology called quantum key distribution to provide ultra-secure levels of communication. For sensitive applications particularly in the fields of healthcare, defines, and finance, this application is very important.

When quantum communications is combined with 6G, augmented and VR applications can be made highly immersive and secure thanks to its extremely low-latency, high-speed data transfer. Then, brain-computer interfaces will allow users to enter into a traditional virtual world and navigate it not by doing point-and-click, but entirely with their thought. Although 6G's ultra-low latency, and AI-enabled networks will be a large upgrade compared to 5G, the combined capabilities of brain-computer interfaces will offer extraordinary new capabilities that many only dream could be possible. Consider how easily one could navigate a virtual world by only using their thoughts; what distinctive virtual or digital experiences could these amazing new technologies enable as innate components to the experience.

11.8.2 Role of 6G in the Metaverse Evolution

The metaverse is a shared immersive digital realm that will be a significant aspect of overall digital life. 6G will serve as the backbone of the metaverse, enabling seamless involvement, real-time rendering, volumetric video streaming, multi-sensory experience across various devices and platforms.

Key benefits of 6G include volumetric video streaming giving users the ability to create avatars and environments that will be realistic, holographic communication providing the opportunity to communicate and collaborate in real-time as if physically present, and cross-platform compatibility allowing users to easily transition between AR, VR, and mixed reality spaces. 6G will allow the metaverse to move beyond entertainment and explore the option of education (such as virtual campuses to host classrooms for students across the globe), commerce (such as immersive digital marketplaces with live AI assistance while you shop), governance (such as virtual town hall meetings) [11] and socializing (such as a place where individuals can come together to network).

AI-enabled 6G networks will lead experiences within the metaverse to be more personalized based on how app users behave with respect to the interface and technology, ultimately giving users a more engaging and intuitive experience. There are also significant policy challenges including data ownership, identity, and consent that need strong policies as the metaverse continues to evolve.

For example, would moving through a virtual environment render possible or impractical adaptations based on one's cognitive or emotional state or provide useful assistive technology for persons experiencing challenges on the mobility spectrum. Certainly there still some navigational processes ahead to successfully integrate 6G with brain-computer interfaces and quantum communications technology.

11.8.3 Policy Frameworks and Global Collaboration

To unlock the potential of 6G-enabled AR and VR ecosystems to its fullest extent, some level of international cooperation and clear policies will be essential. These technologies require standard frameworks, ethical principles, and regulatory management.

The potential policy areas include spectrum for edge computing and terahertz communication, data privacy and data security as it pertains to immersive experiences, and the ethical use of neurotechnology, AI content, and behavioral persuasion. To achieve interoperability across networks, devices, and platforms will require collaboration of many different organizations and efforts, such as actionable plans from the ITU [12] and IEEE as well as the telecommunications authorities in their respective countries.

Public-private partnerships can help create a sense of public accountability but are intended to spur economic impact, where the industry may contribute technical work and expertise, government funding may build physical and social infrastructure and research, and institutions would research the human dimension and develop professionals to work in the new ecological space. Delivering a successful 6G-enabled AR and VR society will require technology and responsible governance, prioritizing user-centric or inclusive design, and to a common long-term vision of sharing the technology to foster cross-connected communities.

11.9 CONCLUSION

The move to 6G in AR and VR shows serious movement toward digital environments that display immersive, intelligent, and intuitive principles. 6G provides ultra-fast terahertz bandwidth, edge computing, AI-enabled networks, and ultra-low latency that push the boundaries of current immersive environments to allow real-time, multi-sensory engagement across sectors and domains including, but not limited to, healthcare, education, smart cities, environmental sensing, and entertainment that provides greater accuracy, collaboration, and scalability to fundamentally change user's experiences and transform industries. However, merely because we have the potential to realize such outcomes does not mean that it will be realized since there are an array of operational challenges that remain, such as developing the necessary underlying infrastructure, protecting data privacy, decreasing energy use, and managing ethical concerns. Equally important to enable equitable access, sustainability, and interoperability will be the collaboration of governments, corporations, research and academia, and standards organizations.

Additionally, the prospects of pairing 6G with additional technologies such as quantum communication and brain-computer interfaces has the potential of extending human-machine interaction even further into the future and the capabilities of next-generation digital societies or metaverse. That said, the potential success of 6G-enabled AR/VR, will not, in and of itself, be enough as its success will not simply become a product of the underlying technology. The outcomes will depend on intentional planning, measures of inclusivity, defining possibilities of human flourishing, and globalism.

REFERENCES

1. M. W. Akhtar, S. A. Hassan, R. Ghaffar, H. Jung, S. Garg, and M. S. Hossain, "The shift to 6G communications: Vision and requirements," *Human-centric Computing and Information Sciences*, vol. 10, no. 1, p. 53, 2020. https://doi.org/10.1186/s13673-020-00258-2.
2. S. Chen, Y. C. Liang, S. Sun, S. Kang, W. Cheng, and M. Peng, "Vision, requirements, and technology trend of 6G: How to tackle the challenges of system coverage, capacity, user data-rate, latency, and mobility," *IEEE Wireless Communications*, vol. 27, no. 2, pp. 218–228, 2020. https://doi.org/10.1109/MWC.001.1900333.
3. M. Giordani, M. Polese, M. Mezzavilla, S. Rangan, and M. Zorzi, "Toward 6G networks: Use cases and technologies," *IEEE Communications Magazine*, vol. 58, no. 3, pp. 55–61, 2020. https://doi.org/10.1109/MCOM.001.1900411.
4. S. Hu, Y. Zhang, and B. Ai, "Terahertz communication for 6G: Applications, challenges, and opportunities," *China Communications*, vol. 19, no. 2, pp. 1–14, 2022. https://doi.org/10.23919/JCC.2022.02.001.
5. J. Kim, S. Choi, and Y. Park, "Augmented and virtual reality applications powered by 6G: Challenges and research directions," *IEEE Access*, vol. 9, pp. 133257–133270, 2021. https://doi.org/10.1109/ACCESS.2021.3115118.
6. M. Latva-aho and K. Leppänen (Eds.), *Key Drivers and Research Challenges for 6G Ubiquitous Wireless Intelligence: 6G Flagship*. University of Oulu, 2019. https://www.6gflagship.com.
7. M. Z. Chowdhury, M. Shahjalal, S. Ahmed, and Y. M., Jang, "6G wireless communication systems: Applications, requirements, technologies, challenges, and research directions," *IEEE Open Journal of the Communications Society*, vol. 1, pp. 957–975, 2020. https://doi.org/10.1109/OJCOMS.2020.3010270.
8. T. V. Nguyen, T. Huynh-The, D. S. Kim, and D. Kim, "Immersive virtual reality for smart cities in 6G wireless networks," *Sensors*, vol. 21, no. 15, p. 5073, 2021. https://doi.org/10.3390/s21155073.
9. L. Chen, H. Zhao, C. Shi, Y. Wu, X. Yu, W. Ren ... and X. Shi, "Enhancing multi-modal perception and interaction: An augmented reality visualization system for complex decision making," *Systems*, vol. 12, no. 1, p. 7, 2023. https://doi.org/10.3390/systems12010007.
10. R. Singh and P. Tyagi, "6G-enabled healthcare: Vision, applications, and challenges," *Journal of Network and Computer Applications*, vol. 205, p. 103451, 2022. https://doi.org/10.1016/j.jnca.2022.103451.
11. P. Yang, Y. Xiao, M. Xiao, and S. Li, "6G wireless communications: Vision and potential techniques," *IEEE Network*, vol. 33, no. 4, pp. 70–75, 2019. https://doi.org/10.1109/MNET.2019.1800418.
12. Z. Zhang, Y. Xiao, Z. Ma, M. Xiao, Z. Ding, X. Lei, G. K. Karagiannidis, and P. Fan, "6G wireless networks: Vision, requirements, architecture, and key technologies," *IEEE Vehicular Technology Magazine*, vol. 14, no. 3, pp. 28–41, 2019. https://doi.org/10.1109/MVT.2019.2921208.

Redefining Reality *The Convergence of Metaverse, 6G, and Edge Intelligence*

12

Abdullah Mohammed Kaleem, Charu Sood, Shubneet, and Anushka Raj Yadav

12.1 INTRODUCTION TO METAVERSE-POWERED 6G EDGE INTELLIGENCE

The rapid evolution of information and communication technologies (ICT) is redefining how digital interactions, collaborations, and immersive experiences are created, experienced, and sustained. At the heart of this transformation lies the convergence of three powerful domains: the Metaverse, sixth-generation (6G) wireless communication, and edge intelligence. Together, these technologies promise to deliver unprecedented levels of fidelity, responsiveness, and scalability for next-generation virtual worlds and augmented environments.

The Metaverse is envisioned as a persistent, interconnected digital realm that combines physical reality with augmented, virtual, and mixed realities. It enables seamless, shared experiences spanning social, professional, educational, and entertainment contexts. The possibility to interact and coexist in hyper-realistic environments, powered by real-time dataflows and intelligent processing, sets new requirements for underlying infrastructure and computational paradigms [1].

6G wireless communication represents the forthcoming leap beyond 5G, harnessing frequencies in the sub-terahertz (THz) range, supporting data rates of up to 1 Tbps, and promising ultra-low latency (sub-millisecond) connectivity. These features are not incremental but transformational; 6G aims to enable continuous, always-on presence and interaction in the Metaverse, with seamless swapping of physical and digital identities. Its robust support for massive machine-type communications and context-aware network optimizations lays the foundation for planetary-scale virtual collaboration [2].

Edge intelligence introduces artificial intelligence (AI) and data analytics directly at the edge of the network—on local devices or proximal servers—reducing dependency on remote cloud infrastructures. By leveraging distributed learning, context-aware decision-making, and adaptive model deployment, edge

DOI: 10.1201/9781003683599-12

intelligence empowers real-time personalization and privacy-preserving data handling at scale. In the context of Metaverse environments, it is the catalyst enabling fluid, context-rich experiences without latency or privacy trade-offs. Key enabling paradigms such as federated learning, multi-agent AI, and secure aggregation are essential for collaborative intelligence while respecting user autonomy and global diversity [1].

12.1.1 Historical Evolution

The journey from 4G and 5G to 6G is marked by dramatic shifts in performance, capabilities, and design philosophy. 4G established mobile broadband and high-speed data transmission, while 5G unlocked ultra-reliable low-latency communications and large-scale IoT integration through millimeter-wave bands and massive MIMO. These technologies transformed mobile internet usage and enabled real-time applications like video streaming and augmented reality. However, as user expectations and application demands rise—with widespread XR devices, digital twins, and fully immersive platforms—the limitations of 5G in terms of bandwidth, energy efficiency, and context-aware adaptation become pronounced [2].

Looking ahead, 6G and edge intelligence respond directly to these gaps by focusing on an ecosystem that moves beyond communication to incorporate sensing, computing, and intelligence. 6G expands spectrum allocation into THz bands, integrates native AI for network management, and supports not just connectivity but intelligent sensing and computing at scale. Edge intelligence enables local adaptation, privacy enhancement, and resilient operations even during network disruptions, allowing the Metaverse to function seamlessly and securely across diverse geographies and user groups [1].

This evolution reflects a paradigm shift from centralized clouds toward distributed, intelligent edge networks, crucial for reducing latency and improving reliability in immersive digital environments. Devices and edge servers will collaboratively manage data and inference, pushing capabilities closer to the user and enabling real-time responsiveness vital for the Metaverse.

12.1.2 Core Motivations and Real-world Impact

The convergence of the Metaverse, 6G, and edge AI is driven by several imperatives. First, ultra-high performance is needed to deliver instant, responsive, and believable experiences in distributed virtual worlds. Immersive applications like augmented reality, virtual reality, and mixed reality require synchronization of audio, video, haptics, and environmental data with minimal delay to avoid user discomfort and motion sickness.

Second, decentralized intelligence ensures privacy and security of user data, mitigating risks associated with centralized data repositories. Edge computing limits the exposure of sensitive information by processing data closer to its source and leveraging privacy-preserving AI techniques.

Third, massive scalability and global accessibility are achieved by harnessing both network and computational resources at the edge, supporting billions of simultaneous users and devices interconnected through 6G networks [1]. The high density of connected devices in urban and remote scenarios mandates intelligent orchestration for resource allocation, fault tolerance, and service continuity.

These technological foundations enable novel applications—from immersive remote collaboration and intelligent healthcare diagnostics to smart cities and hyper-personalized education. At the same time, new challenges arise in terms of interoperability, ethical AI, sustainability, security, and regulatory compliance, necessitating cross-disciplinary innovation that balances user experience with technical feasibility and social responsibility.

12.1.3 Preview of Chapter Structure

This chapter systematically explores:

- The foundational concepts and architectural models of Metaverse-powered 6G edge intelligence.
- Key workflows and ecosystems enabling real-time immersive experiences.

- Intelligent orchestration and AI deployment strategies at the edge.
  - Security, privacy, and trust management challenges.
  - Policy, standards, and regulatory frameworks guiding global deployment.
  - Representative applications, use cases, and deployment lessons.
  - Societal, ethical, and human-centric implications.
  - Practical limits, future directions, and open research challenges.

By addressing these interconnected themes, this chapter provides a comprehensive foundation for understanding the transformative potential and hurdles associated with bringing the Metaverse to life through 6G-powered edge intelligence.

12.2 CORE CONCEPTS AND ARCHITECTURAL FOUNDATIONS

The transformation of the Metaverse through 6G-powered edge intelligence relies on a nuanced understanding of where computing and intelligence reside—across cloud, edge, and device layers—and the unique contributions each makes to immersive, real-time digital experiences.

12.2.1 Cloud, Edge, and Device-Level Intelligence

Traditional computing and data analytics have been predominantly centralized in vast cloud infrastructures. The cloud offers immense computational power, global data storage, and advanced analytics capabilities, which are ideal for handling intensive workloads and archiving historical data. However, the cloud-centric paradigm suffers from inherent limitations such as increased latency due to physical distance and communication bottlenecks, bandwidth congestion during transport of high-volume data streams, and greater exposure to privacy risks, particularly in highly interactive and sensitive Metaverse applications [3].

Edge intelligence—putting AI and analytics on servers, routers, gateways, and other computation nodes at the network's periphery—dramatically shifts operational benefits. By processing data locally, systems achieve ultra-low latency, which is essential for VR/AR feedback cycles, gesture recognition, and real-time collaborative events. Context-awareness increases markedly: edge nodes learn user preferences and environmental cues, adapting content and services accordingly. Furthermore, by filtering, compressing, or abstracting data before transmission, systems can achieve significant bandwidth savings, especially important in scenarios with thousands of real-time sensor feeds or high-resolution video streams.

Device-level intelligence, distributed across user wearables, smartphones, and IoT objects, pushes AI models directly onto hardware-rich endpoints. As lightweight inference engines and efficient neural network architectures advance, these devices increasingly enable on-the-fly, autonomous decisions—reducing dependence on centralized resources and improving the overall reactivity of Metaverse experiences. A key challenge here is balancing compute requirements with energy consumption and hardware constraints, but ongoing improvements in microprocessor design and embedded ML continue to expand device-level capabilities [4]. The ideal architectural approach blends the strengths of all three tiers, creating a cloud-edge-device continuum dynamically balanced according to use-case requirements.

12.2.2 6G Technologies Enabling the Metaverse

6G represents a paradigm-shifting leap, engineered specifically to address the complexity and scale of immersive digital worlds. The most critical features for the Metaverse include:

- **Extreme Data Rates**: 6G is expected to support raw speeds exceeding 1 Tbps, enough to stream multisensory experiences—including volumetric video, spatial audio, holographic data, and unprecedented virtual interactions.
- **Ultra-Low Latency**: Round-trip times below 1 ms are necessary to support high-fidelity XR environments, live tactile feedback, and synchronous collaboration in digital twins or telepresence scenarios.
- **Massive Device Connectivity**: Designed for planetary-scale reach, 6G can connect billions of heterogeneous devices—smart glasses, sensors, autonomous robots, edge servers—supporting densely populated virtual spaces and ubiquitous digital presence.
- **Intelligent Network Slicing**: Using AI-driven network partitioning, 6G can offer tailored connectivity, allocating resources to specific applications based on urgency, bandwidth need, and user priority. This ensures reliable performance for critical Metaverse tasks even under surges of demand [5].

A unique advantage of 6G is its tight integration of sensing, computing, and communication, unlike previous generations where networks served only as data pipes. As a result, the infrastructure actively supports real-time analytics, adaptive security, and context-aware orchestration.

12.2.3 AI at the Edge: Distributed Learning and Collaboration

Edge AI empowers local and distributed nodes to learn, infer, and adapt from real-time data, which is vital for responsive Metaverse services. Resource-constrained inference techniques enable complex AI to run efficiently on edge hardware with limited power and memory—supporting rapid analytics and immediate content personalization.

Federated learning is an especially promising paradigm: here, devices collaboratively train models on local data and aggregate insights without needing to share raw information. This not only preserves privacy but also enables robust knowledge synthesis across diverse environments. Distributed learning architectures facilitate cross-domain collaboration, allowing Metaverse entities—avatars, smart agents, and sensors—to autonomously coordinate to enrich virtual experiences [6].

AI at the edge also facilitates secure adaptation in the face of adversarial threats and environmental changes, supporting dynamic trust models and resilient problem-solving. When edge nodes detect anomalies or context shifts, they can immediately trigger safety protocols or update virtual environments, maintaining a seamless and secure experience for users.

12.2.4 Architectural Models for Metaverse over 6G Edge

Various architectural models organize how intelligence and resources are distributed (Figure 12.1):

- **Cloud-Edge-Device Continuum**: Dynamic task allocation shifts data processing and AI inference where it is most needed—cloud for long-term analytics, edge for real-time synthesis, device for rapid, personal feedback.
- **Digital Twin Ecosystems**: Continuous synchronization between physical and virtual entities allows collaborative diagnostics, real-time simulations, and predictive planning.
- **Multi-Access Edge Computing (MEC)**: Networks of distributed edge resources perform critical Metaverse functions—local XR rendering, biometric authentication, distributed asset management—with seamless backhaul to cloud for coordination.

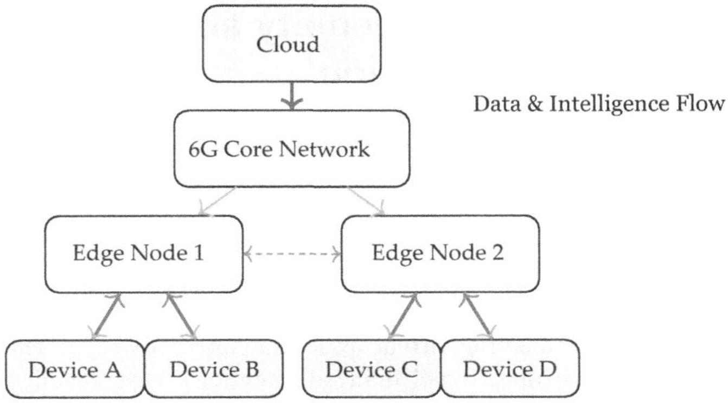

FIGURE 12.1 Integrated Metaverse–6G–edge AI stack.

These flexible models are now augmented with blockchain-based frameworks, cross-layer authentication schemes, and decentralized identity management, empowering user autonomy and verifiability.

12.3 METAVERSE ECOSYSTEM ENABLED BY 6G EDGE INTELLIGENCE

The integration of 6G wireless networks, edge intelligence, and the Metaverse orchestrates a multifaceted ecosystem where immersive digital experiences are powered by real-time data generation, intelligent processing, and adaptive feedback. In this section, we map the end-to-end workflow, examine the handling of heterogeneous data, dissect key architectural modules, and highlight the foundational importance of latency, reliability, and resilience for mission-critical Metaverse functions.

12.3.1 End-to-End Workflow: From Data Generation to Feedback Loops

The Metaverse begins with data generation at the edge: Extended Reality (XR) devices, wearables, smart sensors, and IoT endpoints continuously generate streams of user gestures, spatial data, physiological signals, and context-aware environmental factors. These raw data feeds are locally processed—on-device or at adjacent edge servers—enabling rapid feature extraction, content rendering, biometric authentication, and preliminary inference. This local stage is crucial for minimizing delay, ensuring that virtual reactions are tightly synchronized with real-world interactions [7].

Once processed, data are transmitted using ultra-fast 6G transport. 6G networks support massive bandwidth and sub-millisecond latency, facilitating uninterrupted high-volume communication between edge nodes, devices, and the cloud. MEC nodes perform intermediate processing such as memory caching, real-time analytics, and intelligent routing. Edge inference closes control loops swiftly—rendering virtual content, executing haptic feedback commands, or updating digital twins. This highly responsive system architecture supports a variety of immersive Metaverse applications, ranging from interactive XR gaming and collaborative design platforms to remote surgeries and smart manufacturing environments [8].

12.3.2 Managing Data Heterogeneity and Supporting Personalization

The Metaverse architecture must seamlessly support heterogeneous data originating from various device types—augmented reality headsets, 3D cameras, biosensors, touch-sensitive wearables, smart home appliances, and environmental monitors. Data vary by format, sampling rate, quality, and relevance. Edge intelligence leverages context-aware federated learning and multimodal sensor fusion to integrate these diverse streams, producing enhanced digital twin models that reflect real-time states of physical entities, whether they are people, machines, or entire urban environments [9].

Personalization is achieved by adapting virtual spaces and interactive experiences to individual users based on preferences, biometric feedback, and inferred intentions. Distributed AI models at the edge enable such privacy-preserving, adaptive immersion. For example, virtual avatars can mirror real user movements instantaneously, smart homes respond to environmental shifts, and workplace platforms provide dynamic interfaces tailored to team context and workflow.

12.3.3 Architectural Modules: Interfaces, AI, Privacy, and Orchestration

Key modules drive the Metaverse's robust operation:

- **Device Interfaces**: XR headsets, haptic controllers, and smart sensors serve as data-capture and feedback endpoints, bridging users and their environments to virtual platforms.
- **AI Model Management**: Robust frameworks encompass dynamic AI model training, deployment, and updating at both the edge nodes and central platforms. Models continuously adapt to new data, threats, and behavioral changes.
- **Privacy Layers**: Technologies such as federated learning, differential privacy, secure multi-party aggregation, and blockchain systems protect user data confidentiality during collaborative AI processes.
- **Smart Orchestration**: Automation of resource allocation, load balancing, and fault tolerance is managed by intelligent controllers, adapting to shifts in network congestion, device mobility, or usage surges.
- **Cross-Domain Interoperability**: Open standards and cross-platform protocols facilitate seamless integration across diverse device ecosystems and virtual worlds, ensuring consistent user experiences and secure data exchange.

12.3.4 Latency, Reliability, and Resilience for Critical Metaverse Functions

Achieving ultra-low latency, extreme reliability, and operational resilience is imperative. In remote surgery, even tiny delays or packet losses can compromise safety. Industrial XR relies on persistent digital twins and predictive sensor fusion for operational efficiency. Automated failover mechanisms, AI-based anomaly detection, and adaptive path selection prevent interruptions. Edge intelligence supports redirection, graceful degradation, and rapid recovery during failures, making the Metaverse robust against both network and hardware disruptions [8]. The result is uninterrupted service and trustworthy real-time immersion across mission-critical applications.

12.3.5 Feature Mapping Table

See Table 12.1

TABLE 12.1 Mapping Metaverse components/applications to 6G-edge features and intelligence levels

COMPONENT/ APPLICATION	*6G-EDGE FEATURE*	*REQUIRED INTELLIGENCE*	*PERSONALIZATION/RESILIENCE*
XR avatar interaction	Sub-ms latency	Real-time AI inference	Custom environments; rapid feedback
Digital twin manufacturing	High data rate	Predictive analytics	Fault tolerance; continuous update
Remote surgery	Ultra reliability	Haptic modeling	Strict latency control; zero packet loss
Collaborative industrial XR	MEC + sensor fusion	Distributed learning	Load balancing; adaptive resource allocation
Smart home IoT	Context-aware routing	Lightweight edge AI	User preference adaptation; redundancy
Cross-platform gaming	Network slicing	Real-time coordination	Multi-device sync; resilience to spikes

12.4 INTELLIGENCE AND ORCHESTRATION AT THE EDGE

Edge intelligence forms the backbone of a responsive, scalable Metaverse ecosystem, enabling distributed AI/ML training, rapid inference, dynamic adaptation, and resilient orchestration among heterogeneous devices and platforms. Here, we survey the mechanisms that empower collaborative model training, efficient resource allocation, and failure recovery in live Metaverse environments.

12.4.1 Distributed AI/ML Collaboration at the Edge

Edge nodes in the Metaverse are equipped with computational capabilities and AI models that allow decentralized model training and deployment. Unlike cloud-centric learning, distributed edge AI removes bottlenecks of centralization, supporting faster model updates and real-time reaction to local contexts. Collaboration among edge nodes occurs through techniques such as model aggregation, consensus-driven parameter sharing, and peer-to-peer synchronization. Edge devices locally train models using their own sensor inputs or local user data. Model updates are exchanged with nearby nodes, or via federated servers, and integrated through weighted aggregation to ensure model consistency [10].

This distributed approach enables unique personalization for Metaverse participants. For example, in a multi-user virtual collaboration scenario, each participant's device continuously learns preference, gesture, and activity patterns, refining locally deployed AI models. As participants interact over time, federated updating ensures global coherence of intelligence while maintaining privacy by keeping raw data decentralized.

12.4.2 Task Partitioning, Rapid Inference, and Adaptive Context Integration

Edge-based collaborative intelligence utilizes advanced task partitioning strategies. Complex learning and inference tasks are split into modular jobs or workflows, distributed to different capable edge nodes according to current resource states, proximity to data, and workload demand. Directed acyclic graph (DAG) structures, agent-based workload mapping, or dynamic clustering optimize computation placement for minimum latency and balanced load [11].

For instance, in a virtual manufacturing environment, sensor data is locally processed for equipment status and predictive maintenance, while high-fidelity XR rendering and anomaly detection tasks are routed to MEC nodes with the necessary hardware resources. Edge systems also enable real-time adaptation: models can auto-tune parameters based on environmental variables (lighting, noise, device mobility), user feedback, or workload changes—sustaining user-centric and context-aware service levels.

12.4.3 Federated and Transfer Learning Strategies for Scalable Edge AI

Federated learning at the edge orchestrates collaborative training without needing to share raw user data, preserving privacy and enabling cross-domain synthesis of AI capabilities. Each edge device computes model updates on local data, transmitting only model parameters to a central server or peer aggregator, which merges contributions into a global model. This iterative, privacy-preserving process harmonizes learning across the entire Metaverse while ensuring the security and regulatory compliance needed for sensitive applications (e.g., healthcare, finance, and industrial IoT) [12].

Transfer learning complements federated learning by allowing rapid adaptation of pre-trained models to new domains or user profiles. For example, a global behavior prediction model is transferred and fine-tuned with data from live Metaverse events, so it rapidly gains relevance for emerging activities or demographics within the digital ecosystem. The synergy of federated and transfer learning facilitates scalable, adaptive, and context-rich AI deployment in complex Metaverse scenarios.

12.4.4 Orchestration: Resource Allocation, Load Balancing, Service Migration, and Resilience

Orchestration in edge-enabled Metaverse environments involves systematic management of computational, communication, and data resources to ensure real-time responsiveness and reliability. Automated resource allocation guarantees that high-priority tasks—such as XR rendering, safety monitoring, or synchronized digital twin updates—receive necessary bandwidth and compute cycles based on immediate utility and network congestion metrics.

Load balancing distributes AI workflows and inference requests to avoid overloading any single node and to maximize throughput. Service migration mechanisms enable ongoing model deployments or user sessions to shift dynamically between nodes or data centers as user mobility, network fluctuations, or device failure require. For example, a participant in a multi-user VR event moving between city locations will have their experience seamlessly supported by handover and migration of relevant computation tasks among the nearest edge nodes. Robust failure recovery schemes trigger immediate fallback processes: switching to redundant backup models, rerouting traffic through alternate nodes, and re-instantiating key virtual services to maintain continuity and minimize disruption.

These orchestration tools collectively deliver resilience to the Metaverse environment, allowing multi-user events and mission-critical applications to withstand connectivity challenges, hardware breakdowns, or unexpected surges in demand.

12.5 SECURITY, PRIVACY, AND TRUST MANAGEMENT

The convergence of 6G, edge intelligence, and the Metaverse presents new dimensions of security, privacy, and trust risks—ranging from identity theft in virtual environments to novel adversarial attacks that exploit massive connectivity and distributed environments (Figure 12.2).

In this section, we identify unique security risks, review advanced privacy-preserving strategies, analyze trust management models, and provide a mapping of threats, countermeasures, and privacy-enabled edge intelligence solutions.

12.5.1 Security and Privacy Risks in Metaverse-6G-Edge Integration

Metaverse ecosystems are uniquely vulnerable to a spectrum of security and privacy threats. Key issues include:

- **Identity Theft and Impersonation**: Avatars and digital identities are targets for unauthorized usage, leading to reputational and social harm. Sophisticated phishing or password-guessing attacks have been demonstrated in immersive platforms, enabling attackers to assume control of high-value digital personas.
- **Virtual Asset Theft**: Digital assets, currencies, and property—often represented as NFTs or transferable tokens—are subject to malicious exfiltration, especially in blockchain-enabled environments. Smart contract vulnerabilities and malware can allow attackers to transfer assets illicitly.
- **Adversarial Attacks**: AI models operating at the edge are susceptible to input manipulation, poisoning, and backdoor attacks that degrade model performance or leak private data [13]. For example, adversaries may inject malicious data into federated training or design edge inference queries to reveal private states.

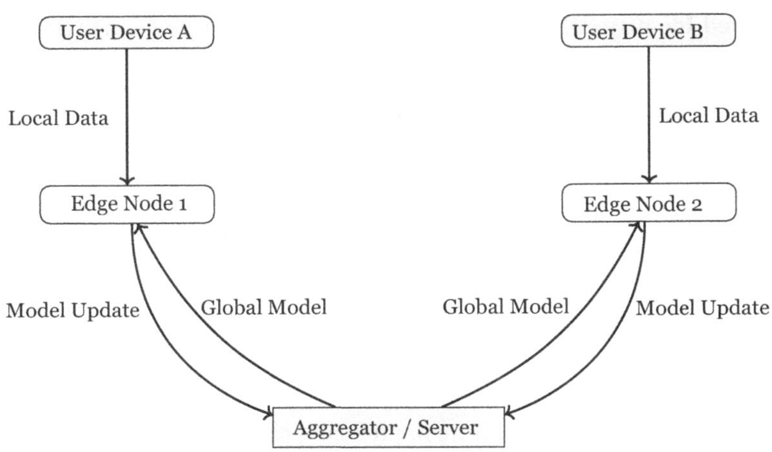

FIGURE 12.2 Workflow for distributed edge model training/inference in a multi-user Metaverse event.

- **Data Interception**: In-transit data between edge nodes, devices, and cloud servers may be intercepted, exposing personal information and operational secrets [14]. The use of public communication channels, coupled with limited device security, amplifies interception risks.
- **Privacy Leakage at the Edge**: Sensitive information, including location, biometric data, and behavioral patterns, is often processed locally. Inadequate privacy measures may allow linkage or reconstruction attacks, letting adversaries infer identities or habits from seemingly innocuous metadata [15].
- **Sybil and Collusion Attacks**: Multiple fake identities are created to manipulate consensus, evade reputation scores, or distort collaborative training. In Metaverse-based voting or crowd moderation, Sybil attacks can change outcomes by swamping systems with fraudulent users.
- **Data Reconstruction/Model Inversion**: Attackers exploit gradients, metadata, or intermediate parameters to reverse-engineer sensitive user data from model updates during federated learning. Membership inference further enables adversaries to identify if a user's data was seen by a model.

12.5.2 Privacy-Preserving Mechanisms

- **Differential Privacy (DP)**: DP techniques inject controlled random noise into AR/VR data, gradients, or query outputs, protecting individuals' identities without substantially degrading utility. Local DP in edge nodes has shown success in protecting privacy-sensitive location streams and personalized recommendation services in Metaverse environments [14]. Advanced mechanisms use adaptive noise schemes, balancing privacy and usability by adjusting noise levels according to user consent or data sensitivity.
- **Secure Aggregation**: During federated learning or distributed analytics, edge nodes send encrypted model parameters to central servers or peer aggregators. Homomorphic encryption and secure multi-party computation (SMPC) allow computations on encrypted data—preserving privacy even during aggregation. Secure aggregation prevents both data reconstruction and adversarial eavesdropping, and supports anonymity among multiple edge nodes.
- **Trusted Execution Environments (TEEs)**: TEEs such as secure enclaves run sensitive computations in isolated environments, protecting code and data from external tampering and inspection. TEEs are increasingly integrated into edge devices, such as AR headsets and smart sensors, enabling local processing of authentication, biometric matching, and private content rendering. In practice, TEEs can guarantee compliance with privacy regulations (such as GDPR) even when working with third-party code or data feeds.
- **Decentralized Identity (DID)**: DID platforms assign self-sovereign digital identities to users, devices, and avatars, managed on distributed ledgers (blockchain). This enables robust authentication, privacy-aware consent management, and cryptographic proof of identity, reducing risks of impersonation. Practical DID deployments in Metaverse gaming and collaborative XR are showing fast, frictionless logins and strong accountability [9].

12.5.3 Trust Models

- **Reputation Scoring**: Edge nodes, avatars, and virtual entities maintain reputation scores based on behavioral consistency, interaction history, and participation in collaborative tasks. Reputation-driven trust management mitigates Sybil and collusion attacks, supports peer vetting for distributed intelligence workflows, and informs automated moderation.
- **Blockchain for Auditability**: Blockchain protocols ensure transparent, tamperproof records of asset transfers, model updates, access events, and security incidents in the Metaverse. They enable decentralized verification and long-term traceability, underpinning digital asset trust

TABLE 12.2 Metaverse threat scenarios, countermeasures, and privacy strategies at the edge

THREAT SCENARIO	*SECURITY COUNTER MEASURE*	*PRIVACY-ENABLED EDGE STRATEGY*
Identity theft, impersonation	Decentralized ID, reputation scoring	Self-sovereign identity, consensus validation
Virtual asset theft	Blockchain audit, smart contracts	Digital provenance, tamperproof logs
Adversarial/model attacks	Gradient obfuscation, secure aggregation, federated learning	DP injection, SMPC, TEEs
Data interception/eavesdropping	Encrypted channels, TEEs, SMPC	Homomorphic encryption, local DP
Privacy leakage, linkage	Local DP, edge anonymization	Adaptive noise, context-aware privacy controls
Sybil/collusion attacks	Reputation analysis, blockchain consensus	Node vetting, audit trails
Model inversion/reconstruction	Differential privacy, secure aggregation, data minimization	Federated learning, encryption-at-rest

and regulatory compliance. For instance, regulatory sandboxes in financial Metaverse platforms use blockchain logs to demonstrate post-event integrity.

- **Zero Trust Networking**: Zero Trust Architecture (ZTA) operates on a "never trust, always verify" principle; all users, devices, and data flows require continuous authentication, dynamic access control, and real-time risk assessment [15]. ZTA is implemented via granular policy rules, context-based authorization, and continuous trust scoring. Practical challenges include balancing user experience with security—frequent authentication requests can impede smooth interaction—but research is advancing unobtrusive, context-aware zero trust protocols for immersive environments (Table 12.2).

12.6 POLICY, STANDARDS, AND REGULATORY PERSPECTIVES

The proliferation of Metaverse ecosystems enabled by 6G and edge intelligence brings new regulatory challenges. Robust policy frameworks, well-defined standards, and deep-rooted organizational compliance guidance are essential to foster safe, fair, and sustainable innovation in these immersive environments.

12.6.1 Global Policy Frameworks

- **Digital Identity Standards**: Digital identity has moved to the forefront in Metaverse spaces, shaping authentication, access control, auditability, and accountability. The landscape is rapidly evolving, with federated identity platforms and decentralized identity (DID) models emerging as best practices. Laws such as the eIDAS Regulation in Europe, along with the proliferation of self-sovereign identity solutions, mandate interoperable, privacy-preserving identity ecosystems and trusted credential exchange [16].
- **Spectrum Allocation for 6G**: Next-generation connectivity—hinging on 6G—is heavily dependent on equitable, interference-free spectrum allocation. International agencies such as the ITU work alongside national regulators to harmonize allocation for ultra-low latency, high-bandwidth Metaverse operations. Coordinated spectrum usage is vital for delivering seamless cross-border services and reliable deployment in densely populated digital regions.

- **Cross-Border Data Sovereignty**: Regulations such as the GDPR (European Union), CCPA (California/USA), and China's Cybersecurity Law dictate stringent data localization, consent management, and transborder data flow standards. Metaverse operators must guarantee data residency and maintain compliance with local and international mandates by supporting lawful cross-territory data sharing, privacy-by-design architectures, and clear user consent mechanisms [17]. These frameworks address increasing concerns about privacy, surveillance, and the misuse of user data in virtual worlds.

12.6.2 Current and Emerging Standards

- **IEEE, 3GPP, and Metaverse Standards Forum**: Leading organizations are working toward unified standards for Metaverse implementations. The IEEE Metaverse Standards Committee and Metaverse Standards Forum (MSF) set frameworks for interoperability, digital identity, content portability, and privacy. Meanwhile, 3GPP is developing 6G technical standards, including URLLC (Ultra Reliable Low Latency Communications) and integration of IoT architectures [16]. These standards are critical to support multi-platform digital economies and seamless cross-domain user experiences.
- **Industrial Edge AI Protocols**: ISO/IEC and other industry groups are standardizing performance, transparency, and security protocols for AI at the network edge. This includes requirements for algorithmic transparency, explainability, robust failure detection, and governance of autonomous systems—a necessity for complex, distributed Metaverse environments [17]. Efforts are ongoing to harmonize protocols and certification frameworks that ensure AI behaves ethically and safely in critical health, financial, and industrial Metaverse deployments.

12.6.3 Organizational Compliance and Best Practices

Organizations deploying Metaverse-6G-edge solutions must implement comprehensive compliance, ethics, and risk management plans (Table 12.3):

- **Transparency and Explainability**: Users and regulators demand visibility into how AI models make decisions. Organizations should offer clear disclosures, policy documentation, and dashboards for monitoring real-time data flows and algorithmic outputs [18].
- **Incident Reporting and Response**: Formal protocols for documenting, investigating, and resolving privacy breaches, cyberattacks, or failures are required. Automated real-time alerting and post-incident forensics provide accountability and help organizations meet legal reporting obligations.
- **Continuous Audits**: Scheduled reviews, internal risk assessments, and third-party audits ensure systems stay compliant and responsive to evolving laws and performance requirements. This supports preventive security and continuous improvement.

TABLE 12.3 Policy, regulatory, and internal standards matrix for Metaverse-6G-edge deployments

DOMAIN	*RELEVANT STANDARD*	*ORGANIZATIONAL PRACTICE*	*COMPLIANCE*
Digital identity	IEEE, W3C, eIDAS, MSF	Decentralized authentication, multi- factor, consent logs	
Spectrum allocation	ITU, national/federal	Usage audits, reporting, interference management	
Data sovereignty	GDPR, CCPA China	Data localization, privacy dash-board, audit trails	
Interoperability	3GPP, MSF, ISO/IEC	API conformance, cross-platform testing, certification	
Edge AI transparency	ISO/IEC AI, IEEE	Explainable algorithms, real-time model introspection	
Incident management	NIST CSF, internal policy	Real-time alerts, structured reporting, root-cause analysis	
Audit and improvement	Third-party, MSI, ISO	Scheduled security/privacy audits, continuous improvement	

- **Consent Management**: Adaptive mechanisms give users clear control over data sharing and virtual engagements—enabling opt-in/opt-out, management of permissions, and auditable consent histories.

12.7 APPLICATIONS AND USE CASES

The fusion of 6G, edge intelligence, and Metaverse environments is reshaping real-world scenarios across industries—including remote work, education, healthcare, smart cities, entertainment, and gaming. This section analyzes how these converging technologies empower immersive experiences and offers deployment lessons for practitioners.

12.7.1 Immersive Remote Work, Education, and Social Platforms

Metaverse-powered communication platforms, built on edge-personalized experiences, allow geographically distributed teams to collaborate in virtual offices, classrooms, and event venues. Edge AI tailors content delivery—avatars, backgrounds, and presentations—based on user activity, device capability, and network health, thus reducing delay and enhancing engagement. Remote meetings happen in XR spaces with instant feedback and adaptive content sharing, while virtual classrooms enable interactive labs and student-driven simulations accessible from anywhere [19]. Social platforms in the Metaverse use distributed intelligence for real-time moderation and sentiment analysis. Each device runs local privacy controls and context-aware customization, ensuring safety and user trust in digital interactions.

12.7.2 Digital Twinning for Smart Cities, Industry, and Healthcare

Digital twins, enabled by real-time 6G data and edge processing, are central to smart city management, industry automation, and advanced healthcare. Smart cities maintain urban digital twins to monitor infrastructure, energy use, population flow, and emergencies. IoT sensors deliver continuous streams about air quality, traffic, or utility consumption, and edge ML algorithms integrate this data to optimize city functions. Planners can simulate crises, predict equipment failures, and visualize resource distribution, all within responsive virtual environments. Industrial digital twins drive predictive maintenance and production optimization. Real-time sensor readings from machines and robots are fed into edge platforms, allowing faults to be detected and addressed before disruptions occur. In healthcare, digital twins represent patient anatomy and vital signs. XR-driven remote surgery relies on millisecond-scale 6G feedback loops, combining haptic controls and live analytics for guided diagnosis and intervention [20]. Telepresence in the Metaverse allows surgeons across continents to participate in collaborative operations, with personalized data and instant communication through edge-enabled streaming.

12.7.3 Entertainment and Gaming: Ultra-Responsive Experiences

Entertainment and gaming in the Metaverse benefit profoundly from ultra-low latency and real-time adaptation. Massive multiplayer online games, XR concerts, and esports events run on distributed architectures that scale to support millions of avatars and participants. Edge inference ensures lag-free action, synchronized event progress, and context-aware gameplay regardless of user location or connection quality. Dynamic resource allocation through edge orchestration helps studios deliver smooth, resilient experiences. Smart streaming technologies handle content delivery and adapt graphics, audio, and user controls on the fly. Beyond gaming,

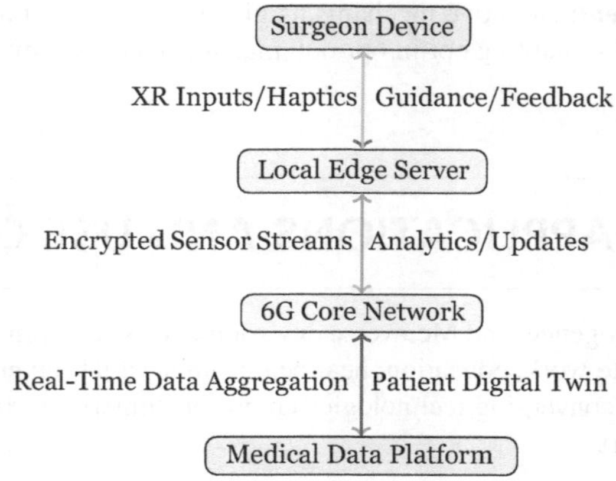

FIGURE 12.3 Workflow for multi-user VR surgery

interactive storytelling and immersive media events leverage edge AI for personalized narratives, adaptive challenges, and collaborative creativity. Multi-user XR concerts or digital festivals allow global participants to join and influence live performances, with edge-based privacy and trust controls managing access and interaction.

12.7.4 Scalability, User Experience, and Integration Lessons

Key deployment lessons have emerged from practice. Scalability is achieved through intelligent orchestration: workloads migrate between edge nodes and cloud whenever demand peaks, supporting millions of simultaneous users without congestion. Complex integration across platforms and standards is addressed by open APIs, federated authentication, and cross-device testing, although legacy system compatibility remains a challenge. User experience is elevated by context-driven adaptation: interfaces adjust to device type and bandwidth, AI models personalize feedback, and real-time analytics detect anomalies or performance drops. Continuous monitoring of network health, resource allocation, and user activity enables rapid troubleshooting, while privacy and trust controls support safe, enjoyable participation in large-scale, persistent environments.

12.7.5 Workflow Example: Multi-User VR Surgery

XR devices send data to edge servers, 6G networks transports sensor streams, and the cloud platform provides digital twin updates for real-time analytics and haptic feedback (Figure 12.3).

12.8 SOCIETAL, ETHICAL, AND HUMAN-CENTRIC IMPLICATIONS

As the 6G-powered Metaverse, driven by distributed AI and edge intelligence, becomes woven into the fabric of daily life, it brings complex societal, ethical, and human considerations. The potential for autonomy, inclusivity, and human agency is balanced by new risks—ranging from algorithmic bias and deepening digital divides to the erosion of privacy and psychosocial well-being.

12.8.1 Autonomy, Inclusivity, Agency, and Accessibility

Immersive environments powered by edge AI aspire to maximize human autonomy and agency. With highly customizable avatars, real-time adaptive interfaces, and personalized digital spaces, users can transcend physical limitations, shape their experiences, and gain new forms of expression [21]. For people with disabilities or mobility constraints, XR devices with haptic and auditory feedback, as well as edge-based assistive algorithms, offer unprecedented accessibility opportunities, enabling participation in education, work, and social life.

Inclusivity is a fundamental aim, fostering environments where diverse languages, cultures, and backgrounds coexist and collaborate. Machine translation, anti-discrimination algorithms, and digital skills training built into Metaverse platforms help democratize access. Nevertheless, design flaws, unconscious bias in algorithms, and unequal tech distribution continue to pose barriers to widespread inclusion.

12.8.2 Risks: Bias, Digital Divide, Privacy, Psychological Impacts

Edge AI, while enhancing responsiveness and personalization, can inadvertently amplify biases existing in data or system design. Personalized recommendations, moderation, and avatar rendering may reinforce stereotypes, exclude minority groups, or perpetuate social inequities. Algorithmic transparency and participatory evaluation are required to address these issues [22].

The digital divide endures and evolves. People in underserved regions often lack reliable connectivity, affordable advanced devices, or digital literacy, restricting their access to immersive Metaverse experiences. As virtual interactions become the norm for work, learning, and healthcare, the socioeconomic gap widens unless targeted investments in infrastructure and public policy are made [23].

Loss of privacy and control is an ever-present challenge. Edge-enabled environments gather vast amounts of user data—biometric identifiers, behavioral patterns, emotional states—which, if inadequately protected, could be exploited for commercial gain, surveillance, or malicious manipulation. Users may lose visibility of who controls their data and how it is used, leading to consent fatigue and potential harm.

Psychological impacts of extended immersion in virtual worlds are both promising and perilous. On the positive side, research highlights increased engagement, opportunities for social connection, improved creative expression, and even therapeutic applications for mental health. Conversely, risks include addiction, social withdrawal, reduced physical activity, cyberbullying, and heightened susceptibility to misinformation. XR environments that blur reality and simulation can affect identity formation, emotional regulation, and real-world relationships, especially among younger users. Ethical design must prioritize user safety, mental health resources, and continuous moderation.

12.8.3 Decentralized Content Moderation and Ownership

Decentralized content creation and moderation in persistent virtual worlds present both promise and peril. Blockchain-based ownership structures enable users to manage assets, intellectual property, and self-expression without gatekeepers, supporting innovative business models and personal empowerment. Community-driven moderation, using DAOs and reputation systems, fosters participatory governance and resistance to censorship.

However, the absence of central oversight can facilitate harmful behaviors and content, ranging from hate speech and harassment to disinformation campaigns. Community governance models may be vulnerable to capture, gaming, or fragmentation, while automated moderation algorithms struggle to balance liberty and accountability. New paradigms in decentralized moderation must marry transparency, auditability, and dispute resolution with robust protection for vulnerable populations.

12.8.4 Scenario Vignette: Societal Consequences in a 6G Metaverse

Fatima, an aspiring entrepreneur in Nairobi, uses the Metaverse to open a digital storefront—serving customers globally through XR-enabled product demos and smart contracts. Local edge servers translate languages and tailor customer experiences. Across town, Musa, a disabled veteran, accesses immersive rehabilitation therapies and remote work via Metaverse platforms with adaptive XR interfaces, reclaiming autonomy and dignity. Meanwhile, Anya, a teenager in rural Siberia, encounters algorithmic bias and harassment in digital classrooms, her access restricted by unreliable 6G coverage and limited device compatibility. These real-life scenarios illustrate the spectrum of potential outcomes: expanded opportunity, improved well-being, and empowerment exist alongside persistent challenges of equity, inclusion, and safety.

The societal, ethical, and human-centric consequences of 6G-edge-driven Metaverse platforms compel ongoing vigilance, proactive policy, and inclusive design. Stakeholder engagement, cross-disciplinary research, and continuous evaluation must underpin technological development to ensure the Metaverse enables flourishing—while countering emergent risks of exclusion, bias, and psychological harm.

12.9 CHALLENGES AND FUTURE DIRECTIONS

Metaverse systems powered by 6G edge intelligence hold enormous promise, yet face practical constraints and evolving research demands. This section critiques current bottlenecks, security-reliability trade-offs, and standards fragmentation, and outlines advances on the horizon, including self-evolving edge AI and quantum acceleration.

12.9.1 Practical Limits and Bottlenecks

Despite rapid technological advances, several challenges hinder widespread and equitable deployment:

- **Edge Computation Bottlenecks**: Real-time tasks such as graphics-intensive XR rendering, multimodal sensor fusion, and deep ML inference demand computational power beyond what many edge devices can deliver. Coordination between heterogeneous nodes introduces additional latency and raises reliability concerns [24].
- **Energy and Thermal Constraints**: Nonstop model updating, distributed analytics, AR/VR content streaming, and real-time data aggregation impose strong energy and cooling demands, especially on mobile and wearables. Battery depletion and thermal overload can disrupt services, limit device autonomy, and increase costs.
- **Real-Time Synchronization**: The need for synchronous and coherent virtual experiences across distributed users and digital twins is challenged by unpredictable network latency, device failures, and clock drift. Achieving seamless real-time adaptation and update propagation under dynamic network conditions remains a technical hurdle.
- **Standards Fragmentation**: Multiple, overlapping standards from IEEE, 3GPP, Metaverse Standards Forum, and various vertical-specific platforms create significant integration complexity. Lack of unified protocols impedes interoperability, slows innovation, and complicates deployment for multi-vendor and cross-border use cases [5].

12.9.2 Security, Reliability, and Explainability

The trade-offs between security, resilience, and explainability are especially critical:

- **Security Versus Performance**: Mechanisms like zero-trust networking, encrypted edge-data exchanges, and privacy-preserving inference introduce computational overhead and may slow real-time responsiveness. Conversely, optimizing for speed or resource efficiency can expose vulnerabilities to data breaches or adversarial attacks.
- **Explainability of Distributed AI**: Black-box models, especially at the edge, complicate transparency and user trust. Explainable AI (XAI) research is focusing on offering interpretable, auditable, and user-friendly explanations for ML-driven actions, balancing insight and performance.
- **Reliability and Robustness**: Distributed Metaverse systems must automatically detect failures, recover rapidly, and adjust to changing user and network contexts. Redundant edge and cloud paths, self-healing orchestration, and flexible service migration are active research areas in building highly resilient platforms.

12.9.3 Advances on the Horizon

Progressive solutions are emerging to tackle these core challenges:

- **Self-Evolving Edge AI**: Techniques like continuous meta-learning, federated evolution, and on-device training support context-aware personalization, anomaly detection, and proactive optimization with minimal cloud dependence.
- **Quantum-Accelerated Edge**: Quantum computing promises transformative gains in parallelism, cryptography, and simulation. When adapted to the edge, it could enable ultra-fast ML inference, robust security protocols, and real-time user modeling for complex, large-scale Metaverse environments.
- **Responsible Orchestration and Sustainability**: Research in carbon-aware scheduling, green migration, and eco-optimized resource balancing is showing promise. Metaverse architects must prioritize sustainable infrastructure, renewable energy, and circular hardware lifecycles to reduce the ecological impact of persistent immersive platforms.
- **Unified Standards and Governance**: The next phase of innovation will see greater international collaboration for standardization of protocols, APIs, and data management methods. Common frameworks will lower barriers to entry, facilitate interoperability, and safeguard security and privacy.

12.10 CONCLUSION

The advent of Metaverse ecosystems empowered by 6G connectivity and edge intelligence signals a transformative era in how humans interact, collaborate, and innovate in digital realms. This chapter has systematically examined the architectural foundations, enabling technologies, security and privacy challenges, regulatory landscapes, practical applications, societal impacts, and future research directions essential for building resilient, scalable, and inclusive Metaverse platforms.

We have highlighted how the integration of cloud, edge, and device-level intelligence—coupled with ultra-fast, ultra-low-latency 6G networks—enables rich, personalized, and context-aware immersive experiences. Distributed AI, federated and transfer learning, and innovative orchestration mechanisms support real-time adaptation, high reliability, and robust privacy even in highly dynamic, multi-domain environments. Digital twins, XR-driven healthcare, remote work, smart city management, and entertainment demonstrate the potential for edge-enabled Metaverse deployments across diverse sectors.

Societal and ethical considerations remain paramount. As immersive environments and virtual communities grow in scale and influence, autonomy, inclusivity, accessibility, and digital equity must be prioritized while mitigating risks such as bias, digital divide, privacy erosion, and psychological impacts. Robust policy frameworks, ongoing standardization efforts, and transparent compliance practices will underpin trust and safety for all stakeholders.

Yet significant challenges persist: computational bottlenecks, security-explainability trade-offs, standards fragmentation, and sustainability concerns must be resolved to unlock the full promise of the Metaverse. Emerging solutions—such as self-evolving AI, quantum-accelerated edge, responsible orchestration, and unified regulation—show promise for the future.

In closing, the journey toward a truly intelligent, secure, and human-centric Metaverse is collaborative and ongoing. By aligning technological innovation with ethical foresight, inclusive design, and proactive governance, researchers, practitioners, and policymakers can ensure that the Metaverse serves as a platform for global empowerment, creativity, and well-being.

REFERENCES

1. Zawish, M., Dharejo, F.A., Khowaja, S.A., Raza, S., Davy, S., Dev, K., and Bellavista, P. AI and 6G into the metaverse: Fundamentals, challenges, and future research trends. *IEEE Open Journal of the Communications Society* 5, 730–778 (2024). https://doi.org/ 10.1109/ojcoms.2024.3349465.
2. Akbar, M.S., Hussain, Z., Ikram, M., Sheng, Q.Z., and Mukhopadhyay, S. On challenges of sixth-generation (6G) wireless networks. *Journal of Network and Computer Applications* 206, 103524 (2025). https://doi.org/10.1016/j.jnca.2024.104040.
3. Baidya, T., Zhang, Q., Gopalakrishnan, V., and Xu, J. Comprehensive survey on resource allocation for edge-computing-enabled metaverse. *Computer Communications* 207, 34–56 (2024). https://doi.org/10.1016/j.comcom.2023.12.009.
4. Fadhel, M.A. Navigating the metaverse: Technologies, applications, and challenges. *Artificial Intelligence Review* 57, 2159–2187 (2024). https://doi.org/10.1007/s10462-024-10881-5.
5. Yi, Z., Qian, Y., Chen, M., Alqahtani, S.A., and Hossain, M.S. Defending edge computing based metaverse AI against adversarial attacks. *Future Generation Computer Systems* 149, 170–184 (2023). https://doi.org/10.1016/j.adhoc.2023.103263.
6. Chen, Y., Huang, S., Gan, W., Huang, G., and Wu, Y. Federated learning for metaverse: A survey. *arXiv* preprint arXiv:2303.17987 (2023).
7. Ahmad, H.F., Rafique, W., Rasool, R.U., Alhumam, A., Anwar, Z., and Qadir, J. Leveraging 6G, extended reality, and IoT big data analytics for healthcare: A review. *Computer Communications* 207, 57–77 (2023). https://doi.org/10.1016/j.comcom.2023.01.007.
8. Kour, R., Karim, R., Venkatesh, S.N., and Kumar, U. Metaverse in industrial contexts: A comprehensive review. *Journal of Industrial Information Integration* 23, 100538 (2025). https://doi.org/10.3389/frvir.2025.1488926.
9. Rafique, W., and Qadir, J. Internet of everything meets the metaverse: Bridging physical and virtual worlds with blockchain. *Computer Science Review* 54, 100678 (2024). https://doi.org/10.1016/j.cosrev.2024.100678.
10. Tang, C., Wu, H., Xiao, Y., Wu, Y., and Wang, Y. A systematic exploration of edge computing-enabled metaverse: Architecture, technologies, and applications. *IEEE Network* 37(6), 15–21 (2023). https://doi.org/10.1109/MNET.2023.3321541.
11. Xu, Y., Feng, D., Zhao, M., Sun, Y., and Xia, X. G. Edge intelligence empowered metaverse: architecture, technologies, and open issues. *IEEE Network* 37(6), 92–100 (2023). https://doi.org/10.1109/MNET.2023.3317477.
12. Vahabi, M., and Fotouhi, H. Federated learning at the edge in Industrial Internet of Things: A review. *Sustainable Computing: Informatics and Systems* 46, 101087 (2025). https://doi.org/10.1016/j.suscom.2025.101087.
13. Chen, C., Li, Y., Wu, Z., Mai, C., Liu, Y., Hu, Y., Kang, J., and Zheng, Z. Privacy computing meets metaverse: Necessity, taxonomy and challenges. *Ad Hoc Networks* 158, 103457 (2024). https://doi.org/10.1016/j.adhoc.2024.103457.

14. Zhang, Y., Sun, Q., and Li, W. Differential privacy in metaverse: Mechanisms and applications. *Computer Standards & Interfaces* 87, 103866 (2024). https://doi.org/10.1016/j.csi.2024.103866.
15. Gambo, M.L., and Almulhem, A. Zero trust architecture: A systematic literature review. *arXiv* preprint arXiv:2403.11659 (2024).
16. Gasser, U. Interoperability of the metaverse: A digital ecosystem perspective. *arXiv* preprint arXiv:2403.05205 (2022).
17. Singh, R., Sharma, G., and Kim, T.W. Edge AI: A survey. *Journal of King Saud University - Computer and Information Sciences* 35, 101593 (2023). https://doi. org/10.1016/j.jksuci.2021.12.019.
18. Ghazinoory, S., Azizi, F., and Abdi, S. Metaverse technology tree: A holistic view. *Computers in Human Behavior* 147, 107882 (2025). https://doi.org/10.1016/j.chb.2024.107882.
19. Adil, M., Kim, S., Shin, S.Y., and Khan, M.K. 5G/6G-enabled metaverse technologies: Taxonomy, applications, and open security challenges with future research directions. *Computer Networks* 252, 110650 (2024). https://doi.org/10.1016/j.comnet.2024.110650.
20. Yu, W., Zhao, X., Zhang, Q., Ding, K., and Wang, N. Comprehensive analysis of digital twins in smart cities. *Artificial Intelligence Review* 57, 3891–3922 (2024). https://doi.org/10.1007/s10462-024-10781-8.
21. Dwivedi, Y.K., Hughes, L., Baabdullah, A.M., Ribeiro-Navarrete, S., Giannakis, M., Al-Debei, M.M., Dennehy, D., Metri, B., Buhalis, D., and Cheung, C.M.K. Meta-verse beyond the hype: Multidisciplinary perspectives on emerging challenges, opportunities, and agenda for research, practice and policy. *International Journal of Information Management* 66, 102542 (2022). https://doi.org/10.1016/j.ijinfomgt.2022.102542.
22. Aboul-Yazeed, R.S., Darwish, A., and Hassanien, A.E. Flourish the mood: A systematic review on the impacts of metaverse on psychological and mental health. *Frontiers in Virtual Reality* 6, 1429038 (2025). https://doi.org/10.3389/frvir.2025.
23. Rockmann, R., Gewald, H., Wiegand, J., and Vidal, E. Bridging digital divides: A literature review and research agenda. *Electronic Markets* 28(2), 211–227 (2018). https://doi.org/10.1007/s12525-018-0296-2.
24. Villar-Rodriguez, E., Pérez, M.A., Torre-Bastida, A.I., Senderos, C.R., and López-de-Armentia, J. Edge intelligence secure frameworks: Current state and future challenges. *Future Generation Computer Systems* 145, 123–145 (2023). https://doi.org/10.1016/j.future.2023.06.002.

Index